From Neurology to Methodology and Back

Natasha Maurits

From Neurology to Methodology and Back

An Introduction to Clinical Neuroengineering

Natasha Maurits
Department of Neurology
University Medical Center Groningen
Groningen, The Netherlands
n.m.maurits@umcg.nl

ISBN 978-1-4614-1131-4 e-ISBN 978-1-4614-1132-1
DOI 10.1007/978-1-4614-1132-1
Springer New York Dordrecht Heidelberg London

Library of Congress Control Number: 2011936974

Printed on acid-free paper

Springer is part of Springer Science+Business Media (www.springer.com)

Preface

There are many textbooks that teach the basics and more advanced concepts of mathematical techniques and engineering approaches or the pathophysiology of neurological disorders. But for some readers, such a specialized treatise is more than bargained for. A medical doctor, who wants to interpret an EEG spectrum, does not need to know all intricacies of Fourier theory. On the other hand, an engineer who wants to employ spectral analysis techniques to improve the differential diagnosis of tremor, does not need a full understanding of tremor pathophysiology. Yet, both for physicians and engineers, it is important to bridge the gap between technology and medicine.

With this book, I hope to provide a guide for those interested in crossing over from the field of medicine – neurology in particular – to the field of technology and vice versa. Neurologists and residents in neurology, medical engineers, medical students, biomedical engineers and students, technical medicine students, or students of other interdisciplinary fields may therefore all find this book interesting and useful.

This text is inspired by a lecture series that I started teaching to residents in neurology back in 2006. The goal of that lecture series was to explain mathematical and physical principles underlying neurological diagnostic techniques to medical doctors. Being a mathematician by training, I noticed that my initial more conventional approach was not optimal; only by introducing the abstract concepts through concrete neurological cases I was able to get my educational message across. This experience motivated me to choose an uncommon approach for this book.

Each chapter starts with an outline of what should be known after studying the chapter. Then the neurological problems of interest are introduced through one or more patient cases. Subsequently, the neurodiagnostic technique that can help establish a diagnosis is outlined and the necessary mathematics, physics, or engineering principles are explained. Finally, the new knowledge is applied to the patient cases and other applications of the neurodiagnostic technique are discussed. To help understand the essentials of each chapter, questions with answers are provided throughout the text. For the interested reader, mathematical details are provided in separate boxes, but they can be easily skipped by those with less background in mathematics. Furthermore, each chapter is completed with many illustrations, a list of easily accessible additional reading material and a glossary. This set-up is intended to make the book suited for self-study, but also for use in classroom lectures.

This book would not have existed without Janne Geraedts and Aaldrik Sillius, who supported me in transforming a long-existing idea into a book proposal as part of a faculty development programme I was participating in 2008. I am also very grateful to Michael Weston, senior editor at Springer, for receiving my book proposal with such great enthusiasm and for his and his assistant's (Eric Farr) continuing support during the writing process. I highly appreciate the critical feedback on the manuscript by my colleagues (Jan Kuks and Han van der Hoeven) and Ph.D. students (Carolien Toxopeus and Esther Smits) at the University Medical Center Groningen (UMCG) and thank my colleagues Han van der Hoeven, Jan Willem Elting, Fiete Lange, and Angela van Loon at the UMCG for providing me with interesting patient cases. My colleague Fleur van Rootselaar at the Academic Medical Center (AMC) in Amsterdam provided a patient case and many valuable comments on the manuscript. My Ph.D. student Marja Broersma provided some photographs and supporting staff members Lukas Dijck, Janneke Sikkema, and Janette Bijmolt helped in obtaining the patient data. I thank you all for being so helpful. I would like to thank my parents for their continuing interest in and loving support for my education, academic training, and work. Last but not least: Johan, thank you for being so positively critical, understanding, and supportive (technically as well as by catering) when I was writing for hours in a row in our study.

Groningen, The Netherlands Natasha Maurits

Contents

Abbreviations

AASM	American Academy of Sleep Medicine
A/D	Analog-digital
AEP	Auditory evoked potential
AMT	Active motor threshold
ANN	Artificial neural network
APB	Abductor pollicis brevis
AUC	Area under the curve
BAEP	Brainstem auditory evoked potential
BMI	Body mass index
BOLD	Blood oxygen-level dependent
BP	Bereitschafts potential/blood pressure
CBF	Cerebral blood flow
CCA	Common carotid artery
CDIP	Chronic demyelinating inflammatory polyneuropathy
CK	Creatine kinase
CMAP	Compound motor action potential
CMCT	Central motor conduction time
CNV	Contingent negative variation
COPD	Chronic obstructive pulmonary disease
CSF	Cerebrospinal fluid
CT	Computed tomography
CTS	Carpal tunnel syndrome
CVA	Cerebrovascular accident
DC	Direct current
DFT	Discrete Fourier transform
DML	Distal motor latency
DSA	Digital subtraction angiography

ECA	External carotid artery
ECD	Equivalent current dipole
ECG	Electrocardiography
EDC	Extensor digitorum communis
EDV	End diastolic velocity
EEG	Electroencephalography
EMG	Electromyography
EOG	Electro-oculography
EP	Evoked potential
EPSP	Excitatory postsynaptic potential
ER	Emergency room
ERP	Event-related potential
ET	Essential tremor
FCMTE	Familial cortical myoclonic tremor with epilepsy
FDI	First dorsal interosseus
FFT	Fast Fourier transform
FIR	Finite impulse response
fMRI	Functional magnetic resonance imaging
FVEP	Flash visual evoked potential
ICA	Independent component analysis/internal carotid artery
ICF	Intracortical facilitation
ICU	Intensive care unit
IIR	Infinite impulse response
IPSP	Inhibitory postsynaptic potential
IQR	Interquartile range
IRF	Impulse response function
ISI	Interstimulus interval
LAURA	Local autoregressive average
LCMV	Linearly constrained minimum variance
LORETA	Low resolution brain electromagnetic tomography
LRP	Lateralized readiness potential
MEG	Magnetoencephalography
MEP	Motor evoked potential
MRC	Medical Research Council
MRCP	Motor-related cortical potential
MRI	Magnetic resonance imaging
MS	Multiple sclerosis
MSA	Multiple system atrophy
MUP	Motor unit potential
MUSIC	Multiple signal classification
NCSE	Non-convulsive status epilepticus

OSAS	Obstructive sleep apnea syndrome
PCA	Principle component analysis
PET	Positron emission tomography
PI	Pulsatility index
PLED	Periodic lateralized epileptic discharge
PML	Proximal motor latency
PRF	Pulse repetition frequency
PSP	Progressive supranuclear palsy
PSV	Peak systolic velocity
PVEP	Pattern visual evoked potential
RAP-MUSIC	Recursively applied and projected multiple signal classification
REM	Rapid eye movement
RMT	Resting motor threshold
rTMS	Repetitive transcranial magnetic stimulation
SAH	Subarachnoidal hemorrhage
SD	Standard deviation
SFAP	Single fiber action potential
SICI	Short-interval intracortical inhibition
sLORETA	Standardized low resolution brain electromagnetic tomography
SMA	Spinal muscular atrophy
SNAP	Sensory nerve action potential
SNR	Signal-to-noise ratio
SEP	Somatosensory evoked potential
SSEP	Steady-state evoked potential
SSVEP	Steady-state visual evoked potential
SVD	Singular value decomposition
TA	Tibialis anterior
TBS	Theta burst stimulation
TCD	Transcranial Doppler
TIA	Transient ischemic attack
TMS	Transcranial magnetic stimulation
UPDRS	Unified Parkinson's Disease Rating Scale
VEP	Visual evoked potential

Chapter 1
Introduction

Today, hospitals are full of advanced technology, ranging from large machines such as MRI scanners, to tiny devices such as implantable electrodes. In addition, patient information systems and wireless communication technologies help doctors provide the best care for their patients. Although most technology is operated by specially trained personnel, medical doctors also transform technical results into clinically relevant knowledge themselves. One of the tasks of engineers is to provide a link from medical techniques to clinical practice, by optimally designing technological devices and software for the hospital environment and advising physicians in their use of technology. Thus, both for physicians and engineers it is important to bridge the gap between technology and medicine. Clinical neuroengineering is one of the fields providing such a bridge between technology and – in this case – neurology. Clinical neuroengineering strives to advance the clinical neurosciences by employing mathematical, physical, and engineering principles. An essential principle of clinical neuroengineering is that inspiration and challenges are provided by clinical neurological practice. After finding a solution for these challenges using mathematical, physical, and engineering methods, this solution is made available to the neurologist. Thus, clinical neuroengineering approaches proceed "from neurology to methodology and back." Important fields in clinical neuroengineering are neurodiagnostics and neurotherapy. Neurodiagnostics is the study and recording of physiological signals reflecting brain, nerve, and muscle activity to determine the function of these systems or to diagnose a neurological disease.

The fundamental topics that are covered in this book range from basic concepts such as sampling and simple statistical measures, via Fourier analysis to highly complex methods such as source localization which involves the solution of an inverse problem. The neurological diseases that are used to provide the clinically relevant context and introduce the technical concepts are as diverse and range from epilepsy, brain tumors, and cerebrovascular diseases to tremor, MS, and neuromuscular diseases. What all topics have in common is that they are presented in a true clinical neuroengineering approach. Each chapter first provides one or more patient cases for inspiration. The diagnostic work-up of these patients is then shown to

N. Maurits, *From Neurology to Methodology and Back*: *An Introduction to Clinical Neuroengineering*, DOI 10.1007/978-1-4614-1132-1_1,

require a particular neurodiagnostic technique (such as electroencephalography, electromyography, or ultrasound measurements) that is discussed next. Next, the mathematical and physical principles underlying these techniques are explained after which we return to the patients. How can the explained technology help provide a diagnosis for the presented patients? By taking this clinical neuro-engineering approach, the bridge between neurology and technology can be crossed from both sides.

Chapter 2
Carpal Tunnel Syndrome, Electroneurography, Electromyography, and Statistics

After reading this chapter you should:

- Know what basic techniques are employed in electroneurography
- Know what basic techniques are employed in electromyography
- Know how electroneurography and electromyography can help diagnose neuromuscular disorders and distinguish neuropathies and myopathies in particular
- Understand why normal values are important in clinical practice
- Be able to describe the statistical considerations that are important for obtaining normal values
- Be able to describe the distribution of a value in a sample by a few statistics

2.1 Patient Cases

Patient 1

For several years, a female 33-year-old patient suffers from numbness and tingling in her right arm and hand, in particular in her index and middle finger. Opening jars has become more difficult, suggesting that the force in her right hand has decreased. Her complaints do not increase when she drives a car or rides a bicycle, nor are the complaints alleviated when she shakes her hand. At night, she incidentally wakes up because of numbness in her fingers. During daytime, the pain in her arm is most prominent. It is already known that she does not suffer from *rheumatoid arthritis*. The neurological examination does not indicate *atrophy* of the hand muscles, nor is the force of the *abductor pollicis brevis* muscle decreased. Her right grasping force is slightly less than normal (*MRC* 4). The somatosensory function of her hand and fingers is normal. Tapping of the nerves at the wrist does not elicit tingling in the fingers (negative Tinel's sign), nor does compression of the *median nerve* in the *carpal tunnel* by flexion of the wrists (negative Phalen's maneuver).

N. Maurits, *From Neurology to Methodology and Back: An Introduction to Clinical Neuroengineering*, DOI 10.1007/978-1-4614-1132-1_2,

Patient 2

A male 61-year-old patient suffers from tingling, numbness, and occasional pain in the thumb, index finger, and middle finger of both hands. His complaints deteriorate at night causing him to wake up sometimes. The numbness and tingling sensations are always present; the patient does not report any movements that improve or deteriorate the complaints. The neurological examination reveals atrophy of the abductor pollicis brevis muscle in both hands, with preserved force (MRC 5). In both hands, the three affected fingers are *hypesthetic*. Both Tinel's sign and Phalen's maneuver are negative.

2.2 Electroneurography: Assessing Nerve Function

The complaints that patients 1 and 2 suffer from are typical for carpal tunnel syndrome (CTS), a group of symptoms caused by compression of the median nerve in the carpal tunnel. The median nerve is a mixed nerve, containing both motor and sensory nerve fibers. Thus, when the nerve is damaged due to compression, this can lead to both motor (muscle weakness and atrophy) and sensory problems (numbness, tingling sensations). The location of the complaints is directly related to the motor and sensory innervation of the median nerve *distal* to the site of compression, i.e., the motor problems will occur in the *thenar* muscles whereas the sensory problems will occur in the thumb, index, and middle finger and the *radial* side of the ring finger. The motor problems mostly do not occur until a later stage of CTS. Different body positions at night can cause an increase in complaints, whereas shaking of the hand can alleviate them, by decreasing or increasing the available space in the carpal tunnel. Phalen's maneuver also decreases the space in the carpal tunnel, but is not very sensitive for detecting CTS. Causes of CTS are variable (congenital narrow carpal tunnel, hormonal (pregnancy), trauma, rheumatoid arthritis) and cannot always be identified in an individual patient. The test that is most sensitive for diagnosing CTS involves sensory nerve conduction velocity assessment.

2.2.1 *Nerve Physiology and Investigation*

Nerves consist of several nerve fibers (axons) that are oriented in parallel. When the nerves are myelinated, a fatty myelin sheath covers the axon. This myelin sheath has insulating properties, enabling faster signaling along the nerve. When a motor nerve fiber is stimulated, it will conduct the locally elicited action potential along its membrane via the *neuromuscular junction* to the *motor end plate* of the muscle where it will cause muscle fibers to contract. Sensory nerve fibers conduct the action potential to other sensory fibers in the central nervous system, enabling sensations, or to motor neurons, evoking reflexes. To assess the speed of nerve

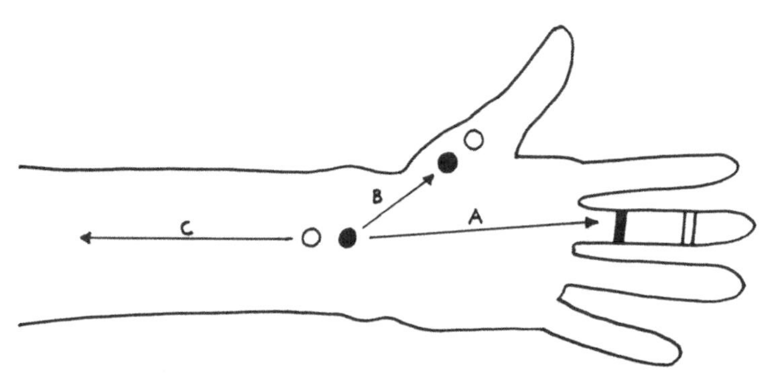

Fig. 2.1 For stimulation at the wrist A indicates antidromic conduction in sensory nerve fibers, B orthodromic conduction in motor nerve fibers, and C orthodromic conduction for sensory nerve fibers and antidromic conduction for motor nerve fibers. *Circles* and *strips* indicate stimulation (at the wrist) and recording electrodes, where the dark electrodes are the so-called active electrodes. Distances between active electrodes determine latencies and velocities

signaling (conduction velocity) or the number of normally functioning fibers in a nerve, electroneurography can be used.

To investigate nerve conduction velocity, the nerve is stimulated and the time it takes for the evoked potential to arrive at another point along the nerve's trajectory is recorded. Usually, electrical stimulation is used, because it can be well quantified and all nerve fibers can be stimulated at virtually the same moment. The artificial stimulation typically takes place somewhere along the nerve, thereby evoking a traveling potential in two directions: the natural direction (orthodromic conduction) and the opposite direction (antidromic conduction) (see Fig. 2.1).

Stimulation can be performed with needle or surface *electrodes*; the latter can only be used for superficially located nerves. Both surface and needle electrodes can be used to record the evoked potential at a location distant from the stimulation site. By stimulating a mixed nerve, both motor and sensory functioning can be evaluated.

2.2.1.1 Motor Nerve Functioning

Electrical stimulation of a motor nerve evokes orthodromic action potentials that activate the neuromuscular junction and, secondarily, cause muscle fibers to contract. The sum of all evoked single fiber action potentials (SFAPs) – the compound muscle action potential (CMAP) – can be recorded from the skin using surface electrodes. When evaluating the CMAP, it is essential that the stimulation is strong enough to activate all muscle fibers (supramaximal stimulation). If the CMAP amplitude is maximal, it can be concluded that stimulation was supramaximal. The time between stimulation and CMAP onset is the motor latency. Because latency is determined by CMAP onset (and not by, e.g., its peak), the fastest conducting nerve fibers determine its value. Note that the latency is not solely determined by the nerve conduction velocity; it also includes the time taken to cross

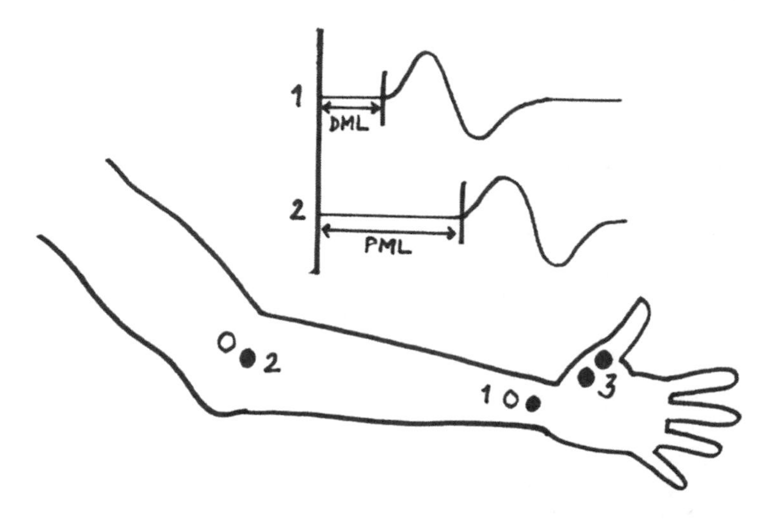

Fig. 2.2 CMAPs are derived from surface EMG recorded over the thenar muscle (3) by distal (1) and proximal stimulation (2), generating the distal and proximal motor latencies (DML and PML), which allow to derive motor nerve conduction velocity

the neuromuscular junction and to travel from the junction to the electrode overlying the muscle belly. Therefore, nerve conduction velocity is determined by stimulating at two different (distal and *proximal*) locations along the nerve while recording from the same position (see Fig. 2.2).

Question 2.1 How can the distal and proximal latencies be used to derive the conduction velocity of the nerve segment between the two stimulation positions? Remember that velocity is calculated as (distance traveled)/(time traveled).

For motor nerve conduction velocity studies of the main nerves in the arm (median and ulnar nerve), the distal stimulation site is usually at the wrist while the proximal stimulation point is at the elbow. For the main nerves in the leg (tibial and peroneal nerve), the distal stimulation site is usually at the ankle and the proximal stimulation site at the knee. Depending on the clinical question, more proximal stimulation sites may also be used.

The CMAP amplitude, measured from top to baseline or from top to top, gives information about the number of (supramaximally) activated muscle fibers. By comparing the CMAPs between distal and proximal stimulation and between both sides of the body, important diagnostic information can be derived. In a normally functioning nerve, both distal and proximal stimulation should stimulate all axons going to the muscle that the CMAP is recorded from. Because the difference in arrival time at the recording site of action potentials in the fastest conducting fibers

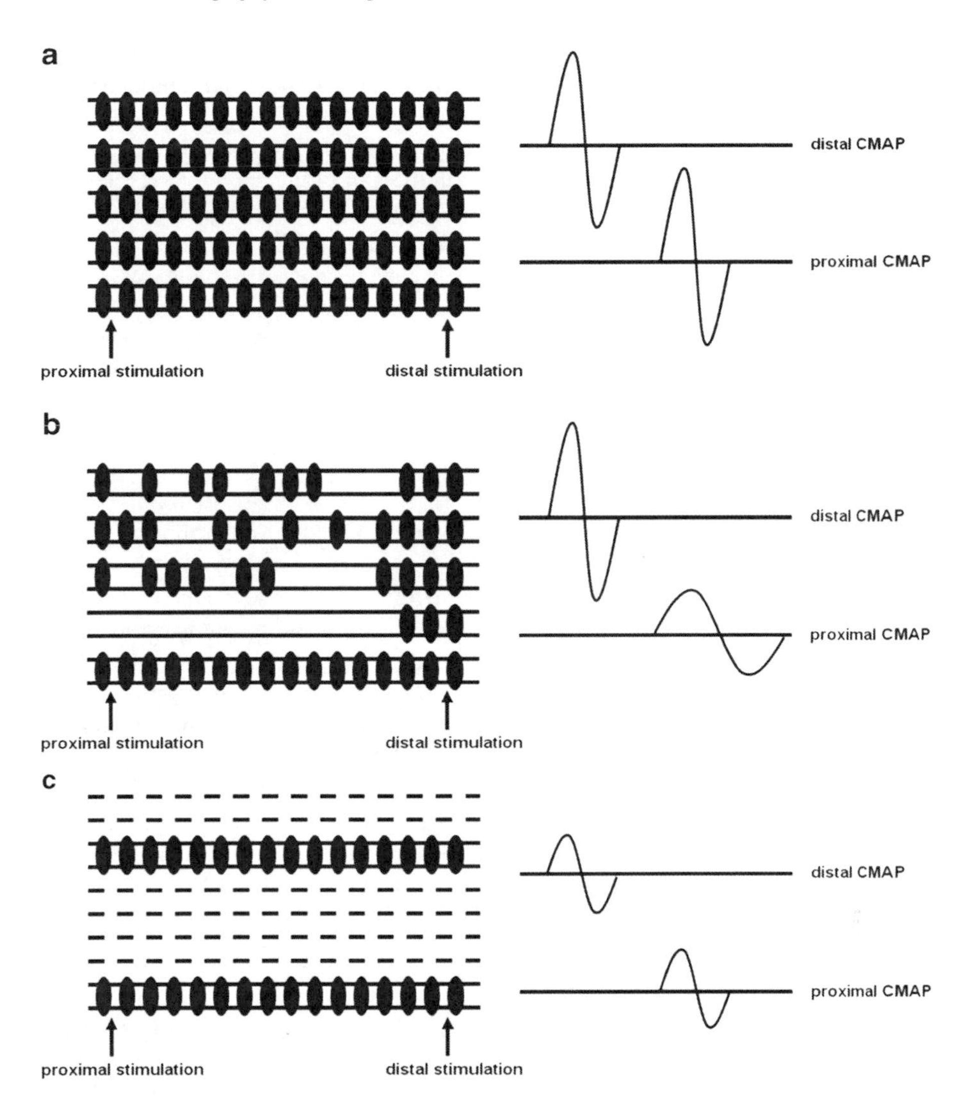

Fig. 2.3 Schematic overview of a nerve consisting of five myelinated axons. (**a**) Normal nerve, with intact myelin sheath. The CMAP morphology is similar between proximal and distal stimulation. (**b**) Segmental demyelination: the myelin sheath has been damaged multifocally. The distal CMAP is normal, but the proximal CMAP has lower amplitude and becomes broader, because the conduction velocity varies strongly between axons. (**c**) Axonal degeneration: several axons have completely disappeared. Functioning of the remaining axons has been preserved resulting in normal conduction velocities but lower amplitude CMAPs

compared to those in the slowest conducting fibers will be larger for proximal stimulation, leading to more *temporal dispersion*, the amplitude of the proximally stimulated CMAP will be slightly lower and the peak slightly broader. In normal motor nerves, these differences are minimal however.

There are three pathological mechanisms that can underlie disturbed nerve conduction: axonal degeneration, segmental demyelination, and conduction block (see Fig. 2.3).

Axonal degeneration entails the functional loss of some motor axons, typically over the entire trajectory of the nerve, inducing a decreased CMAP amplitude for proximal as well as distal supramaximal stimulation. Depending on which axons are affected, the latency may be normal or increased. Segmental demyelination causes decreased nerve conduction velocity in the part of the axon that is affected. If all axons in a nerve are affected, nerve conduction velocity can become severely diminished. Since the axons themselves are still intact, the CMAP amplitude should be preserved. However, the CMAP amplitude can be decreased (and the CMAP can have a broader peak) because of increased temporal dispersion of nerve action potentials along the different axons. Finally, conduction block involves complete cessation of conduction, with an intact axon. If the conduction block is located between the proximal and distal stimulation points, the proximal CMAP amplitude will be decreased whereas the distal CMAP amplitude will be normal.

F-Responses

When a peripheral (motor) nerve is stimulated supramaximally, potentials are conducted orthodromically evoking a muscle response (M-response). At the same time, potentials are also conducted antidromically and arrive at the *motor neurons* in the *anterior horn* of the spinal cord. Some of these motor neurons are then depolarized and send potentials back through the motor nerve fibers (orthodromically again), yielding the later F-response. It is named after its first recording in a foot muscle. The number and type of motor neurons that depolarize differs from stimulus to stimulus, changing the latency, amplitude, and shape of the F-response. Often, the minimum F-response latency is recorded, which reflects only the fastest motor axon contributing to the F-response. This value can be normal in pathology, however. Therefore, the dispersion in F-response latency is more helpful in diagnosing disorders of the proximal segments of motor nerves. Note that F-latencies depend on height and it is thus better to interpret F conduction velocities rather than latencies. For longitudinal recordings, latencies are very useful, however.

2.2.1.2 Sensory Nerve Functioning

As mentioned before, both sensory and motor nerve functioning can be assessed by stimulating a mixed nerve. When a peripheral nerve is damaged, often both sensory and motor functioning are affected; although when damage to the nerve is still limited, there may be sensory dysfunction while motor functioning is still normal. Assessing sensory nerve function is also important for those diseases in which sensory nerves are selectively involved.

When a sensory nerve is or sensory nerve fibers are stimulated by surface electrodes, the (compound) sensory nerve action potential (SNAP) can be obtained from another location along the nerve, by surface or needle electrodes. Both orthodromic and antidromic methods are in use. The sensory nerve conduction velocity can be determined similar to the motor nerve conduction velocity, i.e., by stimulating distally and proximally along the nerve, recording from a more distal site along the nerve, and comparing the distal and proximal sensory latencies (DSL and PSL). A difference with motor conduction velocity is that the DSL can also be directly used to derive the sensory nerve conduction velocity in that trajectory, since the SNAP is directly recorded from the nerve, in contrast to the CMAP which is recorded from the muscle.

Question 2.2 What distance needs to be determined to derive the sensory nerve conduction velocity based on the DSL?

2.2.1.3 Factors Influencing Nerve Conduction

As all things move slower when it is cold, it is not surprising that nerve conduction velocity is reduced at colder temperatures. At the same time SNAP amplitudes are increased at colder temperatures. These temperature effects mainly result from slower functioning of the sodium and potassium channels in the nerve cell membrane. The changes in nerve conduction with temperature have two important practical consequences. First of all, any reported (motor or sensory) nerve conduction velocity should include the (skin) temperature at the recording site (and preferably also at the stimulation site). Otherwise, a second measurement in the same subject cannot be interpreted well. Second, it is even better to always record nerve conduction velocities at the same (skin) temperature, so that values can be compared to normal values or to values obtained at another occasion in the same patient, without having to resort to conversion formulas. This can be achieved in practice by warming up the patient with warm blankets, infrared lamps, or warm water baths until the required temperature is reached (preferably 32°C at the skin).

Another factor influencing nerve conduction is age, which is especially important for recordings in young children. After the age of 3–5 years, nerve conduction velocities have typically increased to normal adult values. The increase in conduction velocity with age in young children largely results from an increase in nerve fiber diameter (the thicker the nerve fiber the faster it conducts) and from increased myelination. In older adults nerve conduction velocity gradually decreases again. Kimura (2001) has reported normal values for different nerves and age groups.

2.2.2 Diagnosing Carpal Tunnel Syndrome

To understand the logic of the electroneurographic investigation in CTS, it is important to know the trajectories of the main nerves in the lower arm and hand with respect to the carpal tunnel (see Fig. 2.4) on the side of the palm of the hand. The median nerve passes through the carpal tunnel whereas the other two nerves do not. Furthermore, the radial side of the ring finger is innervated by the median

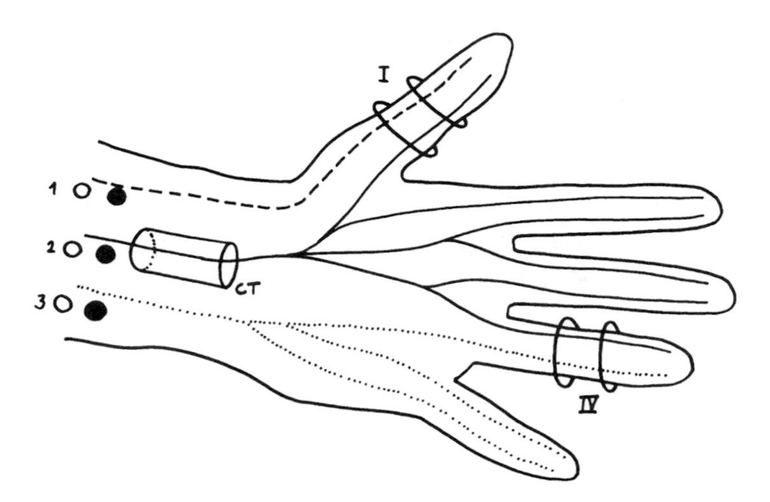

Fig. 2.4 Schematic overview of the trajectories of the radial (*dashed*), median (*drawn*), and ulnar (*dotted*) nerves and some of their branches in the wrist and hand. See text for explanation of the numbers

nerve, while its *ulnar* side is innervated by the ulnar nerve. Similarly, the thumb is innervated by the median nerve on the side of the palm of the hand and by the radial nerve on the other side.

The diagnosis of CTS relies on two investigations: the motor conduction of the median nerve and the sensory conduction of the median nerve. If the median nerve is compressed in the carpal tunnel, it is expected that conduction velocity is decreased, causing the DML to increase, due to focal demyelination. The conduction velocity itself cannot be determined for the most distal part of the motor nerve because of the unknown delay at the neuromuscular junction, only the DML can be reported. To be able to directly compare the DML in a patient to values obtained in healthy subjects, it is important that the same distance (usually 7 cm) between stimulation and recording sites is used. Furthermore, in slight cases of CTS the motor conduction velocity in the lower arm will still be normal, but when CTS is severe the median nerve may demyelinate causing a decreased motor conduction velocity. In addition, axonal damage may occur.

To determine if the sensory nerve conduction velocity of the median nerve is affected, SNAPs are recorded from the fingers that are innervated by the median nerve and one of the other nerves that do not pass through the carpal tunnel, i.e., from the thumb (I in Fig. 2.4) for the radial nerve and from the ring finger (IV in Fig. 2.4) for the ulnar nerve. By stimulating each of the nerves at the wrist proximal from the carpal tunnel (1, 2, and 3 for the radial, median and ulnar nerves and CT for the carpal tunnel in Fig. 2.4) at the same distance from the recording electrode, SNAP latencies can be compared between the possibly trapped median nerve and the other nerves.

If the latency difference is larger than 0.4 ms, conduction of the sensory median nerve fibers is disturbed, which is assumed to be a sign of CTS. In principle, to exclude other causes of the CTS symptoms, it may be necessary to additionally investigate motor conduction of the ulnar nerve, or the function of one of the muscles innervated by the ulnar nerve by (needle) electromyography.

2.3 Electromyography: Assessing Muscle Function

Needle electrodes can also be used to assess muscle fiber functioning by recording their electrical activity from a close distance (needle electromyography). When muscle fibers are activated at the motor end plate, action potentials start to propagate in two directions, similar to what happens in activated nerve fibers. These single fiber action potentials (SFAPs) induce contraction of the muscle fiber by a so-called electromechanical coupling mechanism. Electromyography only allows studying the electrical and not the mechanical phenomena of muscle contraction. A needle electrode, although very thin, is too large to record individual SFAPs, except in pathological circumstances in which muscle fibers can discharge spontaneously and independently. A needle electrode will typically record the activity of multiple muscle fibers belonging to the same or multiple activated *motor units*. During needle EMG the muscle is investigated at rest, during weak contraction to evaluate individual action potentials, and during strong contraction to evaluate the full contraction pattern. For a more detailed account of electromyographic investigations please see the references. Here, only some basic concepts are discussed.

An EMG needle has a very small field of view, allowing picking up the activity of a limited number of muscle fibers directly and of more fibers at a larger distance, depending on needle type. All SFAPs that develop within the field of view of the electrode when a muscle is contracted add up to the motor unit potential (MUP) that is actually recorded. The SFAPs at a larger distance contribute only partially to the MUP, because of volume conduction effects that act as a low-pass filter (see Sect. 4.3.4). The number of fibers within the field of view of the electrode, the distance of the active fibers to the electrode, and the timing between the SFAPs all contribute to the shape of the MUP. Normally, the timing of SFAPs contributing to the MUP is highly synchronized, invoking only 3–4 peaks in the MUP. An MUP that has more than four peaks (phases) is called polyphasic. The presence of too many polyphasic MUPs can indicate *neuropathies* (mostly large strongly polyphasic MUPs) or *myopathies* (small polyphasic MUPs). Note that these characteristics are different for recent neuropathies and long-existing myopathies. Besides MUP evaluation, another important aspect of needle electromyography is the assessment of muscle contraction patterns (see Fig. 2.5).

During weak contraction only a few motor units should be activated, resulting in a pattern with only a couple of separately distinguishable MUPs (single pattern). When the contraction level is increased, more and faster firing motor units should be recruited, resulting in a full contraction pattern in which the individual MUPs can no longer be distinguished (interference pattern). The maximum amplitude in

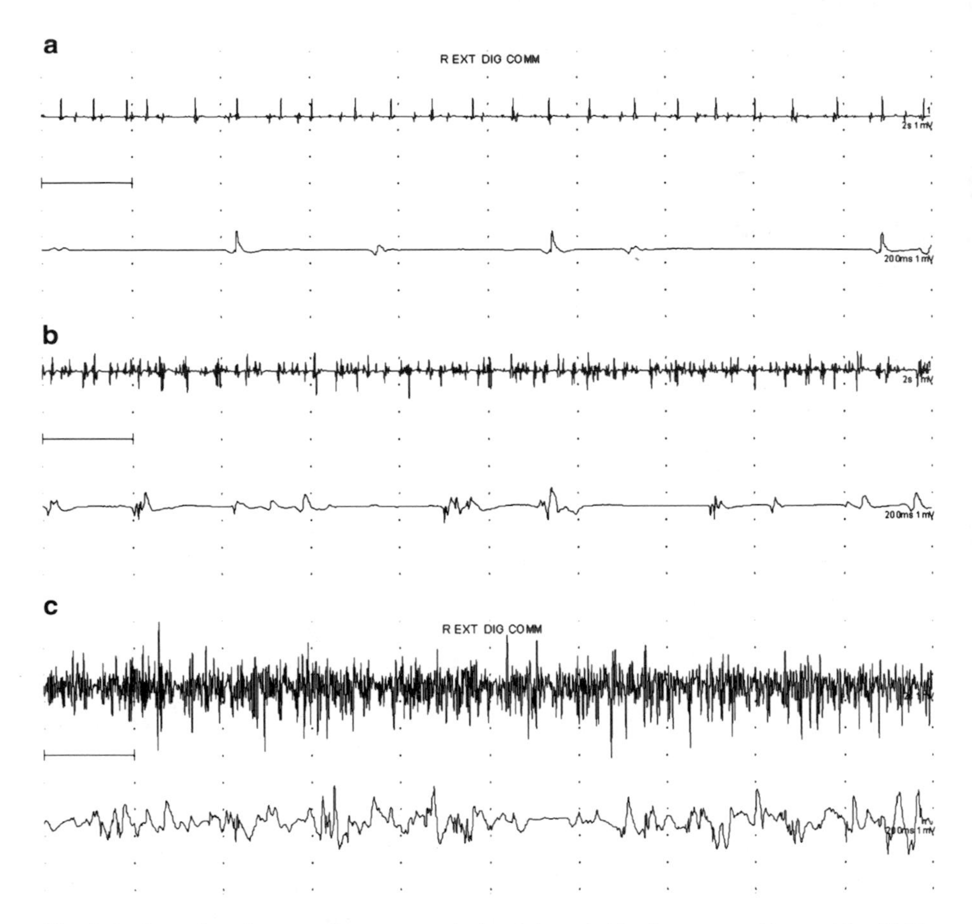

Fig. 2.5 Examples of contraction patterns during (**a**) weak (single pattern with two identifiable motor units), (**b**) moderate (mixed pattern), and (**c**) strong (interference pattern) contraction. The duration is 2 s for every top signal and 200 ms for every bottom signal. Signals were obtained from needle EMG of the m. extensor digitorum communis in the lower arm

the interference pattern may be increased in case of a chronic neuropathy and decreased in case of a myopathy, due to the presence or absence of large MUPs. In between the single and interference pattern a mixed pattern exists, which may occur when a healthy muscle cannot be contracted optimally due to pain or fear, or as a result of pathology. A (single or mixed) contraction pattern can be analyzed automatically to extract the different involved motor units. Normally, during rest, there should be no muscle activity. However, needle insertion or repositioning can result in activity due to mechanical stimulation of the muscle fibers. In addition, when the needle is inserted close to the motor endplate zone, spontaneous miniature endplate potentials, not resulting in action potential propagation, may be recorded. Other potentials occurring during rest, such as fibrillations and positive sharp waves, indicate pathology. Fibrillations have two or three phases, last 1–5 ms and start with a positive phase. They are discharges of individual muscle fibers and indicate changes in membrane properties usually due to axonal degeneration in the nerve innervating the muscle fiber (denervation). Positive sharp waves occur under similar circumstances as fibrillations, start with a short positive phase followed by a long negative phase and last approximately 10 ms. Spontaneous motor nerve discharges can also lead to MUPs during rest. Examples are fasciculations and spontaneous rhythmic MUPs. Fasciculations occur spontaneously and completely irregularly and can be observed through the skin as muscle twitches when they are superficially located. They can be both benign and pathological. Spontaneous rhythmic MUPs are purely rhythmical (in contrast to contraction patterns due to volitional contraction) at 3–9 Hz. They can be seen in nerve compression syndromes such as CTS. The electromyogram can also be recorded using surface electrodes, as for recording of the CMAP. Surface recordings prohibit the evaluation of individual MUPs, however.

2.4 Diagnostic Measures: What Is (Ab)normal?

As referred to in the previous sections, for reliable and consistent interpretation of electroneurographical and electromyographical measures (or any other measure obtained in a patient for that matter), it must be possible to determine if a patient's value falls within the normal range or not. This normal range is also referred to as reference range, reference values, or normal values.

2.4.1 Sampling the Population

To obtain normal values for a certain laboratory test, a representative sample of the general population needs to be taken and the test value must be obtained in this population under the same circumstances as will be used clinically. This implies that normal values should ideally be obtained for each laboratory and each machine again, although some test values differ so little between laboratories that normal values can be adopted from another laboratory. A representative sample of the population must have the same characteristics as the group of patients that they will be compared with, in the same proportion. In addition the sample must be chosen randomly to avoid bias. When the sample is large enough and selectively chosen, it is usually representative. If it is known beforehand that certain parameters that describe population characteristics have a great influence on the test value (e.g., age, sex, ethnicity), they can be taken into account for the composition of the population sample. In practice 30–40 subjects may be measured to obtain normal values, or ten per subcategory if a distinction is made on the basis of, e.g., age or weight. However, the exact number of subjects needed to represent all possible outcomes of the test value in the normal population depends strongly on the characteristics of the test value's distribution and of course, the larger the sample the better the estimate.

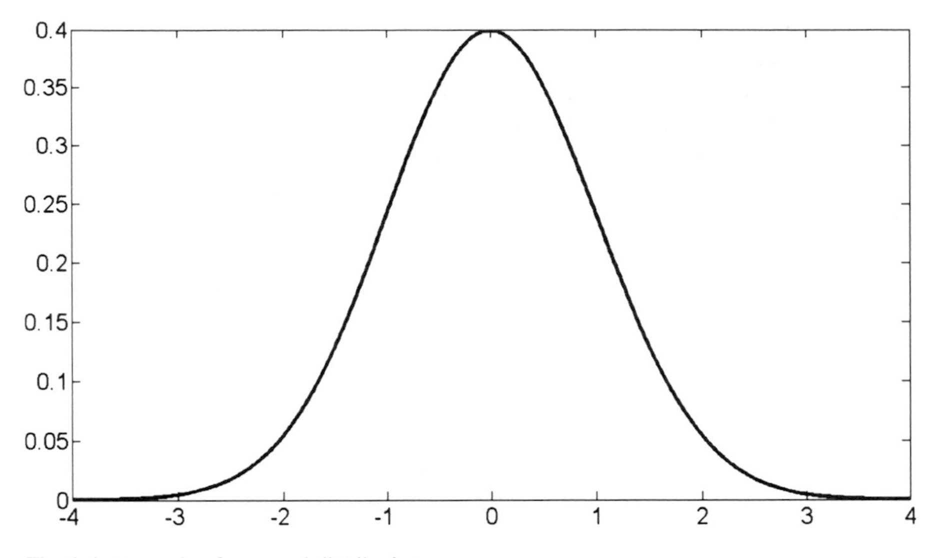

Fig. 2.6 Example of a normal distribution

2.4.2 Normal Distributions

For test values that depend on many independent variables (such as for most medical tests), it is often the case that the histogram of values obtained in a large population of different subjects will have a characteristic bell-shaped distribution known as the Gaussian or normal distribution (see Fig. 2.6).

When a test value in a population sample follows a normal distribution, it is usually assumed that the test value will follow the same distribution in the entire population. The beauty of a normal distribution is that its graph can be completely described by two parameters, the mean μ and the standard deviation σ, as follows:

$$f(x) = \frac{1}{\sqrt{2\pi\sigma^2}} e^{\frac{-(x-\mu)^2}{2\sigma^2}} \tag{2.1}$$

The mean μ is the center of the distribution and this value is mostly found in the population. When σ is small, the distribution is closely centered around the mean, whereas when σ is large, the distribution is very widespread. In the standard normal distribution, $\mu = 0$ and $\sigma = 1$. Furthermore, independent of the exact values for μ and σ, exactly 95% of the values in the distribution lies between $\mu - 1.96\sigma$ and $\mu + 1.96\sigma$ and 99% of the values lies between $\mu - 2.58\sigma$ and $\mu + 2.58\sigma$. In practice the two parameters μ and σ are estimated from a population sample with size N and values x_i as follows:

$$\bar{x} = \frac{1}{N} \sum_{i=1}^{N} x_i \tag{2.2}$$

$$s = \sqrt{\frac{\sum_{i=1}^{N} (x_i - \bar{x})}{N - 1}} \tag{2.3}$$

Here, the estimates for μ and σ are indicated by $\bar{x}$ and s. By replacing μ and σ by $\bar{x}$ and s, it can be seen that 95% of the values in the population sample lies between $\bar{x} - 1.96s$ and $\bar{x} + 1.96s$, if it is normally distributed. This interval can be used as an estimate of the reference interval for a particular laboratory test that can then be used for diagnostic purposes. However, instead of the factor 1.96, the factor 2.58 may be used (99% of the values) or, for ease of use, the factors 2.5 and 3 are also common. In these cases, values are thus said to be abnormal if they are higher (or lower) than the mean + (or −) 2.5 or 3 standard deviations. The higher this factor the fewer false positives a laboratory test will find in pathological cases (higher sensitivity). If it is more important to have a high specificity (fewer false negatives), this factor may be chosen lower.

2.4.3 Descriptive Statistics

The mean and standard deviation as calculated in (2.2) and (2.3) are examples of descriptive statistics. Formally (and rather abstractly), a statistic is a single measure of some aspect of a sample, which can be calculated by applying some formula on the individual values in a sample (or dataset). In other words: a statistic summarizes an important property of a dataset in a single number. Statistics for datasets can be roughly divided into two categories: measures of location and measures of variability.

Measures of Location

The most well known and often used measure of location is the (arithmetic) mean as defined in (2.2). The advantage of the mean is that it uses all values in a dataset and can help to characterize the distribution of the data as in the normal distribution. The main disadvantage of the mean is that it is very vulnerable to outliers: single values that, when excluded, have a large influence on the results. Outliers cannot be simply removed from the dataset, unless there are very good reasons to do so, such as known measurement errors or a value that actually belongs to a different population than the rest of the sample. An alternative measure of location, that is less sensitive to outliers, is the median. The median is the middle value (or the mean of the middle two values if there is an even number of values in the sample) if all values in the sample are ordered in size. The disadvantage of the median of course is that it does not use all values in the sample. A third measure of location is the mode, the most frequently occurring value in the sample. It is not used very much, except for measures such as modal income. In case of a normal distribution, mean, median, and mode have the same value. Other measures of location are quartiles (lower, median, and upper) and percentiles. The quartiles are the values that divide the data into four equal parts, each cutting off another 25% of the data values. The quartiles can be found by ordering the values again and determining the value that has 25% of the other values below it and 75% of the other values above it for the first quartile, etc. The median is equal to the second quartile.

Question 2.3 What are the mean, median, mode, and first quartile for the following dataset: {1, 3, 2, 4, 5, 2, 2, 6, 1, 2, 3, 2, 4, 1, 2, 6, 4, 5, 3, 2, 2, 3, 4, 5, 4, 2, 3, 1}?

Percentiles are the values that have a certain percentage of the values below them when ordered. The tenth percentile thus has 10% of the values in the dataset below it and 90% of the values above it. The first or lower quartile is thus equal to the 25th percentile, the second quartile (median) to the 50th percentile, and the third or upper quartile equals the 75th percentile.

Measures of Variability
The standard deviation (SD) as calculated in (2.3) is a measure of variability. Its square is the variance. From (2.3) it can be seen that s would be zero when all x_i would be the same, i.e., when there would be no variability in the sample. On the other hand, when the x_i are all very different from the mean $\bar{x}$, s would be very large. As mentioned earlier, about 95% of the values in a normally distributed sample lies within two standard deviations from the mean. When data is not normally distributed, there are more appropriate measures of variability that not automatically assume symmetry of the distribution. The range is defined as the difference between the smallest and largest samples in the value. The interquartile range (IQR) is given as the difference between the first and third quartiles and is less sensitive to outliers than the range.

2.4.4 Correlations

Normal values can only be defined on the basis of a representative sample when it is homogeneous, i.e., when there are no properties on which individual subjects differ and that, when the group would be split on that property, would give different distributions in the subgroups. For example, length depends on age and sex; clothes size depends on weight, age, and sex; and nerve conduction velocities depend on temperature. Thus, normal values for length cannot be derived from a sample that contains widely different ages or both men and women, but subcategories should be made for each of these variables. To determine if normal values should be given for subcategories, it should be determined whether the value under consideration is associated with (depends on) the value of another independent variable. This can be achieved by correlation analysis when the two variables under consideration are both continuous (vary smoothly). The correlation coefficient is a statistic that summarizes the strength of the relationship between two variables and can vary between −1 (strong negative correlation) and +1 (strong positive correlation). A correlation coefficient of 0 indicates that there is no correlation. Its calculation is explained in Box 2.1.

Box 2.1 The Correlation Coefficient

When a linear relationship between two variables x and y is expected, Pearson's correlation coefficient can be calculated for a sample of N measurements of these variables, given as x_i and y_i, as follows:

$$r_{xy} = \frac{1}{N-1} \frac{\sum_{i=1}^{N} (x_i - \bar{x})(y_i - \bar{y})}{s_x s_y} \quad (1)$$

Here $\bar{x}$ and $\bar{y}$ are the sample means and s_x and s_y the sample standard deviations of x and y as calculated in (2.2) and (2.3), respectively.

Pearson's correlation coefficient is not sensitive to nonlinear relationships. In that case a rank correlation coefficient (Kendall's or Spearman's) can be calculated, which measures the extent to which one variable increases/decreases as the other variable increases/decreases, without assuming that this increase is explained by a linear relationship.

Note that Pearson's correlation coefficient only indicates the presence of a linear relationship between variables. If there is a strong but nonlinear relationship between variables, Pearson's correlation coefficient can be close to zero. Examples are given in Fig. 2.7.

The correlation coefficient is closer to 0 when the data is noisier, positive for positive slopes, and negative for negative slopes of the linear relation. Yet, the value of the correlation coefficient does not reveal the slope of the linear relation itself. A common misconception about correlation is that a high value necessarily reflects a causal relationship between variables, although it can be indicative of one. Other approaches are needed to establish a causal relationship.

A correlation analysis can indicate whether normal values need to be given for subcategories. For example, in one of our own studies (Maurits et al. 2004), we determined normal values for muscle ultrasound parameters in children. We first calculated correlations between muscle ultrasound parameters and all *independent* variables (length, weight, *BMI*, and age) for each gender (48 boys and 57 girls between the ages of 45 and 156 months were included). Although normal values in children are often given per age group, we first determined which of the variables that correlated with the muscle ultrasound parameters would predict them most strongly by linear regression analysis (see references for details). More sensitive

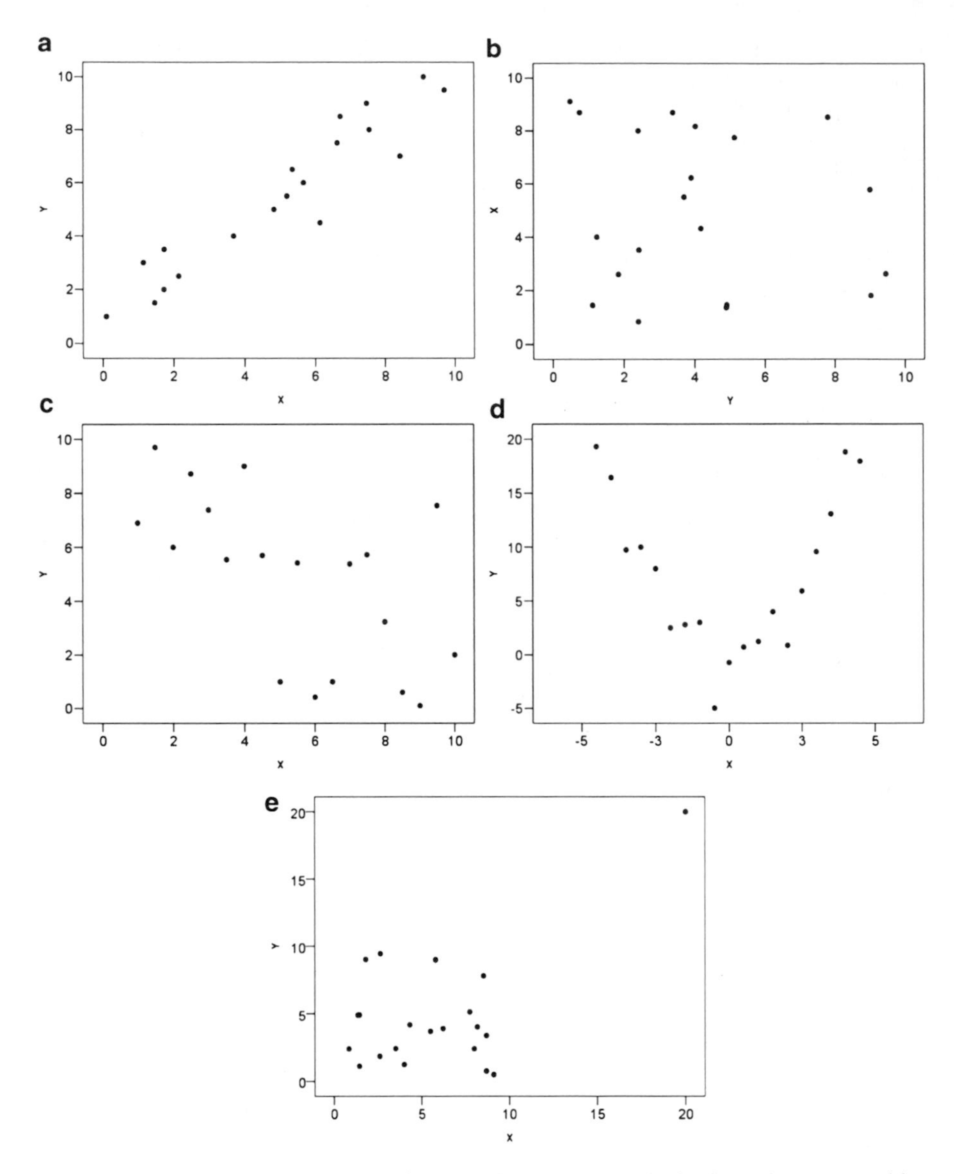

Fig. 2.7 Examples of scatter plots of two variables x and y indicating (**a**) strong positive correlation, (**b**) no correlation, and (**c**) weak negative correlation. An example where the correlation coefficient is inappropriate to describe the relationship between x and y is given in (**d**) for a quadratic relationship and in (**e**) for an outlier

parameters for clinical practice are obtained if classification is performed according to the most predictive variable. The results showed that normal values for subcutaneous tissue thicknesses should be given as a function of BMI, for muscle thicknesses as a function of weight and all other muscle parameters actually did not depend on length, weight, BMI, or age. Note that none of the muscle parameters were eventually given as a function of age.

2.4.5 *Tips and Tricks When Using Descriptive Statistics*

Although descriptive statistics and correlations are not difficult to calculate, there are some practical guidelines that should be considered when using these summarizing measures.

2.4.5.1 Normality

In the previous sections, it was indicated that it is important to know whether the distribution of a value in a sample is normal, to determine which descriptive statistics can be used to describe the distribution. There are different ways to assess normality of a distribution. The simplest is to plot a histogram of the values in the distribution. If its shape is Gaussian (looks like a symmetric bell; see Fig. 2.6), it is likely that the distribution is normal. An important aspect of normal distributions is that they are symmetric around the mean. If the distribution is not symmetric, but has a longer tail to the left (lower values), it is left skewed; if the longer tail is to the right (higher values), it is right skewed (see Fig. 2.8).

If a distribution is skewed, the median and IQR are more appropriate summary measures than the mean and SD which are sensitive to the skewness. A measure of skewness is the third moment around the mean:

$$\sqrt{\frac{\sum_{i=1}^{N}(x_i - \bar{x})^3}{s^3}} \tag{2.4}$$

A simpler expression for skewness exploits the fact that in a normal distribution, the mean and median are close:

$$3\left(\frac{\text{Mean} - \text{Median}}{s}\right) \tag{2.5}$$

> *Question 2.4* Will skewness according to (2.5) be positive or negative for right-skewed distributions?

In some cases, appropriate transformations of the data can make non-normal distributions normal (see Box 2.2).

Other ways to assess normality of distributions is by exploring the so-called Q–Q plots. A Q–Q plot (where Q stands for quantile) allows comparing two distributions (e.g., a theoretical normal distribution with the distribution in a sample) by plotting

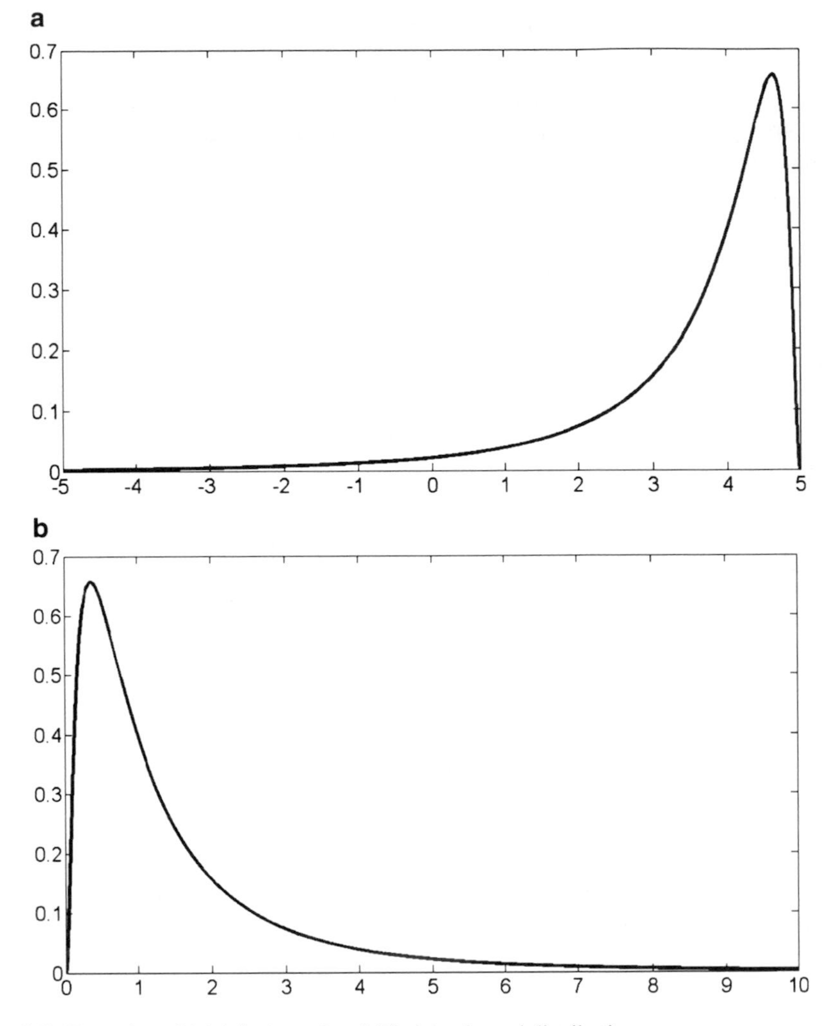

Fig. 2.8 Examples of (**a**) left-skewed and (**b**) right-skewed distributions

their quantiles against each other. If the pattern of points in the plot resembles a straight line, the distributions are highly similar. Examples of quantiles are the 2-quantile (the median), 4-quantiles (the quartiles), and 100-quantiles (the percentiles). Finally, some statistical tests can be used to assess normality of a distribution (e.g., the Shapiro–Wilk or Kolmogorov–Smirnov test).

When the distribution of the values for a laboratory test in a population sample is not normal, and cannot be made normal by a transformation, defining normal values on the basis of the SD is not appropriate. Before resorting to other methods, it is first important to check whether the distribution is not normal because it is bi- or multimodal, i.e., has multiple peaks. This may indicate that two or more different populations (e.g., males and females) underlie the distribution and the total sample

Box 2.2 Normalizing Skewed Distributions

To normalize a right-skewed distribution, larger values in the distribution must be decreased more than low values in the distribution. To accomplish this, we need a transformation that has relatively low output for high input values and relatively high output for low input values. A suitable transformation to achieve this is the (natural or 10-base) logarithm (Fig. 2.9a).

Similarly, to normalize a left-skewed distribution, we need a transformation that has relatively high output for high input values and relatively low output for low input values. A suitable transformation for left-skewed distributions is a quadratic function (Fig. 2.9b).

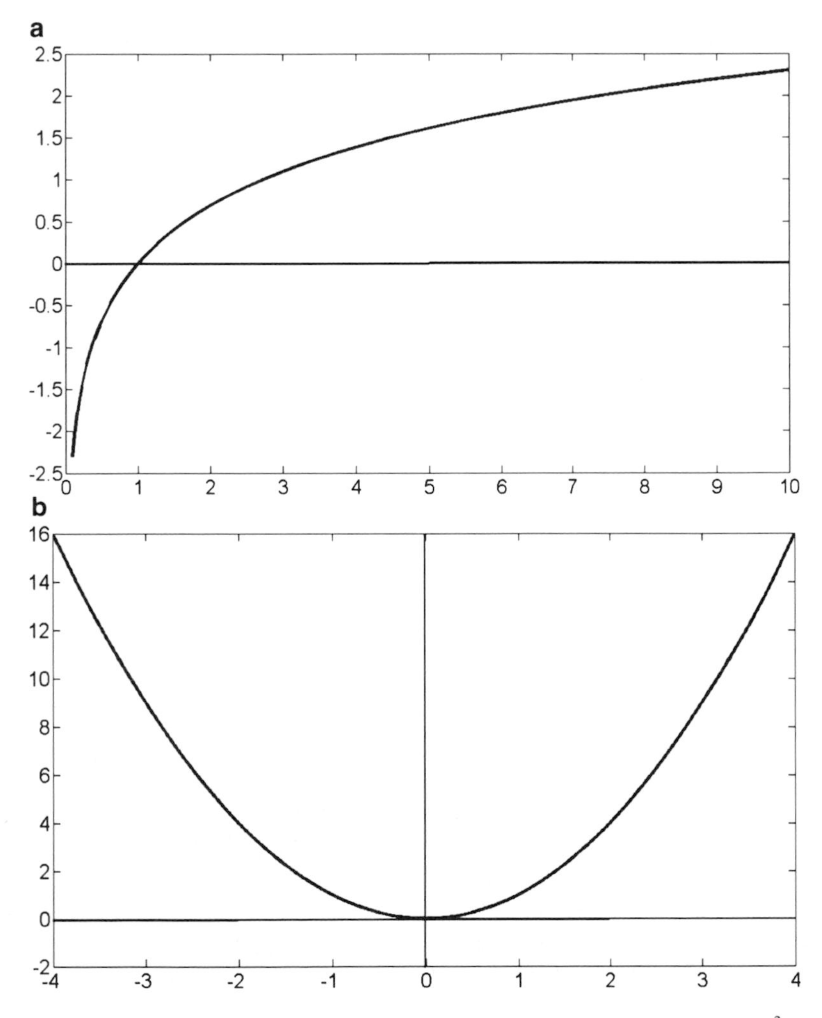

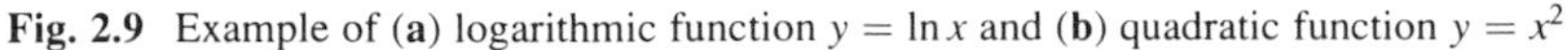

Fig. 2.9 Example of (**a**) logarithmic function $y = \ln x$ and (**b**) quadratic function $y = x^2$

should then first be split. If this is not the case, percentiles (see Sect. 2.4.3) may be used to define normal values. If low values on the laboratory test are abnormal, the 1st, 5th, or 10th percentile may be used as a cut-off value; if high values on the laboratory test are abnormal, the 90th, 95th, or 99th percentile may be used. Finally, the most extreme value seen in healthy subjects may also be used as a limit for the normal range.

2.4.5.2 Plotting Before Calculating

There are at least two reasons for plotting data before calculating descriptive statistics or correlations. As mentioned in Sect. 2.4.3, some descriptive statistics are sensitive to outliers. Outliers are most easily detected by plotting the data in an appropriate manner (in a dot plot, histogram, or boxplot; see Fig. 2.10).

As indicated in Sect. 2.4.4, correlations between two variables can only be calculated if a linear relationship is expected between the two. To investigate if a nonlinear relationship is more likely, the two variables can first be plotted against each other in a scatter plot (see Fig. 2.7). Thus, plotting before calculating prohibits making some elementary mistakes in calculating statistics.

2.4.5.3 Reliability of Sample Statistics

Although it was mentioned earlier (Sect. 2.4.1) that a sample that is large enough and selectively chosen, is usually representative for the entire population, there is still a need to know how reliable sample statistics are. In other words: if we would take another representative sample of the population, how likely is it that we would find a similar estimate for the sample statistic? The precision with which a mean is estimated can be inferred from the standard deviation of the mean, more commonly referred to as the standard error of the mean (SE). It is calculated by dividing the standard deviation of the sample by the square root of the number of subjects making up the sample:

$$\mathrm{SE} = \frac{s}{\sqrt{N}} \tag{2.6}$$

This definition formally confirms the more intuitive notion that the sample mean becomes more reliable (i.e., SE becomes smaller) when the number of subjects in the sample is larger. To be clear, the difference between s and SE is that s provides a measure of the variability between individuals related to the laboratory test value, whereas SE provides a measure of the variability in the mean, derived from the individual values, from sample to sample. Note that because SE is always (much) smaller than s, some authors display SE in figures instead of s. This is a very deceptive practice, because it suggests less variability in individual test values than are actually present in the sample.

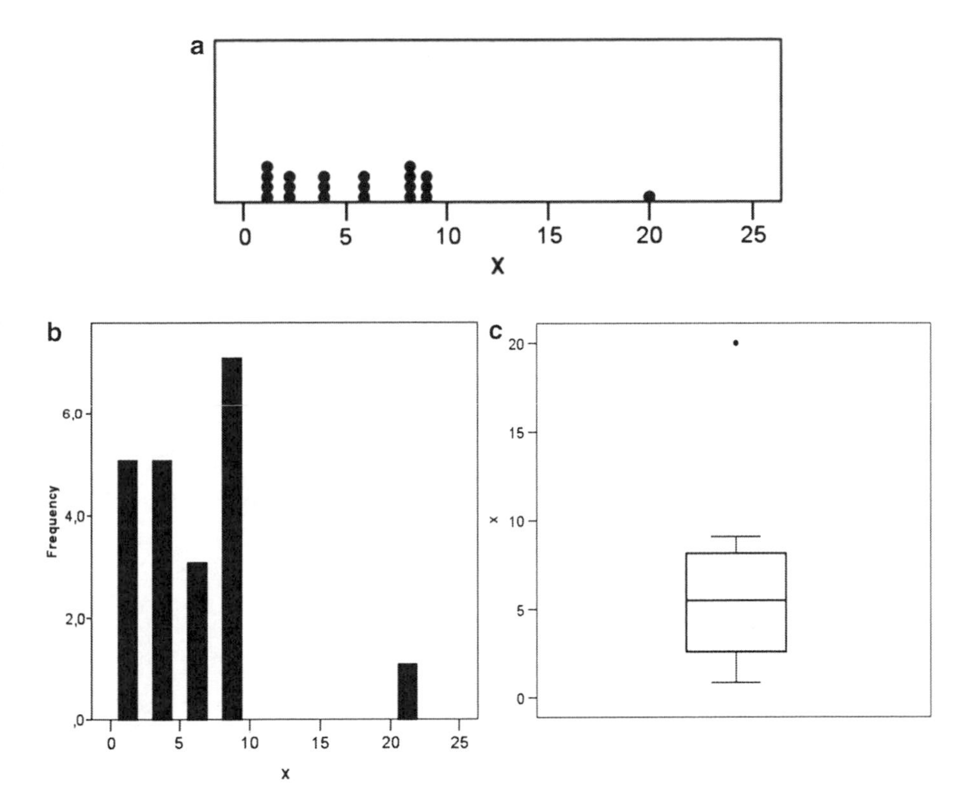

Fig. 2.10 Examples of (**a**) dot plot, (**b**) histogram, and (**c**) boxplot

2.4.5.4 Statistical Group Differences and Individual Patients

Normal values are employed to distinguish patients from healthy subjects. In literature many studies can be found that compare groups of patients to groups of healthy controls finding statistically significant differences between groups. It is important to realize that this does not imply that the measure under investigation is necessarily a sensitive measure to distinguish one individual patient from healthy subjects. For example, even when there is a lot of overlap in the distributions of the test value for patients and healthy subjects, a statistically significant difference may still be present (when the groups are large enough). Yet, in that case, an individual patient is highly likely to be within normal range and cannot be distinguished from healthy subjects. Better measures to predict the applicability of a measure for diagnostic purposes are sensitivity and specificity. Sensitivity is the percentage of patients that is correctly identified as being patients and specificity is the percentage of healthy subjects that is correctly identified as being healthy.

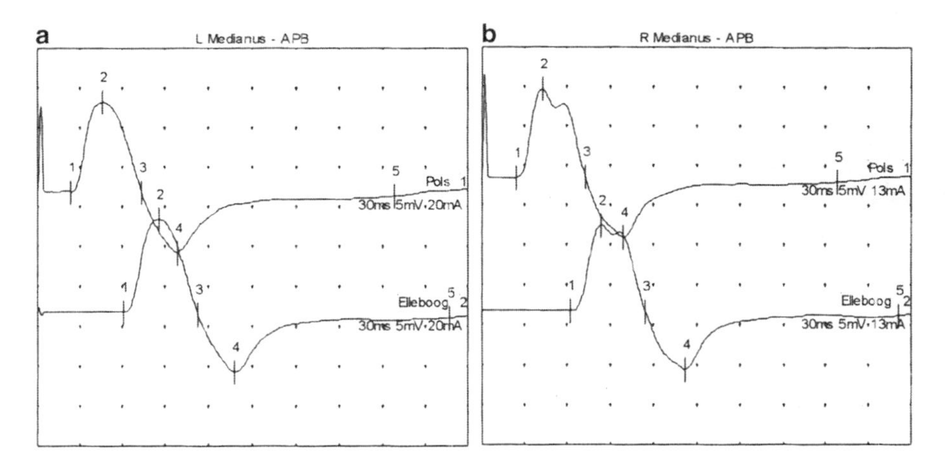

Fig. 2.11 CMAPs as obtained from the abductor pollicis brevis muscle after (**a**) left and (**b**) right distal (*top*) and proximal (*bottom*) stimulation of the median nerve in patient 1

2.5 Electroneurography in Individual Patients

Both patients 1 and 2 were investigated after warming up their lower arms in a warm water bath. Subsequently, motor and sensory conduction velocities were determined through electroneurography. Normal values were the same as those given by Kimura (2001). Deviations from the normal mean are given in SD.

Patient 1

The median nerve motor conduction velocity in the lower arm was determined by stimulating at the wrist and elbow and recording from the abductor pollicis brevis muscle. For the left arm, the DML was 2.35 ms (−3.4 SD, i.e., shorter (better) than the mean normal latency) and the PML was 6.05 ms, resulting in a conduction velocity of 65 m/s over a distance of 24.5 cm (+1.4 SD, i.e., faster (better) than the mean normal velocity) over the lower arm. For the right arm, the DML was 2.4 ms (−3.2 SD) and the PML was 6.25 ms, giving a conduction velocity of 64 m/s over a distance of 24 cm (+1.1 SD) over the lower arm. Thus, the median nerve motor conduction velocity was normal in both arms. In Fig. 2.11 the left and right CMAPs resulting from distal and proximal stimulation are shown.

In addition, to assess nerve conduction in the most distal part of the median nerve, the DML was compared between the median nerve and the ulnar nerve, by stimulating at the wrist proximal from the carpal tunnel, at the same distance (8 cm) from the recording electrodes over the second *lumbrical* muscle (innervated by the median nerve) and the second palmar *interosseus* muscle (innervated by the ulnar nerve). The DML was 2.70/2.75 ms for the left/right median nerve and 2.75/2.80 ms for the left/right ulnar nerve. The differences between the median and ulnar nerve were (far) smaller than 0.5 ms, implicating a normal DML for the median nerve.

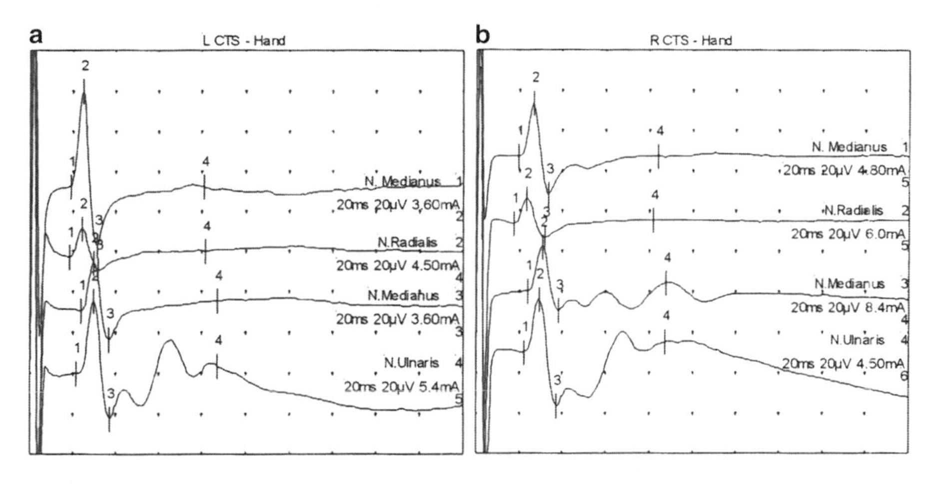

Fig. 2.12 (**a**) Left and (**b**) right SNAPs as obtained from the thumb (first and second trace) and ringfinger (third and fourth trace) after stimulation of the median nerve (first and third trace), the radial nerve (second trace) and the ulnar nerve (fourth trace) in patient 1

Table 2.1 Findings of sensory nerve conduction velocity investigation in patient 1

		Latency (ms)	Distance (mm)	Velocity (m/s)
Left				
Thumb	Median nerve	1.85	90	48.6
	Radial nerve	1.80	90	50.0
Ringfinger	Median nerve	2.35	130	55.3
	Ulnar nerve	2.15	130	60.5
Right				
Thumb	Median nerve	1.95	80	41.0
	Radial nerve	1.70	80	47.1
Ringfinger	Median nerve	2.35	120	51.1
	Ulnar nerve	2.20	120	54.5

Finally, sensory nerve conduction velocities were determined by stimulating the nerves at the wrist proximal from the carpal tunnel and recording from the thumb (for a comparison between median nerve and radial nerve conduction velocities) and from the ring finger (for a comparison between median nerve and ulnar nerve conduction velocities) using ring electrodes. The SNAPs are illustrated in Fig. 2.12, and Table 2.1 gives a numerical overview of the findings.

In conclusion, motor conduction velocities for the median nerve were normal in both lower arms, as were the DMLs. Sensory conduction velocities were normal as well. Together, motor and sensory conduction studies do not give any indications for CTS in this patient, despite her motor and sensory problems. Her problems (pain and numbness) were finally thought to be of *tendomyogenic* origin.

Patient 2

The investigation proceeded along similar lines in patient 2 as in patient 1. For the left arm, the DML was 8.80 ms (+15.6 SD, i.e., longer (worse) than the mean normal latency) and the PML was 14.20 ms, resulting in a conduction velocity

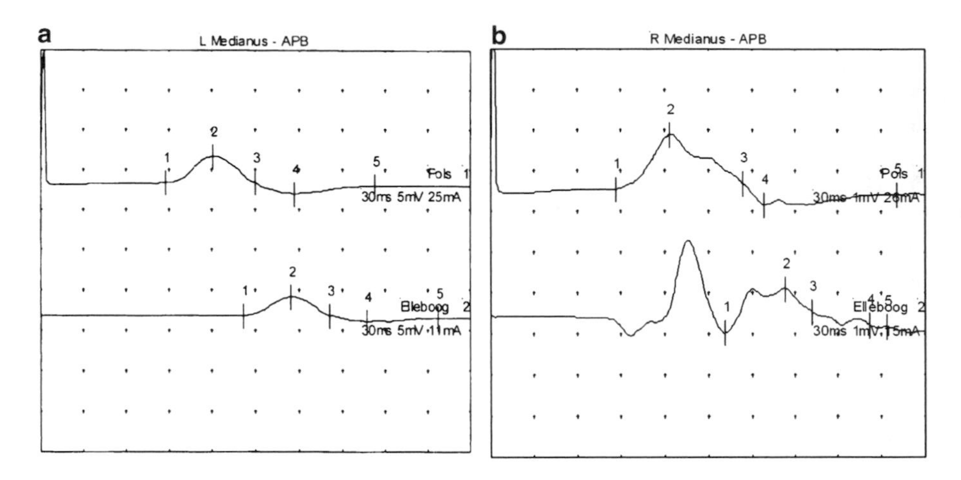

Fig. 2.13 CMAPs as obtained from the abductor pollicis brevis muscle after (**a**) left and (**b**) right distal (*top*) and proximal (*bottom*) stimulation of the median nerve in patient 2. The proximal stimulation of the right median nerve probably costimulated ulnar nerve fibers, causing the first large positive deflection in the bottom right trace. The second positive deflection is the median nerve CMAP, comparable to the one in the top right trace. Note that the vertical scale differs between the two figures (**a**) 5 mV per division and (**b**) 1 mV per division

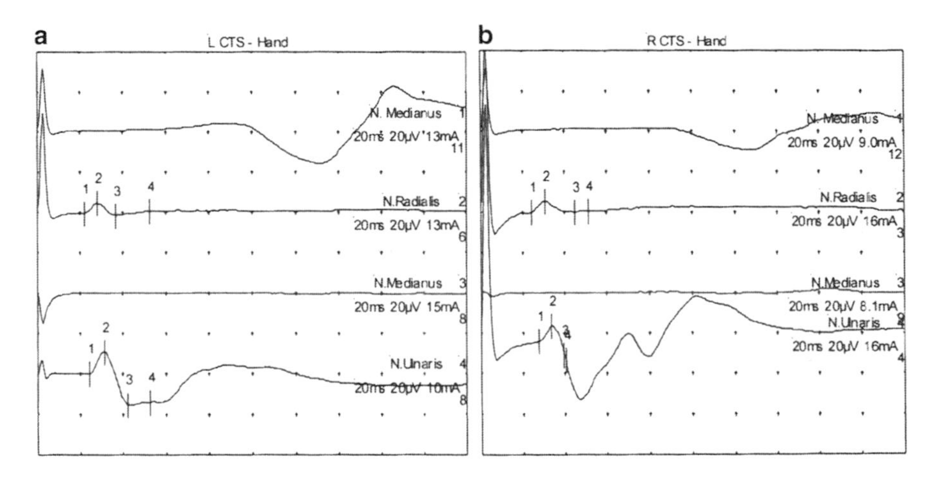

Fig. 2.14 (**a**) Left and (**b**) right SNAPs as obtained from the thumb (first and second trace) and ringfinger (third and fourth trace) after stimulation of the median nerve (first and third trace), the radial nerve (second trace) and the ulnar nerve (fourth trace) in patient 2. Note that the vertical scale is only 20 μV per division

of 44 m/s over a distance of 24 cm (−3.0 SD, i.e., slower (worse) than the mean normal velocity) over the lower arm. For the right arm, the DML was 8.65 ms (+15.2 SD) and the PML was 16.15 ms, giving a conduction velocity of 36 m/s over a distance of 27 cm (−4.9 SD) over the lower arm. Thus, the median nerve motor

Table 2.2 Findings of sensory nerve conduction velocity investigation in patient 2

		Latency (ms)	Distance (mm)	Velocity (m/s)
Left				
Thumb	Median nerve	–	100	–
	Radial nerve	2.20	100	45.5
Ringfinger	Median nerve	–	130	–
	Ulnar nerve	2.40	130	54.2
Right				
Thumb	Median nerve	–	100	–
	Radial nerve	2.40	100	41.7
Ringfinger	Median nerve	–	140	–
	Ulnar nerve	2.75	140	50.9

conduction velocity was abnormal in both arms. In Fig. 2.13 the left and right CMAPs resulting from distal and proximal stimulation are shown.

When comparing the CMAPs from patient 2 with those in patient 1, it can be clearly seen that the amplitudes are also much lower in patient 2 and the waveform is more dispersed in time, possibly reflecting axonal degeneration and segmental demyelination (see Sect. 2.2.1.1).

When stimulating at the wrist proximal from the carpal tunnel, at the same distance (left: 9 and right: 10 cm) from the recording electrodes over the second lumbrical muscle and the second palmar interosseus muscle, the DML was 9.65/7.45 ms for the left/right median nerve and 3.25/3.95 ms for the left/right ulnar nerve. The differences between the median and ulnar nerve were (far) larger than 0.5 ms, implicating a strongly abnormal DML for the median nerve as well.

In this patient, F-responses (see Sect. 2.2.1.1) were additionally obtained in the right hand by stimulating the median nerve at the wrist and recording from the abductor pollicis brevis muscle. The F-wave conduction velocity was found to be 47 m/s (−5.3 SD).

In this patient, sensory nerve conduction velocities were also obtained: SNAPs are illustrated in Fig. 2.14, and Table 2.2 gives a numerical overview of the findings in patient 2.

Note that SNAPs could not be obtained at all when stimulating the median nerve and SNAP amplitudes for the other nerves were normal but low.

To summarize, motor conduction velocities were abnormal in both lower arms, DMLs were strongly increased bilaterally, and the DML difference for the median nerve to the lumbrical muscle compared to the ulnar nerve to the interosseus muscle was considerable. CMAP amplitudes were low to very low. The F-response indicated an increased DML, but a surprisingly normal proximal motor conduction velocity. Sensory conduction velocities could not be established for the median nerve in either hand, because SNAPs could not be obtained in the thumb, middle, or ring finger, not even when stimulating at the hand palm. The radial and ulnar nerve had normal sensory conduction. Together, these findings are in accordance with severe CTS in both hands, with considerable axonal degeneration.

Because of the severity of his problems, patient 2 was referred to the neurosurgeon for surgery. During this procedure, the transverse carpal ligament that forms the roof of the carpal tunnel will be cut in two, thereby relieving the pressure on the median nerve. If his problems would have been less severe, localized corticosteroid injections could have been given. These injections with anti-inflammatory drugs only temporarily relieve the symptoms of CTS, but this may be sufficient when strategic adaptations to how the hands are used can prevent recurrence of the problems.

Although patients 1 and 2 had similar mostly sensory (numbness, tingling, pain) problems, the electroneurographical investigation clearly indicated that CTS was highly unlikely in patient 1 and very likely in patient 2. In this sense, electroneurography can be very helpful in diagnosing neuropathies. Note that for these patients, the clinical tests for CTS (Tinel's sign and Phalen's maneuver) were not contributing to the diagnosis.

2.6 Other Applications of Electromyography in Neurology

In Sect. 2.2.1.1, it was shown that surface EMG can be used to assess the CMAP. The CMAP is also recorded in the context of transcranial magnetic stimulation to assess central motor conduction times, i.e., to assess the integrity of the *pyramidal tract* (see Chap. 11). The CMAP is then called a motor evoked potential (MEP).

Since surface EMG records the electrical activity of a contracting muscle, it can also be used more generally (1) to determine the (relative) timing of activity in (multiple) muscles, (2) to estimate the force delivered by the muscle, and (3) to determine the rate at which a muscle fatigues. Using EMG to estimate muscle force can only be done for *isometric* contractions: under those circumstances *EMG power* is linearly related to force. Muscle fatigue is reflected in the EMG by a decrease in the frequency components of the EMG signal, typically by a drop in the center frequency. Section 3.3.3 describes how frequency components can be obtained from any signal by Fourier transform.

The applications (2) and (3) of surface EMG are mostly employed to investigate fundamentals of healthy muscle physiology and to study muscle functioning in populations of patients in whom (motor) fatigue or muscle weakness is part of their symptoms (e.g., MS, chronic fatigue, neuromuscular disorders) but is less relevant for diagnosis of neurological disorders. Application (1) is discussed extensively in Chap. 3, in which EMG recordings of multiple muscles are shown to be relevant for differential diagnosis of *tremor*.

2.7 Answers to Questions

Answer 2.1 The difference between the proximal and distal latencies is the latency (in ms) over the part of the nerve between the two stimulation sites. The distance (in mm) between the two stimulation sites can be measured on the skin, after which the conduction velocity can be calculated by dividing the distance by the latency. The velocity is then obtained in mm/ms and has the same value in m/s.
Answer 2.2 To obtain the sensory nerve conduction velocity based on the DSL, the distance between the stimulating electrode and the recording electrode needs to be determined.
Answer 2.3 The mean and median are both 3, the mode and first quartile are both 2.
Answer 2.4 In a right-skewed distribution, the mean will be larger than the median and skewness will be positive according to (2.5).

Glossary

Abductor pollicis brevis Muscle used to move the thumb away from the palm of the hand.
Adduction Movement toward the plane dividing the body in a left and right half.
Anterior horn (Frontal) grey matter in the spinal cord.
Atrophy Decrease in muscle volume.
BMI Body mass index: weight corrected for length as weight/length^2
Carpal tunnel Canal on the inside of the wrist, through which several tendons of lower arm muscles and the median nerve pass on their way from the lower arm to the hand palm.
Distal Far(ther) from the trunk.
Electrode Usually a small metal (silver, gold, tin) plate or ring that can be used to record electrical activity. Other forms are needle and sticker electrodes.
EMG power See power spectrum in Sect. 3.3.3.
Extension Opposite of flexion: stretching a joint.
Flexion Bending a joint using flexor muscles.
Hypesthetic Reduced sense of touch or sensation.
Independent Here: sampled variables that are already known (e.g., age, length, weight). In contrast, dependent variables depend on independent variables.
Interosseus Small muscles in the hand palm, used for *adducting* the fingers toward the middle finger.
Isometric Muscle contraction which keeps muscle length constant.
Lumbrical Small muscles in the hand palm, used for *flexing* and *extending* hand joints.
Median nerve One of the main nerves of the arm, another one is the ulnar nerve.
Motor end plate Region of the muscle membrane involved in initiating muscle fiber action potentials.

Motor neuron Neuron located in the central nervous system (CNS) that projects its axon outside the CNS to control muscles.

Motor unit Motor neuron including all muscle fibers it innervates.

MRC Manual muscle force assessment scale, established by the Medical Research Council in the UK. 0 = no contraction, 1 = flicker or trace of contraction, 2 = active movement with gravity eliminated, 3 = active movement against gravity, 4 = active movement against gravity and resistance, and 5 = normal power. Grade 4 is sometimes divided in 4−, 4, and 4+ to indicate movement against slight, moderate, and strong resistance.

Myopathy Disease in which (part of) the muscle is not functioning normally, resulting in muscle weakness, cramps, stiffness, and/or spasms.

Neuromuscular junction Junction of the axon of a motor nerve and the motor end plate.

Neuropathy Disease in which (part of) the nerve is not functioning normally, resulting in sensory and/or motor disturbances, depending on the damage and the type of nerve (sensory, motor, or mixed).

Proximal Close(r) to the trunk.

Pyramidal tract Motor pathway from the cortex to the spinal cord, involved in voluntary skilled movement in particular.

Radial On the side of the thumb.

Rheumatoid arthritis Chronic inflammatory disorder, particularly attacking joints.

Temporal dispersion Here: arrival of action potentials at different moments in time.

Tendomyogenic Originating from the tendons and muscles.

Tremor Oscillating movement of one or more body parts.

Thenar muscles Group of three muscles in the palm of the hand at the base of the thumb.

Ulnar On the side of the pink.

References

Online Sources of Information

http://en.wikipedia.org/wiki/Carpal_tunnel_syndrome. Extensive overview of all aspects related to CTS (anatomy, symptoms, diagnosis, treatment etc)

http://en.wikipedia.org/wiki/Electromyography. Overview of EMG methods, (ab)normal results etc

http://www.teleemg.com. Educational site providing detailed guides and information on electroneurography and electromyography and anatomy of nerves and muscles

http://en.wikipedia.org/wiki/Normal_values#Standard_definition. Short description of important aspects in determining normal values

http://en.wikipedia.org/wiki/Correlation. Some mathematics of correlation, includes examples of data sets with high and low correlations

http://en.wikipedia.org/wiki/Regression_analysis. Overview of concepts and mathematics involved in regression analysis

Books

Blum AS, Rutkove AB (eds) (2007) The clinical neurophysiology primer. Humana Press, Totowa (available on books.google.co.uk.)
Campbell MJ, Machin D (1999) Medical statistics. A commonsense approach. Wiley, New York
Kimura J (2001) Electrodiagnosis in diseases of nerve and muscle: principles and practice. Oxford University Press, Oxford (available on books.google.co.uk. Section 5.6 in particular)
Saunders WB (2000) Aids to the examination of the peripheral nervous system. Elsevier, Philadelphia (available on books.google.co.uk. Publication by MRC)
Weiss L, Silver J, Weiss J (eds) (2004) Easy EMG. A guide to performing nerve conduction studies and electromyography. Elsevier, Amsterdam

Papers

Maurits NM, Beenakker EA, van Schaik DE, Fock JM, van der Hoeven JH (2004) Muscle ultrasound in children: normal values and application to neuromuscular disorders. Ultrasound Med Biol 30:1017–27

Chapter 3
Tremor, Polymyography, and Spectral Analysis

After reading this chapter you should know:

- Why spectral analysis can help differentiate between different types of tremor
- How spectral features are related to clinical features of tremor
- How spectra are generated
- How spectra can be interpreted
- How Fourier series, Fourier analysis, and Discrete and Fast Fourier Transforms are related
- How spectral analysis can be used to assess other neurological disorders
- The basic mathematics of goniometry and spectral analysis (additional material)

3.1 Patient Cases

Patient 1

A male 40-year-old patient is suffering from a progressive tremor in his hands since 20 years, right more than left, which was diagnosed as *essential tremor* 5 years earlier. The tremor is most obvious when picking up objects, drinking or eating. Alcohol diminishes the tremor. Writing has become almost impossible. The clinical neurological investigation shows an evident, coarse tremor, increasing on intention, right more than left, but no other neurological abnormalities. Several tremor medications were tried, but to no avail. Because several types of tremors can be suppressed by an intracranial operation whereas others cannot, it is important to analyze the tremor.

N. Maurits, *From Neurology to Methodology and Back: An Introduction to Clinical Neuroengineering*, DOI 10.1007/978-1-4614-1132-1_3,

Patient 2

At age 16 a female, now 28-year-old patient, fell from her horse which resulted in a pelvic fracture. Since a surgical treatment for this fracture 3 years earlier, she suffers from tremor in both arms, hands, and legs. This tremor does not increase when paid attention to, nor with relaxation or stress. There is no family history of tremor. At examination, there was a fast tremor with small amplitude which appeared at *extension* of the arms and hands. There was no tremor at rest and no *intentional component*. Further clinical neurological examination revealed no abnormalities. From this clinical examination, the type of tremor cannot be diagnosed with certainty.

Patient 3

A 54-year-old male patient is suffering from a tremor since many years. He tells that his mother and aunt also had a tremor since the age of 40 years which was really a handicap. However, gait and balance were not affected as they were still ambulant in their 70s. His tremor started in the left hand but has progressed to the right hand and has become increasingly severe over the last few months, now making his job as a truck driver unachievable. The tremor is present both at rest and during action. Drinking is difficult and eating soup impossible. Alcohol use does not improve his complaints. The clinical neurological examination reveals a distal tremor in the arms, left more than right, present at rest and increasing during action, but no other abnormalities. Because most medications have been ineffective in this patient, an intracranial operation is now considered. In these cases, it is standard practice to first analyze the tremor.

3.2 Tremor Recording: Polymyography and Accelerometry

The above examples show that although a particular diagnosis of tremor can be made more likely or unlikely by clinical examination of a patient, it may be necessary to further analyze the tremor to have more certainty about the diagnosis. Tremor is actually found in every person. Usually, this is a barely visible tremor that may be observed when extending the arms or when moving with great precision and that may increase when the muscles become fatigued. This is called physiological tremor. Pathological tremor occurs in a number of conditions, such as Parkinson's disease in which other signs and symptoms accompany the tremor or essential tremor, in which the tremor typically is the only symptom. For tremor diagnosis, the clinical observations in which attention is paid to the circumstances under which the tremor develops, *amplitude* and frequency (the number of times the movement repeats itself in 1 s) of the tremor, are most important for diagnosis, but not always sufficient as indicated in the previous paragraph. Although all tremors exhibit involuntary oscillatory motion, their underlying pathophysiology varies and is not completely understood. However, in all cases there are specific disturbances in the balance between *agonist* and *antagonist* muscle activity that is normally required to execute a movement smoothly or to maintain a position steadily. This is why a polymyographic recording can help to achieve a more detailed tremor analysis. In a polymyogram (see Fig. 3.1), muscle activity is recorded by surface electrodes on the skin above the muscles of interest. Surface electromyography allows getting an overview of muscle activity quickly, noninvasively, and without pain. Usually, activity in lower and upper arm musculature, but also in lower leg muscles, is examined. Depending on the movement that needs to be investigated, head and neck muscles may be added to the registration. Often, *accelerometers* attached to each limb are additionally used to record their movements, providing the so-called *kinematic* information (e.g., position, velocity, acceleration). The accelerometer recording helps to assess the amplitude and frequency of the tremor quantitatively. A distinguishing feature of tremor is that it is a very regular, rhythmical movement around a state of equilibrium. Polymyography with subsequent spectral analysis is particularly suited to establish rhythmicity of movement, as will be shown in the next section.

A polymyogram gives information about the presence or absence of the tremor during execution of different movements and during rest, on the extent and regularity of the tremor, its symmetry (is it present in the same amount on both sides of the body), on the involved muscles, and on its frequency. Furthermore, bursts of EMG activity can be quantified (duration, amplitude, frequency) and studied for synchronous or alternating occurrence between agonist and antagonist muscles. This information is sometimes difficult to obtain from clinical examination alone and can help to distinguish between different types of tremors. The most often encountered tremors with some of their distinguishing characteristics are described in Table 3.1.

Some of these distinguishing characteristics can be observed clinically, such as when the tremor is most prominent or the effect of load (adding a weight to the limb). Other characteristics can clinically only be assessed *qualitatively*, such as frequency. Finally, characteristics such as burst pattern can clinically not be assessed at all. The observation that most tremors have a relatively fixed frequency was already made by Gordon Holmes in 1904, and it is this specific property of tremors that makes them especially suited for spectral analysis.

3.3 Spectral Analysis

3.3.1 Tremor Frequency and Period

Spectral analysis, which is also known as Fourier or harmonic analysis, can be used to derive the frequency of a tremor from an EMG recording of an involved muscle. Frequency is expressed in Hertz (Hz) and here refers to the number of times the basic tremulous movement is repeated per second. If a hand tremor has a frequency of 4 Hz, it means that the entire movement of, e.g., wrist *flexion*, *extension*, and back to flexion again is executed four times per second. Most people cannot intentionally move faster than at a frequency of 4–7 Hz. The tremor period is the duration of one of

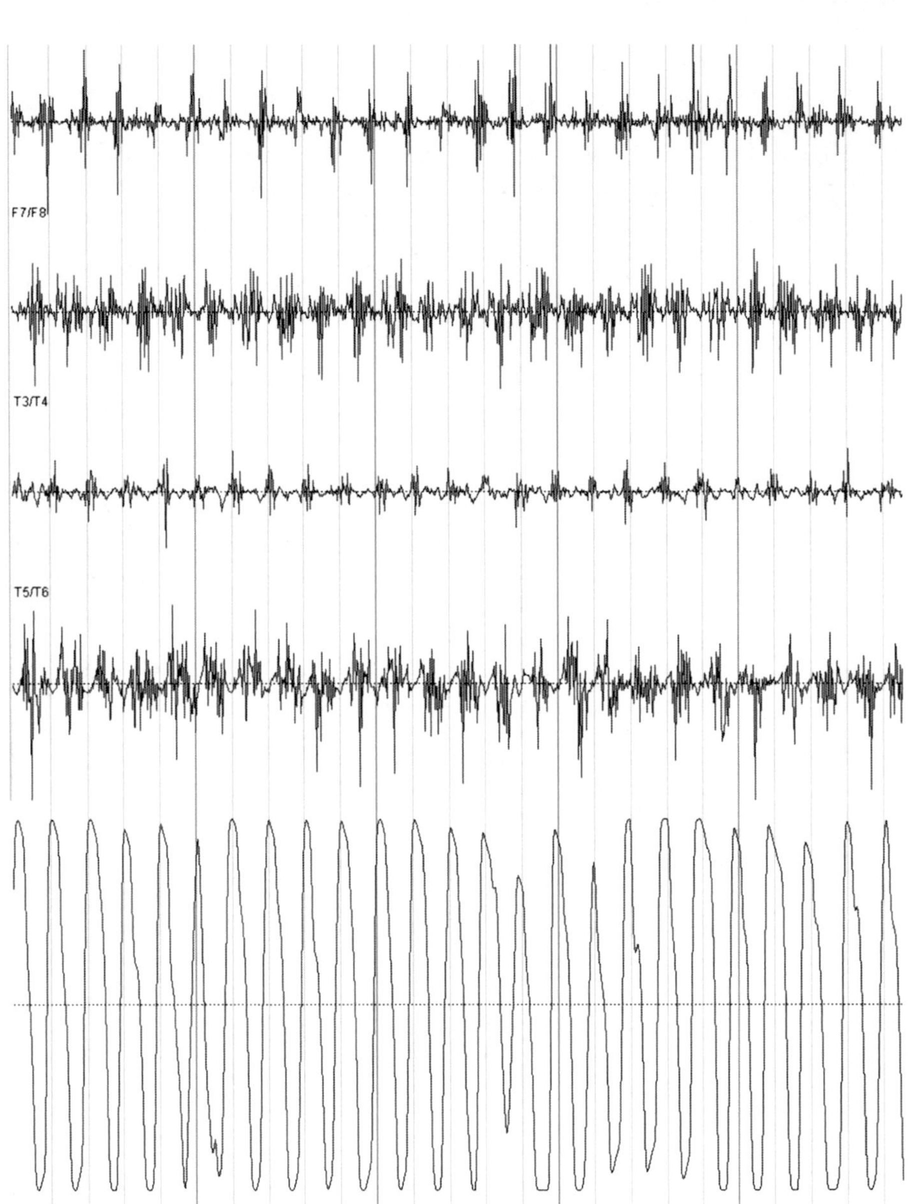

Fig. 3.1 Example of polymyographic recording extended with accelerometry (lowest channel). This recording illustrates the muscle activity during 5 s when a tremor patient extends the arm and fingers. The EMG was recorded from (*top* to *bottom*) the right flexor carpi radialis and extensor digitorum communis in the lower arm and the biceps brachii and triceps in the upper arm. The accelerometer was attached to the dorsal (*upper*) side of the right hand

Table 3.1 Frequently diagnosed tremors and their typical distinguishing characteristics. Frequencies may depend on the (part of the) limb involved and various sources provide different values

Tremor	Most prominent during	Frequency (Hz)	Effect of load	(Ant)agonist burst pattern
Enhanced physiological	Posture	5–12	Lower frequency	Synchronous
Essential	Posture	4–12	None	Variable
Parkinson	Rest	4–7	None	Alternating
Orthostatic	Standing	13–18	None	Synchronous
Psychogenic	Variable	Variable, typically 4–7	Lower amplitude	Synchronous

these tremulous movements. It is expressed in seconds and can be calculated as 1 per frequency. To obtain a rough estimate of tremor frequency from an EMG or accelerometer recording, you can count the number of repeating events per second, i.e., in the EMG the number of bursts and in the accelerometer recording the number of peaks (see Fig. 3.2). Higher frequencies are thus reflected in more closely spaced waves with shorter periods in the EMG or accelerometer recording.

Question 3.1 What are typical frequencies of a beating adult human heart?

3.3.2 *Toward Spectral Analysis of Tremor by Computer: Sampling*

The EMG and accelerometer signals are originally analog, which means that the signal has a value at the exit of the *amplifier* at any moment in time. Such a signal is also called continuous. Unfortunately, computers cannot work with continuous signals and thus, before a computer can be used to store a recording and subsequently use software programs for further analysis, the signals must be transformed to series of distinct numbers. This transformation is called analog-to-digital (A/D) conversion, discretization, or sampling. During this process, the amplitude of the signal is stored at regular time intervals. The number of times per second that an amplitude value is stored is called the sampling frequency, which is, again, expressed in Hz.

As a rule, when using commercially available amplifiers and software, the best sampling frequency for a particular physiological signal is already taken care of. However, some systems allow to adapt the sampling frequency of a signal, in which

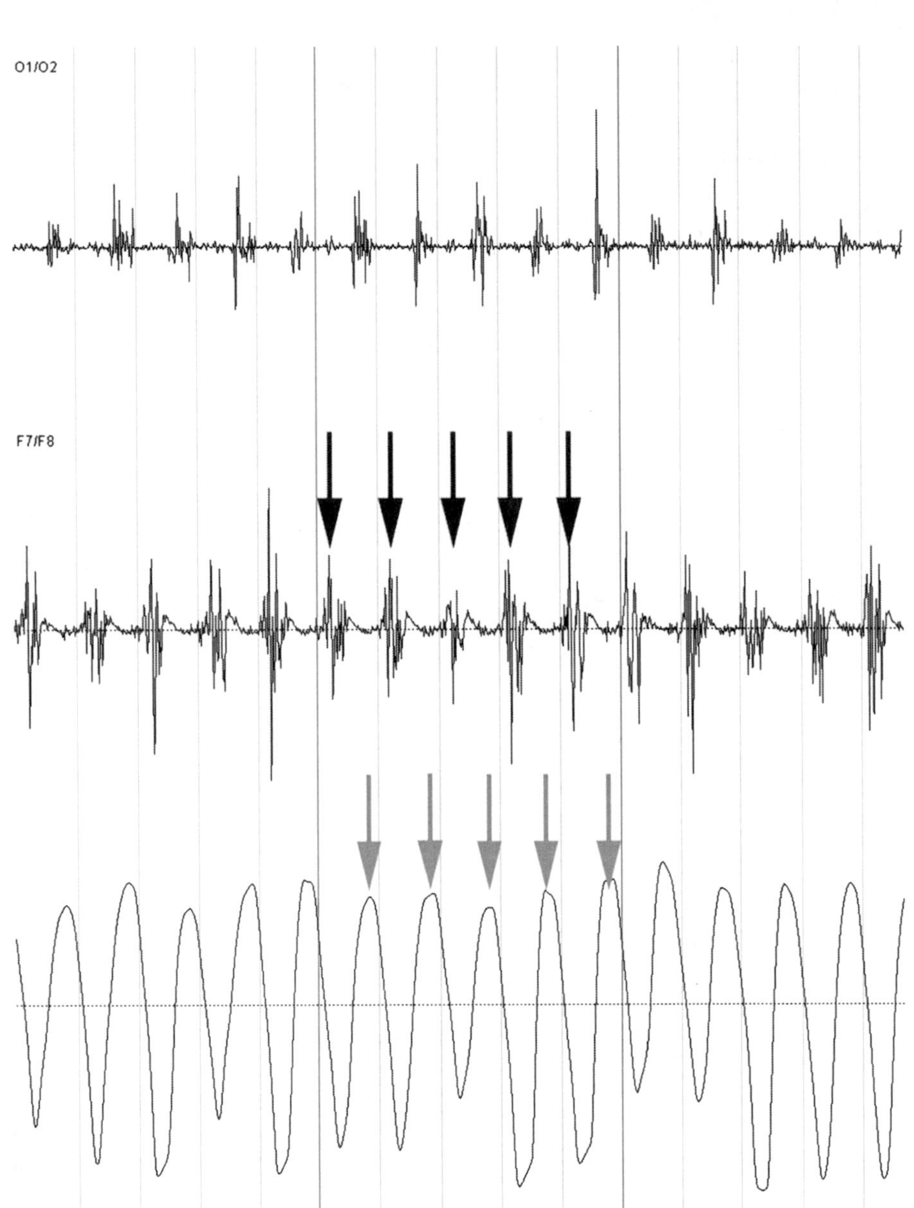

Fig. 3.2 Example of estimating tremor frequency by counting bursts in the EMG recording (*black arrows*) or peaks in the accelerometer recording (*gray arrows*). The time between two vertical lines is 1 s. Here, the tremor frequency can thus be estimated to be approximately 5 Hz

case care must be taken to choose an adequate value, since both undersampling (sampling at a too low frequency) and oversampling (sampling at a too high frequency) are possible. When sampling a rapidly changing signal at, e.g., only 1 Hz (i.e., once per second), it is not hard to imagine that information in the signal is

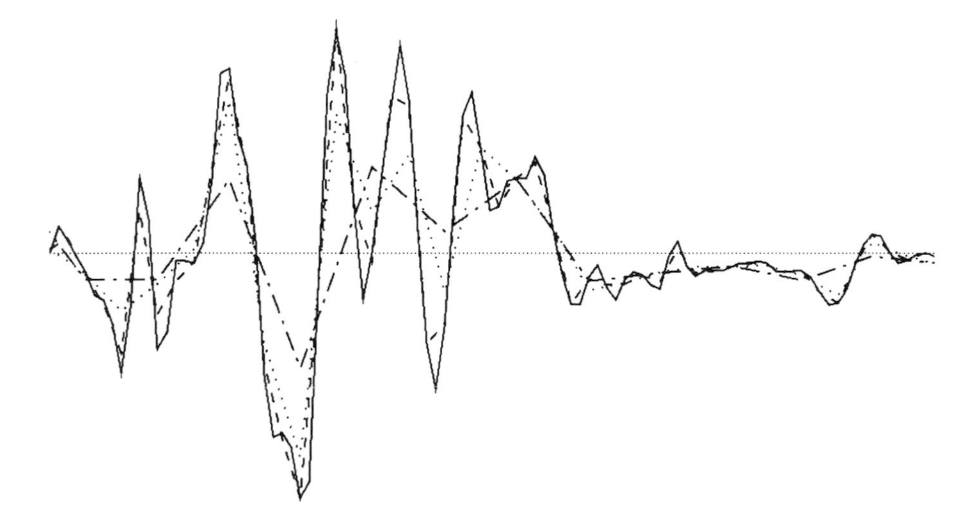

Fig. 3.3 Examples of the effect of sampling frequency on a digitized EMG signal: 100 ms of EMG is displayed, with one EMG burst. *Solid line*: original sampling frequency of 1,000 Hz, *dashed*: 500 Hz, *dotted*: 250 Hz, *dash-dotted*: 125 Hz. Although theoretically a sampling frequency of 500 Hz should suffice for the EMG which contains frequencies up to 250 Hz, it can be clearly observed that EMG quality already deteriorates

lost. Depending on the characteristics of the signal, sampling always causes more or less loss of information. Therefore, to adequately represent an analog signal digitally, it is important to use a minimal sampling frequency. From a theoretical point of view, sampling twice as fast as the highest frequency in the original signal suffices. This highest frequency that is still adequately represented is the so-called Nyquist frequency, which is half of the sampling frequency. Thus, if an EMG signal is sampled at 2,000 Hz, the highest frequency that is still adequately represented is 1,000 Hz. Or, the other way around, if you want to adequately represent signals with frequencies up to 50 Hz, you have to sample the signal at least at 100 Hz. Here, adequately represented means one sampling point at each peak and at each trough of a wave, i.e., two sampling points per period. When you think about it, these two points are what you really need to distinguish two waves of different frequencies from each other. Yet, two sampling points per period are insufficient to represent the characteristics of the signal in time. The examples in Fig. 3.3 will make this more clear.

Furthermore, undersampling of a signal results in aliasing, which is also referred to as back-folding. As an effect of aliasing, signals of different frequencies become indistinguishable (aliases) from one another. For example, when sampling a signal with a frequency of 75 Hz at 50 Hz, a digital signal of 25 Hz results (see Fig. 3.4).

This aliasing effect may have unexpected consequences: suppose that you are sampling an EEG signal at 70 Hz, because frequencies higher than 35 Hz are not expected. But, somehow, maybe because of noise, there is actually a 100 Hz component in the original analog signal. This 100 Hz component will then result in a 30 Hz component in the digitized signal, which is artificial. Since it is

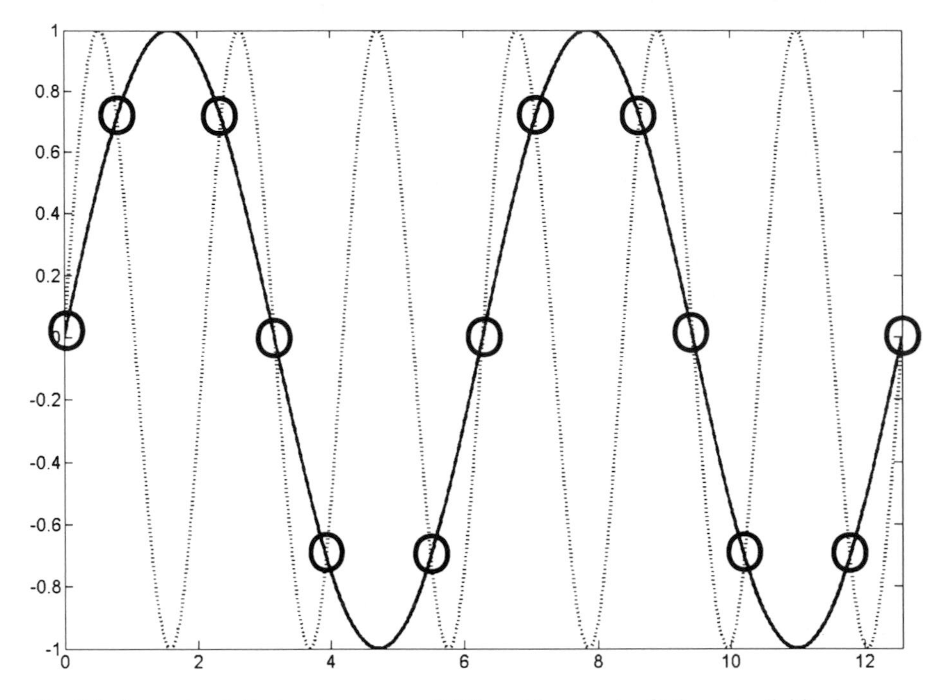

Fig. 3.4 Example of aliasing: two waves of different frequencies fit the same digitized samples (*circles*) and can no longer be distinguished. The higher frequency will be aliased on the lower frequency

practically impossible to know all frequencies in the analog signal, anti-aliasing filters are typically applied before sampling. Anti-aliasing filters cut off all frequencies higher than the Nyquist-frequency in the analog signal before sampling, thereby providing an effective solution to the aliasing problem without the need to sample at very high frequencies.

On the other hand, why not simply sample at the maximum frequency the amplifier allows? The most important reason not to do this is that sampling at a higher frequency leads to larger data files: a 16-channel EMG recording of an hour that is sampled at 2,000 Hz will yield a file of approximately ($16 \times 3{,}600 \times 2{,}000=$)115 MB. To balance between minimal and optimal sampling requirements, the typical sampling frequency is chosen to be 5–10 times as high as the highest frequency in the analog signal. For surface EMG, which may contain frequencies up to 250 Hz, the sampling frequency thus needs to be quite high, preferentially between 1,500 and 3,000 Hz. For scalp EEG, which contains frequencies up to 35–50 Hz, a sampling frequency of 150–250 Hz suffices in practice. Note that the sampling frequency automatically determines the accuracy of subsequent measurements in time; when sampling a signal at 1,000 Hz, events can never be measured at a higher accuracy than 1 ms.

Question 3.2 What frequency should theoretically be used to sample an ECG signal, which typically contains frequencies up to 40 Hz? And what frequency would you use in practice, taking the shape of an ECG signal into account?

3.3.3 Obtaining Frequency Content from a Digitized Signal: Fourier Transform

To gain some understanding of the methods that are used to calculate the frequency content of a digitized signal, it is important to realize that signals can be represented in different ways. The most obvious way to think about a signal is according to the way in which it is recorded, as a variable amplitude over time. This representation can be referred to as a "time domain" representation. However, although less intuitive, a signal can just as well be characterized by its frequency content. This representation can be referred to as a "frequency domain" representation. The time and frequency domain representations do not change the signal, but only describe it in a different way. This has some similarity to using different properties to describe the same object: a football can be described in terms of its shape (round), but also in terms of its movement (rolling).

A common way to describe a signal in the frequency domain is a frequency spectrum. This can be derived from a signal in the time domain using the so-called spectral analysis (also named frequency analysis or Fourier analysis). Spectral analysis employs the decomposition of a signal into basic wave functions: sines and cosines of different frequencies. As a reminder, a brief update on these goniometric functions and goniometric concepts is given in Box 3.1.

Box 3.1 Update on Goniometry

Suppose a point P moves at a constant speed over a circle of radius 1 (see Fig. 3.5). The projection of the position of P on the vertical (y-)axis yields the sine of the accompanying angle α with the horizontal axis. Similarly, the projection of the position of P on the horizontal (x-)axis yields the cosine of α. The angle α can be expressed in degrees or radians: once around the circle (the period of the circular movement) equals 360° or 2π radians.

When the movement continues, the projections repeat themselves. This is why sines and cosines are also called periodic functions. The number of rotations per second of the point P is the frequency of the accompanying movement.

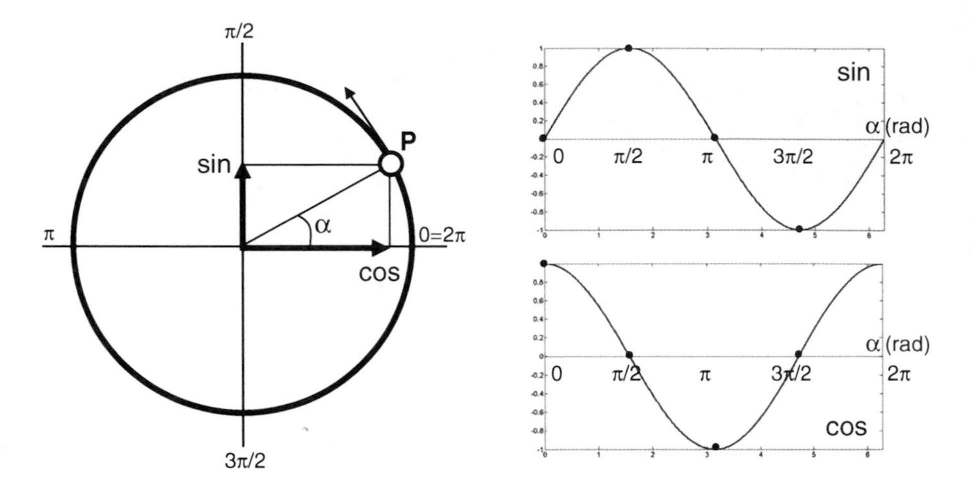

Fig. 3.5 Graphic representation of sines (sin) and cosines (cos) as projections of circular movement

Around 1800 the French mathematician Fourier showed that any periodic signal can be accurately expressed as an infinite sum of sines and cosines of different frequencies. This expression is called a Fourier series. The frequencies in this series are not random, but are rather determined by the period T of the signal under investigation. The lowest frequency in the series is $1/T$ and higher frequencies are simply multiples of this lowest frequency: i.e., $2/T$, $3/T$, $4/T$, etc. The lowest frequency is called the base, ground, or fundamental frequency and the multiples are called harmonic frequencies. This is why this type of analysis is sometimes also referred to as harmonic analysis. Thus, when a signal with a period of 10 s (i.e., the signal repeats itself every 10 s) is expressed as a Fourier series, the base frequency is $1/10 = 0.1$ Hz and the harmonic frequencies are 0.2, 0.3, 0.4 Hz, etc. This does however not imply that each of these frequencies is actually present in the series (see Fig. 3.6). Furthermore, if the signal is *symmetric* around time 0 only cosines are needed in the Fourier series, whereas when the signal is *antisymmetric* around zero (as in Fig. 3.6) only sines are needed to represent the signal. The more terms are added to the series, the more accurate the approximation of the original signal will be. In practice, only a few (up to 10) terms are often enough for a reasonable approximation. Calculating the coefficients (weights) for a Fourier series requires quite some mathematics but can be done (see Box 3.2). Unfortunately, physiological signals such as EMG, EEG, or ECG are never periodic, making it impossible to determine a Fourier series. These signals may be almost periodic over short periods of time, but physiological variation over time is inevitable. Therefore, Fourier series do not provide a practical approach to get information on the frequency content of physiological signals.

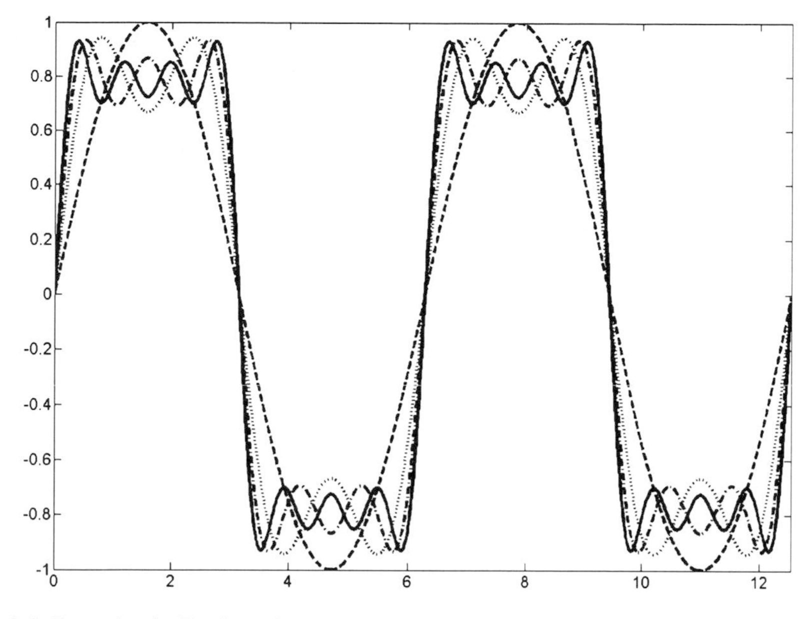

Fig. 3.6 Example of a Fourier series approximation for a periodic block wave signal (period 2π). The Fourier series approximation of the signal is displayed when more and more terms are added. *Dashed line*: first approximation [sin(x)], *dotted line*: second approximation [sin(x) + 1/3 sin($3x$)], *dash-dotted line*: third approximation [sin(x) + 1/3 sin($3x$) + 1/5 sin($5x$)], *solid line*: fourth approximation [sin(x) + 1/3 sin($3x$) + 1/5 sin($5x$) + 1/7 sin($7x$)]. Note that not all harmonic frequencies are present and only sines are needed to represent the original antisymmetric signal

Box 3.2 Summary of the Mathematics of Fourier series, Fourier transform and DFT

This summary is not intended to explain the details of spectral analysis (many excellent books exist on this topic), but only as a reminder of the most important formulas.

Fourier series

Any periodic function $x(t)$ of period T can be expressed as a Fourier series as follows:

$$x(t) = \sum_n a_n \sin\left(\frac{2\pi nt}{T}\right) + b_n \cos\left(\frac{2\pi nt}{T}\right), \quad n = 0, 1, 2, \ldots \qquad (1)$$

Here, the frequency of each sine and cosine is n/T. Given $x(t)$, the coefficients a_n and b_n can be calculated by integrations over one period of $x(t)$:

$$a_n = \frac{2}{T}\int_0^T x(t)\sin\left(\frac{2\pi nt}{T}\right)dt, \quad n{>}0,\ a_0 = 0, \qquad (2)$$

(continued)

Box 3.2 (continued)

$$b_n = \frac{2}{T}\int_0^T x(t)\cos\left(\frac{2\pi nt}{T}\right)dt, \quad n>0, \quad b_0 = \frac{1}{T}\int_0^T x(t)dt. \tag{3}$$

Other ways to calculate a Fourier series representation of a signal, e.g., using only sines or only cosines, exist but are not shown here.

Fourier transform

The Fourier transform is an extension of the Fourier series, but for nonperiodic signals. A Fourier transform requires a functional description of the signal $x(t)$ and also results in a functional (complex-valued) expression:

$$X(f) = \int_{-\infty}^{\infty} x(t)e^{-2\pi ift}dt. \tag{4}$$

Here, the sum in (1) has become an integral and the sines and cosines have become infinitely small and are found back as $e^{-2\pi ift} = \cos 2\pi ift - i\sin 2\pi ift$. Here $i = \sqrt{-1}$. In this case the spectrum usually displays the magnitude $|X(f)|$.

Discrete Fourier Transform (DFT)

DFTs can be applied to discretized physiological data that is neither periodic nor expressed as a function. The DFT was especially designed to analyze the frequency content of discretized signals of finite length. N samples of the signal x in the time domain are transformed to N samples of the DFT X in the frequency domain as follows:

$$X[n] = \sum_{k=0}^{N-1} x[k]e^{-2\pi ikn/N}, \tag{5}$$

where both k (in the time domain) and n (in the frequency domain) run from 0 to $N-1$. The samples $x[k]$ correspond to the values of $x(t)$ at $t = k/f_{\text{sampling}}$, where f_{sampling} is the sampling frequency. In that case, $X[n]$ corresponds to the values of $X(f)$ at $f = f_{\text{sampling}}\, n/N$.

The Fourier transform can be used to assess the frequency content of *non-periodic* signals and is, in that sense, a generalization of the Fourier series. Still, this transform is not suited for frequency analysis of sampled signals, because the Fourier transform needs a mathematical expression (function) for the signal which is not available for sampled, digitized signals. Fortunately, there is a method to apply Fourier transforms to digitized signals: the discrete Fourier transform (DFT). To actually perform the DFT using a computer, the transform has been implemented in several software programs as a fast Fourier transform (FFT), making spectral analysis quite easy in practice. The FFT is a fast implementation of the DFT, which is commonly used. To be fast, most FFT implementations assume that the signal length N is a power of 2. If that is the case the FFT algorithm will return exactly the same value(s) as a DFT. A very brief summary of the mathematics of Fourier transforms is given in Box 3.2 for the interested reader.

To be able to interpret the results of a DFT/FFT, it is necessary to understand some details of the transform. In a DFT of a signal of T seconds and consisting of N samples, for all N sines and cosines with frequencies 0, $1/T$, $2/T$, $3/T$, $4/T$, ..., $N - 1/T$ the extent to which they are present in the original signal is determined by a mathematical expression. Evaluation of this expression results in two coefficients or weights for each frequency: one for the sine and the other for the cosine. Note that these coefficients can be zero, implying that the related frequency is not present in the signal. The result of this calculation is often displayed as a *spectrum* (see Fig. 3.7 for an example), with frequency on the horizontal axis. Some form of the coefficients for each frequency is set out vertically; for a power spectrum the sum of the squared coefficients is displayed, for an amplitude spectrum the square root of the values in the power spectrum is presented.

Often, the values in a spectrum are displayed on a logarithmic scale. When a $^{10}\log$ is used, 1,000 is displayed as 3, 100 as 2, 10 as 1, 1 as 0, 0.1 as -1, etc. This allows better comparison of very large and very small values with each other in one graph. On the horizontal axis, values are typically displayed until the Nyquist frequency: this is the $N/2$th frequency value. As an example, suppose that you have 512 samples of a signal that is sampled at 256 Hz, i.e., 2 s of data. In this case the base frequency is $1/2 = 0.5$ Hz and the Nyquist frequency is $256/2 = 128$ Hz. The frequencies on the frequency axis will then be 0, 0.5, 1.0, 1.5, ..., 127.5 Hz. In this case the spectral resolution (the distance between neighboring values on the frequency axis) is 0.5 Hz.

Question 3.3 What is the expected maximum frequency in the spectrum that is calculated for a piece of EMG of 5 s which is sampled at 4,000 Hz? And what is the spectral resolution?

3.3.4 Tips and Tricks when Interpreting Spectra

To understand some of the intricacies of spectra, it is important to keep in mind that the spectral resolution of an FFT is determined by the length of the signal in seconds. The maximum frequency in the spectrum is the Nyquist frequency, which is determined by the sampling frequency. This implies that if a short

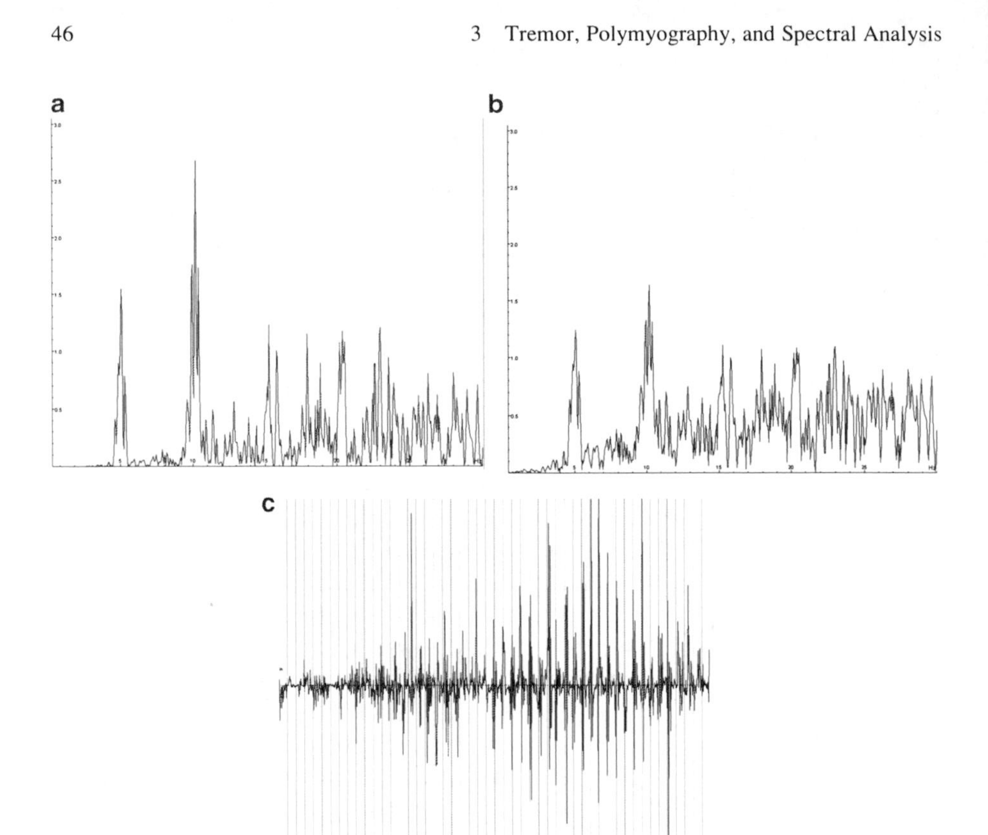

Fig. 3.7 Examples of both (**a**) the power spectrum (in μV^2) and (**b**) amplitude spectrum (in μV) of (**c**) a piece of EMG signal, calculated by FFT. Notice that the peaks are relatively more clear in a power spectrum, which is therefore used more often in practice. However, to evaluate the relative contribution of different frequencies to the signal, an amplitude spectrum is more suitable

piece of signal is selected for an FFT, the result is a rather coarse spectrum with broad peaks and troughs (Fig. 3.8b). Thus, a longer piece of signal results in a higher frequency resolution (finer detail in the representation of frequencies; Fig. 3.8a).

3.3.4.1 Avoiding Spectral Leakage

When using an FFT, it is important to realize that the algorithm assumes that the piece of signal that is transformed is representative for the entire signal. More specifically, it assumes that the signal is periodic over its length. As an example, when considering a 10-s piece of EMG, the FFT actually treats the signal as consisting of infinite repetitions of the 10-s piece that are stuck together. When not treated properly, this assumption leads to so-called "spectral leakage,"

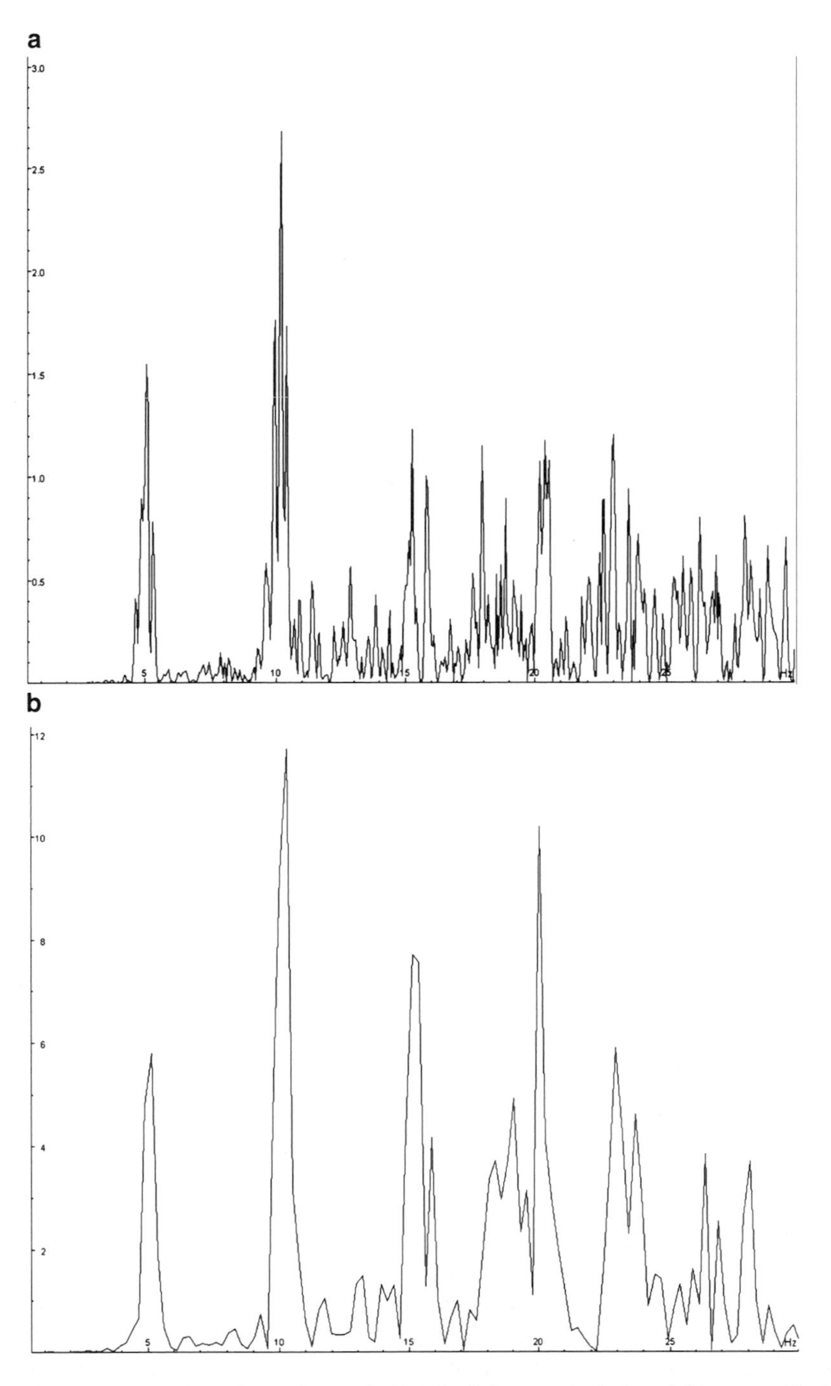

Fig. 3.8 (**a**) Fine (obtained from 10 s EMG; high frequency resolution) and (**b**) coarse (obtained from 2 s EMG; low frequency resolution) spectrum

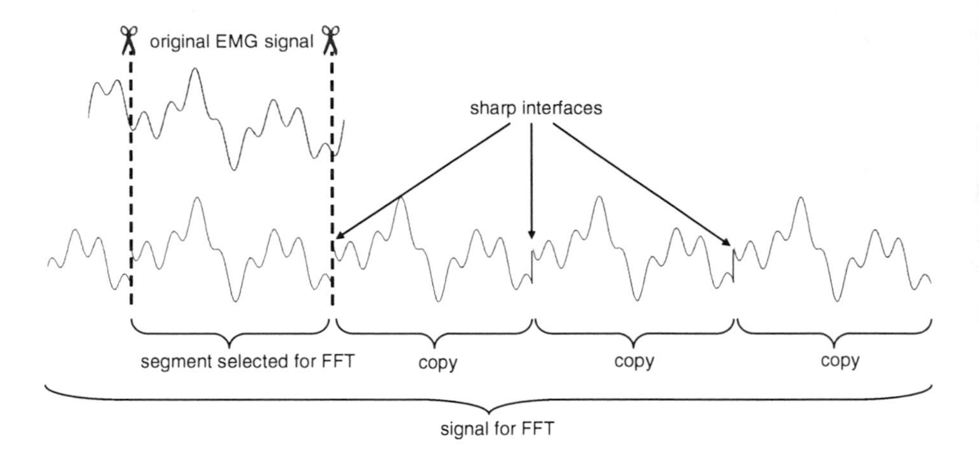

Fig. 3.9 Illustration of the cause of spectral leakage for an artificial EMG signal

Fig. 3.10 The effect of a window on the signal for FFT in Fig. 3.9

artificial power in the spectrum at all but especially at high frequencies. The spectral leakage results from the sharp interfaces that arise in the signal when pieces of signal are stuck together, since fast changes in signal amplitude contain many, including high frequencies (see Fig. 3.9).

There is a practical solution to the problem of spectral leakage: the so-called Hann (or Hanning) or Hamming windows can be applied to the signal before FFT. These windows set the signal to zero at the beginning and end of the selection and provide a smooth transition over a short period of time (e.g., 10% of the length of the signal selection) to the original value of the signal. Application of one of these filters has the effect that the sharp interfaces in Fig. 3.9 are replaced by smooth interfaces (Fig. 3.10). As a result spectral leakage disappears (Fig. 3.11). Depending on the software that is used to calculate the FFT, one of these windows is often automatically applied or has to be chosen optionally. Besides Hann and Hamming windows, many other types of windows exist (e.g., cosine or Gaussian windows). Application to the data has a similar effect for all of them.

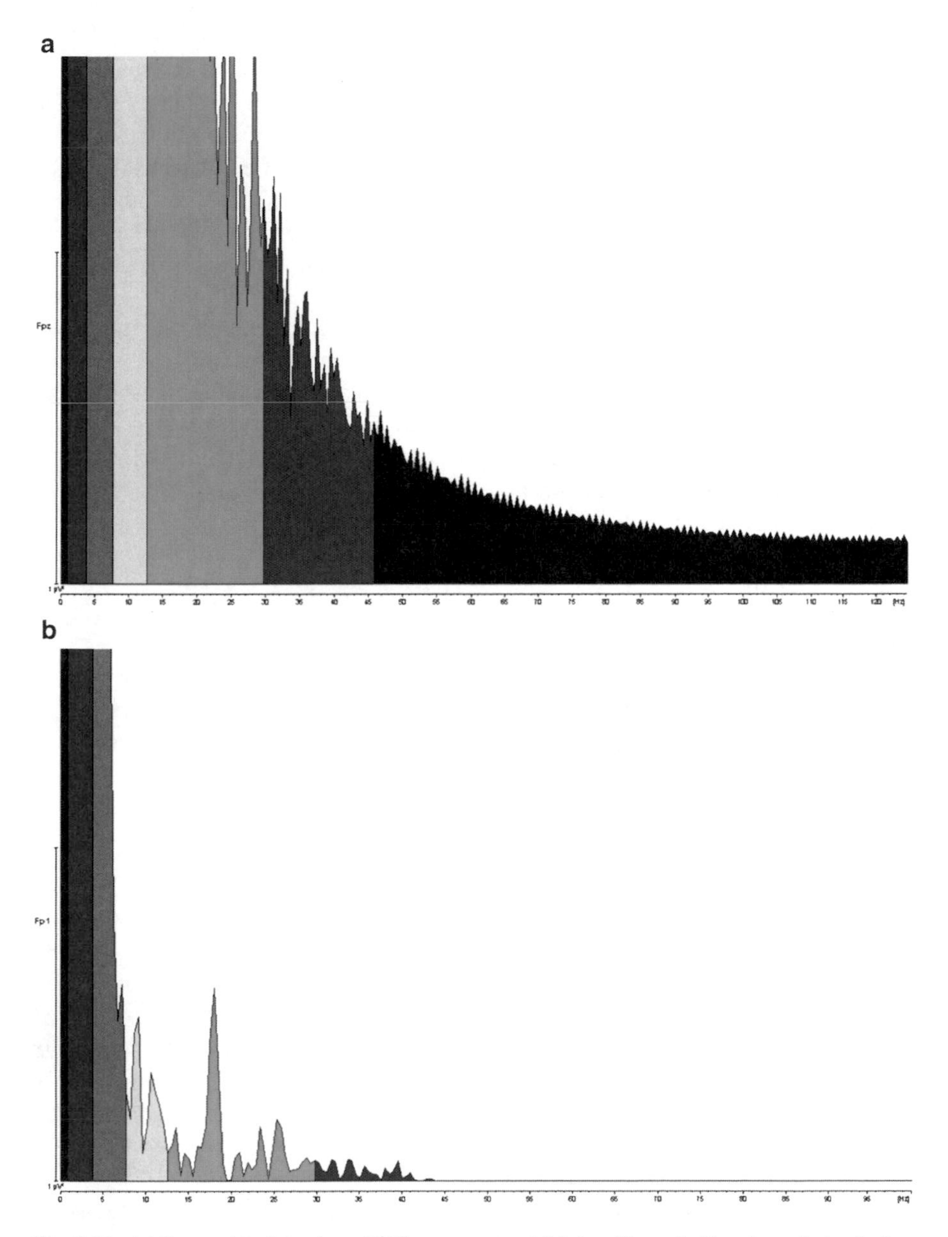

Fig. 3.11 (**a**) Spectral leakage in an EEG spectrum and (**b**) the effect of a Hanning window before FFT on the same signal (axes have the same scale). Note that this EEG was sampled at 250 Hz and subsequently a low-pass filter of 30 Hz (keeping only frequencies below 30 Hz) was applied. This implies that there is no (or hardly any) actual EEG signal above 30 Hz and the additional power in the spectrum in (**a**) must really be due to spectral leakage. These two spectra display frequencies up to 100 Hz (the Nyquist frequency is 125 Hz)

3.3.4.2 Avoiding Zero-Padding

The most common implementation of the FFT assumes that the signal length is a power of 2 (128, 256, 512, 1,024, etc.). Calculating the FFT of a discretized signal that has a different length is possible, but does have some effects that may not be expected. Usually, the FFT algorithm adds zeroes to the signal until the new length is a power of 2 and then does the calculation. Depending on how close the initial length is to a power of 2, the effect of this so-called "zero-padding" may be quite large. When the original signal consists of 513 data points, 511 zeroes are added to reach the next power of two being 1,024, resulting in a completely different signal. Furthermore, zero-padding has an effect on the frequency resolution. Suppose the FFT of a signal of 6 s sampled at 250 Hz needs to be calculated. This signal consists of $6 \times 250 = 1{,}500$ data points. This is not a power of 2. The nearest power of two is 2,048. But 2,048 data points actually represent $2{,}048/250 = 8.192$ s of data. Thus, instead of a frequency resolution of $1/6 = 0.166$ Hz, a resolution of $1/8.192 = 0.122$ Hz results. This is an artificially high frequency resolution, resulting only because of zero-padding. It will come as no surprise that zero-padding is easily avoided by choosing the signal length as a power of 2.

Question 3.4 If a signal is sampled at 100 Hz, how long should your selection of the signal be to avoid zero-padding?

3.3.4.3 Improving Spectral Reliability

The spectrum that results from the FFT of just one selection of a few seconds is not very reliable for a nonperiodic signal. Undoubtedly, the spectrum that results from the FFT of another part of the signal will be different (Fig. 3.12). One way to improve upon this unreliability is to use averaging. Using this approach the phrase "spectral estimation" is now more appropriate than "spectral calculation." Averaging can be done by averaging within the spectrum (spectral smoothing) or by averaging over spectra (spectral averaging). Spectral smoothing makes the spectrum more reliable by calculating a weighted average of the spectral value itself and its neighbors in the spectrum [e.g., ½(the spectral value) + ¼(left and right neighboring spectral values)]. By assuring that the sum of the weights is one, no spectral power is lost.

Because physiological signals are usually not stationary (i.e., do not have a stable amplitude and frequency distribution), it also makes sense to determine a spectrum as the average of spectra calculated from disjoint (Bartlett's method) or overlapping (Welch's method, also includes windowing) pieces of signal. Typically, these pieces will have a length of 1–10 s. Often, spectral averaging is applied first, after which spectral smoothing is additionally used.

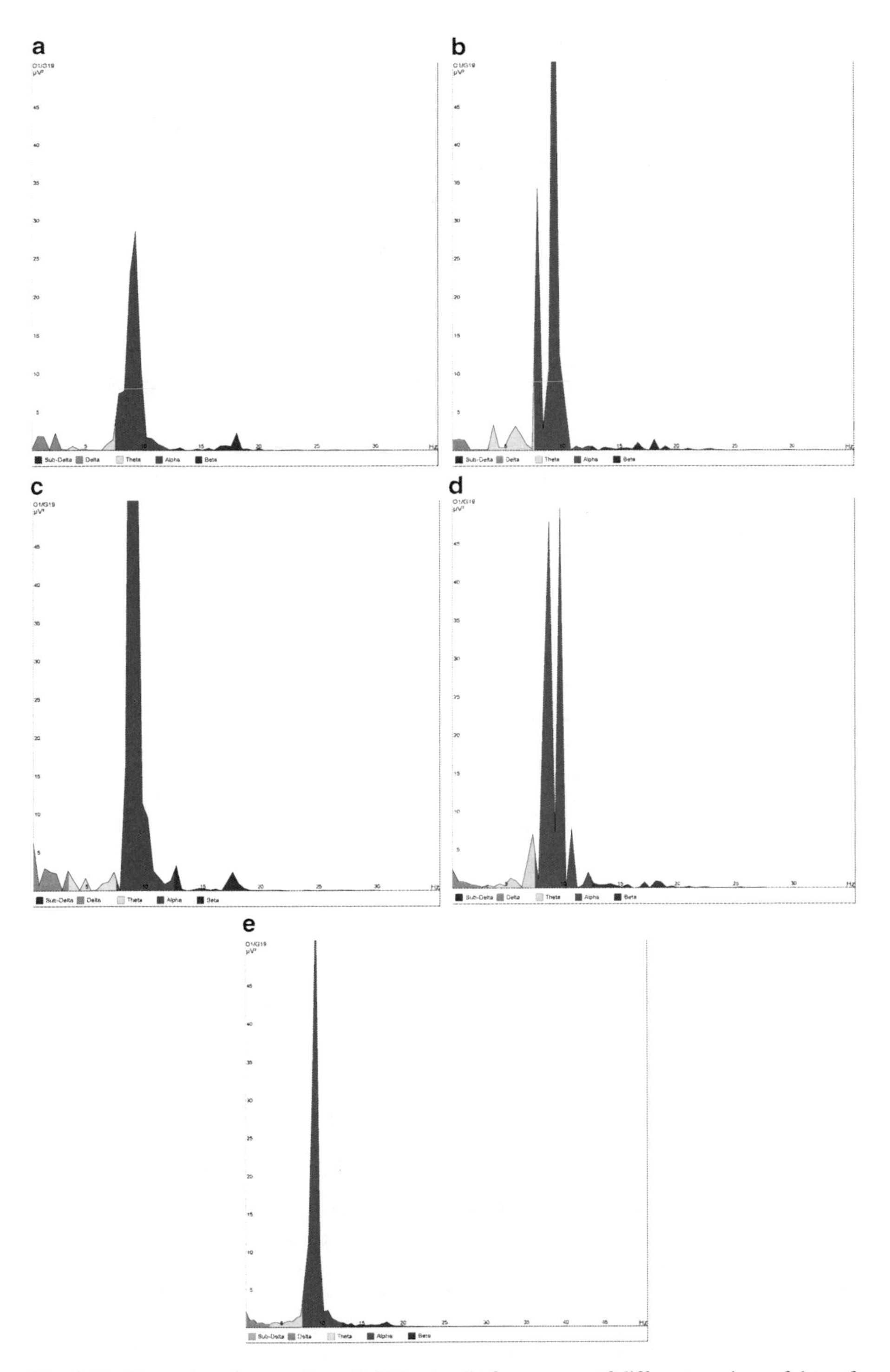

Fig. 3.12 Illustration of spectral unreliability (**a–d**), four spectra of different portions of data of the same length from one EEG recording. (**e**) Average of 24 spectra of the same length from the same dataset

Note that an FFT spectrum does not provide information on the temporal occurrence of particular frequencies. A peak in the spectrum can result both from persistent activity as well as from a short-lasting burst of activity at that particular frequency. Other analysis techniques such as *wavelet analysis* provide a means to deal with this difficulty, but will not be explained here.

3.4 Tremor Analysis in Individual Patients

Now that the generation and interpretation of spectra has been explained; it can be shown how these spectra can help to differentiate between different types of tremor and how spectral properties are related to clinical neurological features in the patients introduced in Sect. 3.1. Each tremor investigation proceeds according to the same protocol, which may vary slightly between hospitals. The EMG and accelerometer signals are evaluated during different actions, to evaluate whether the tremor is mostly present during rest, posture, or movement (see also Table 3.1) and to determine the variability of the tremor frequency. These actions can be, but are not limited to, rest, hand extension, hand relaxation, finger extension, forward and sideward arm extension, moving of the index finger to a position just in front of the nose (top-nose test), pointing the right and left index finger toward each other at a position in front of the nose (top-top test), contralateral distraction (e.g., by using *diadochokinesis* of the left hand while extending the right arm and vice versa), load application (a weight is attached to the arm), standing, lying down in a *supine* position, extending the hand while lying down, and finally, during entrainment (e.g., by extending one hand while tapping with the other). Other actions may be added for untypical tremors (e.g., a tremor that only occurs when executing one specific daily life activity, such as eating).

Patient 1

In this patient who suffered from a progressive tremor for 20 years, the EMG was recorded from upper and lower arm muscles, bilaterally. Accelerometers were attached to both hands. All tests were executed with the right arm and hand. No tremor was observed in the EMG during rest, nor during hand extension, although the lower arm muscles were active during this movement. During full relaxation of the hand, a tremor developed, initially at a frequency around 5.0 Hz which quickly slowed down to 4.5 Hz (Fig. 3.13a, b). During finger extension, a coarse tremor developed accompanied by tremor bursts in upper and lower arm muscles. EMG bursts alternated in agonist and antagonist muscles and the tremor had a frequency of 5.0 Hz, both according to the accelerometer as well as to the EMG signal. In the accelerometer spectrum, a harmonic peak was visible at twice the base frequency. Hereafter, the tremor continued unchanged in rest. During forward arm extension, an initial tremor quickly disappeared to return again at a frequency of 5.0 Hz after the fingers were additionally extended. Hereafter, the tremor first continued in rest, but then suddenly disappeared. During sideward arm extension, during the top-nose test (Fig. 3.13c, d) and during the top–top test, a tremor was visible at frequencies

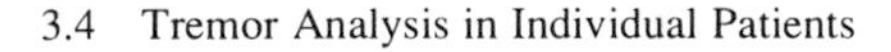

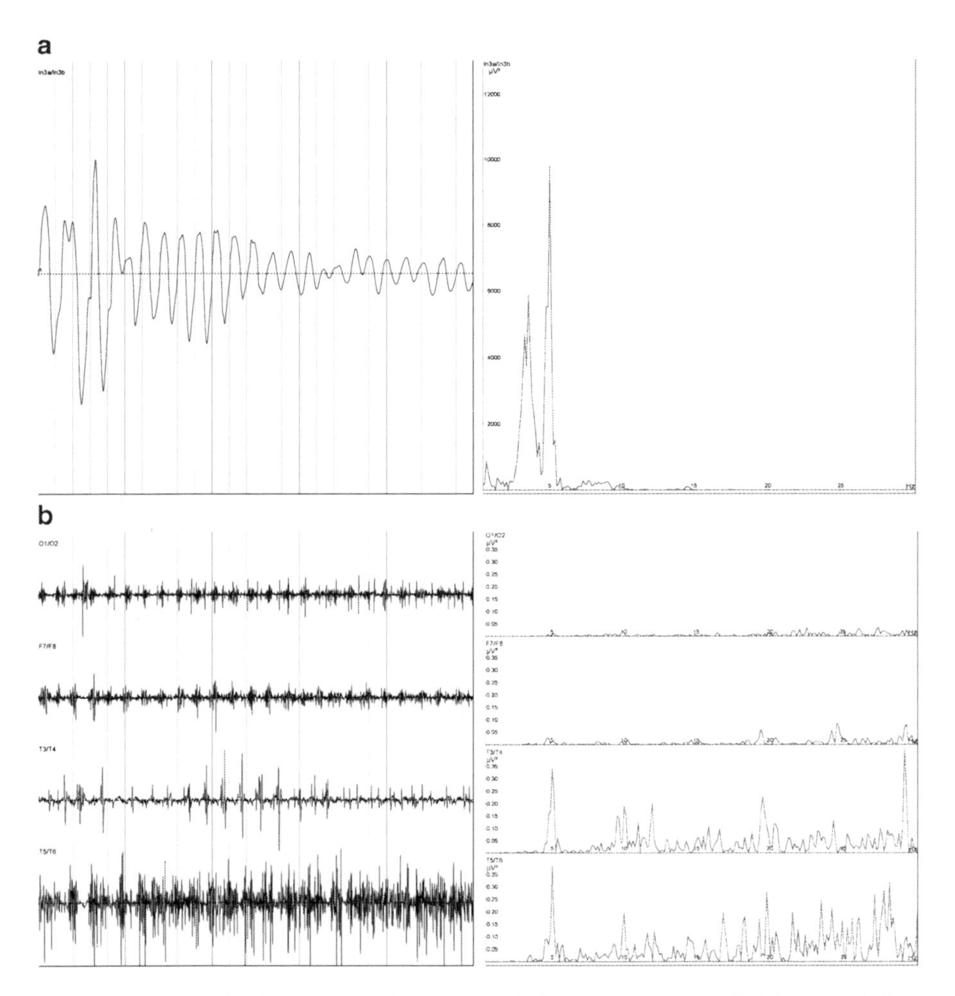

Fig. 3.13 Results of polymyography in patient 1. *Left*: Accelerometer or EMG traces (4–5 s), *right*: FFT results of selected data. EMG from (*top* to *bottom*) flexor carpi radialis, extensor digitorum communis, biceps brachii, and triceps. The spectrum is displayed from 0 to 30 Hz. Vertical axes have been adapted individually for each figure. (**a**) Accelerometer results during hand relaxation: tremor in right hand at 5.0 Hz. (**b**) EMG results (same time interval as **a**) during hand relaxation: tremor in the right hand at 5.0 Hz. Tremor bursts are visible in the EMG. (**c**) Accelerometer results during top-nose test: coarse tremor in right hand at 4.9 Hz. Slight tremor at the same frequency in the left hand. Note the triple harmonic peak in the spectrum at 14.7 Hz. (**d**) EMG results (same time interval as **c**) during top-nose test: tremor in the right hand at 4.9 Hz. Harmonics are visible in the spectra at two, three, and multiple times this base frequency. Clear EMG bursts visible. (**e**) Accelerometer results during diadochokinesis with the left hand (*top signal*) at 1.2 Hz. A tremor develops in the right hand (*bottom signal*) at approximately the double frequency (2.6 Hz). (**f**) EMG results (same time interval as **e**) during diadochokinesis with the left hand at 1.2 Hz. The tremor that is observed in the accelerometer recording in **e** is not visible in the right hand EMG (spectrum) because of its irregularity

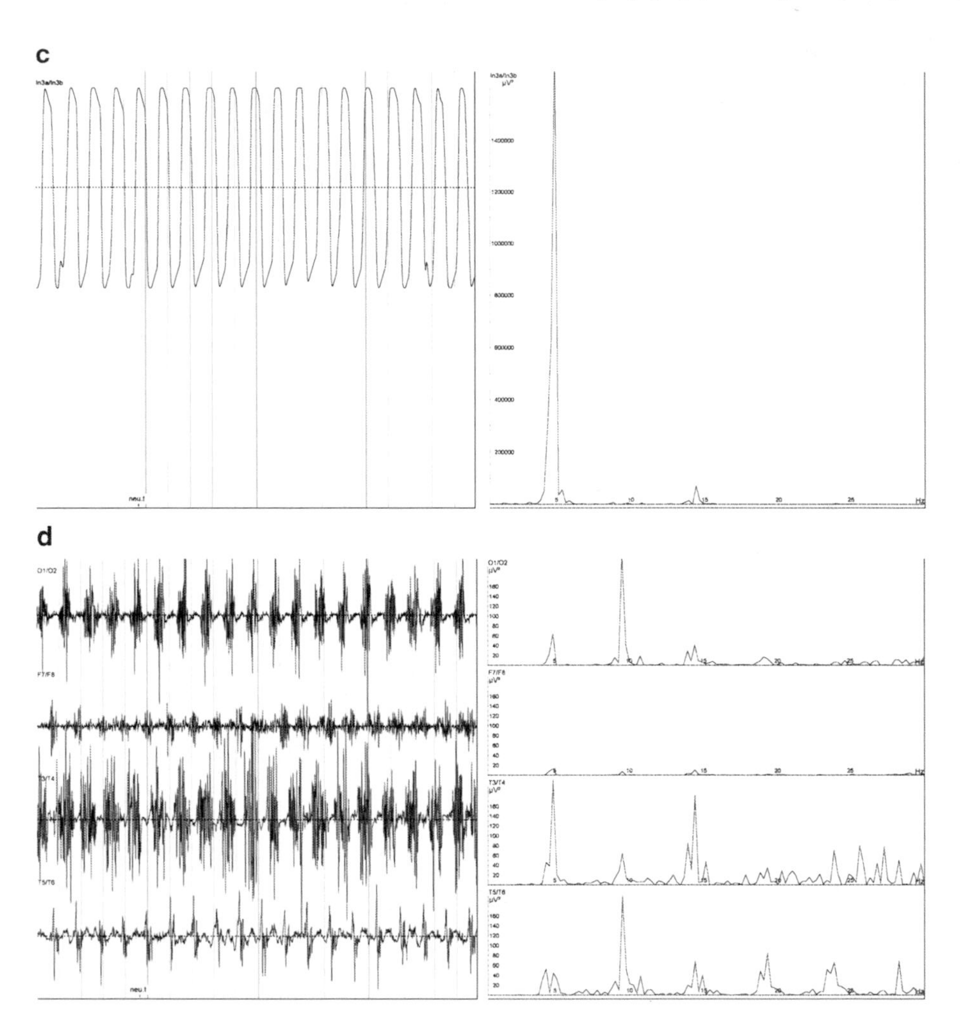

Fig. 3.13 (continued)

varying between 4.6 and 5.0 Hz. When performing diadochokinesis with the left hand at 1.2 Hz, a tremor at the double frequency developed in the accelerometer recording, which was not clear in the EMG however (Fig. 3.13e, f). When a weight was attached to the right arm, the tremor frequency slowed down to 3.8 Hz during finger extension. When the arm was then extended forward, the tremor frequency increased to 5.1 Hz again. At the end of this position, the tremor frequency suddenly slowed down to 3.9 Hz. When the fingers were then extended, the tremor changed in appearance with many more *pronation* and *supination* movements than before.

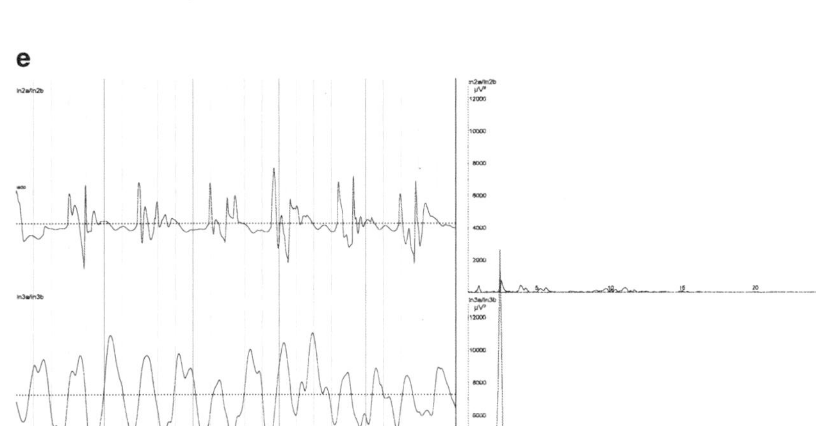

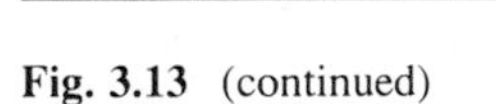

Fig. 3.13 (continued)

When standing, the tremor completely disappeared, as well as when lying down. When the hand was extended while lying down, a tremor appeared at a frequency of 5.6 Hz. During entrainment (tapping with the left hand), the tremor in the right hand completely disappeared. No tremor appeared at a different frequency. This effect of entrainment was also found in the other hand. When switching, both hands sometimes shortly moved in unison. This entrainment was reproducible.

Summary:

- The tremor in this patient was found during action and posture, as well as during rest.
- The tremor frequency varied between 3.8 and 5.6 Hz.

- The tremor frequency slowed down to 3.8 Hz during loading.
- The tremor disappeared during contralateral distraction and during entrainment.

Altogether, when looking at the typical distinguishing characteristics of the different types of tremor in Table 3.1, the results of the polymyographic recording and subsequent spectral analysis are most consistent with a psychogenic tremor. The tremor was present during different actions, varied considerably in frequency, slowed down in frequency when the arm was loaded (which could be consistent with an enhanced physiological tremor, but the other properties of the tremor do not fit), and the tremor completely disappeared when the patient was distracted or during entrainment (which is highly suspect for a psychogenic tremor). In this particular case, the findings of the polymyography thus strongly contradicted the clinical diagnosis of essential tremor and this patient did not have an intracranial operation. Instead, other ways of treating the patient were explored.

Patient 2

In this patient who fell from her horse 12 years ago, the EMG was recorded from upper and lower arm muscles, bilaterally. Accelerometers were attached to both hands. All tests were executed with the right arm and hand. During rest, standing, and lying down, no tremor was observed. When the fingers were extended (Fig. 3.14a, b), a very slight and irregular tremor at 8–9 Hz could be observed, but with no clear peaks in the spectrum. Agonist and antagonist muscles did not show alternating activity. A slight and irregular tremor developed at frequencies between 9 and 10 Hz during the top-nose (Fig. 3.14c, d) and top–top tests, as well. Again, no clear peaks were visible in the spectra. When a weight was attached to the arm, the tremor frequency slowed down slightly, to 8–9 Hz, but no other changes in the tremor were apparent.

Summary:

- The tremor in this patient was very fine, irregular, and had a high frequency between 8 and 10 Hz.
- There was no clear alternation between agonist and antagonist contraction, nor were any clear peaks found in the spectra.
- On loading, the tremor frequency slowed down slightly but noticeably.

Altogether, when looking at the typical distinguishing characteristics of the different types of tremor in Table 3.1, the results of the polymyographic recording and subsequent spectral analysis are most consistent with a postural tremor, most likely an enhanced physiological tremor. There are no indications for an essential tremor; it is highly unlikely that the tremor frequency would decrease on loading in case of essential tremor, and the agonist–antagonist burst pattern would be more variable. Typically, enhanced physiological tremor is considered a benign condition for which no further treatment is necessary.

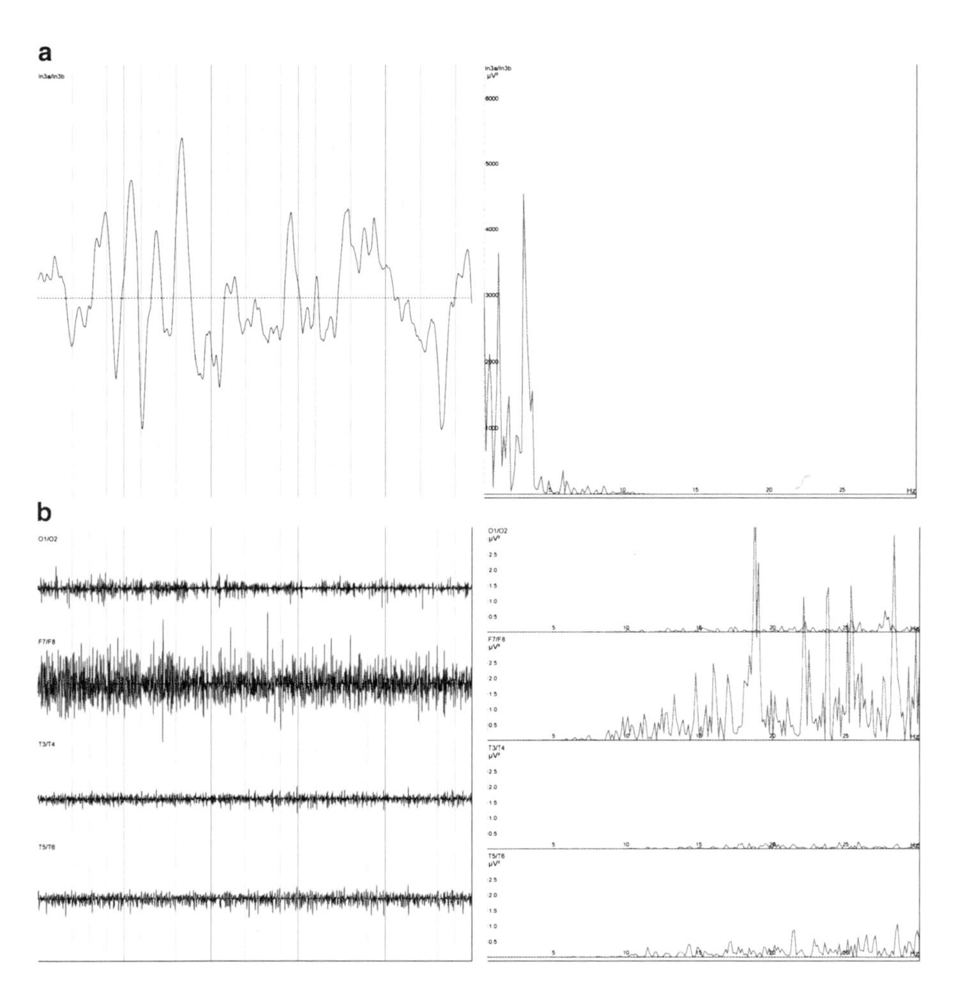

Fig 3.14 Results of polymyography in patient 2. *Left*: Accelerometer or EMG traces (4–5 s), *right*: FFT results of selected data. EMG from (*top* to *bottom*) flexor carpi radialis, extensor digitorum communis, biceps brachii, and triceps. The spectrum is displayed from 0 to 30 Hz. Vertical axes have been adapted individually for each figure. (**a**) Accelerometer results during finger extension. There are peaks in the spectrum, but they are not very clear. (**b**) EMG results (same time interval as **a**) during finger extension. No clear EMG bursts or spectral peaks. (**c**) Accelerometer results during top-nose test. No spectral peaks. (**d**) EMG results (same time interval as **c**) during top-nose test. No clear EMG bursts

Patient 3

In this 54-year-old truck driver, the EMG was recorded from upper and lower flexor and extensor muscles, bilaterally. Accelerometers were attached to both hands. All tests were executed with the left arm and hand. During rest, a tremor was present at a frequency of 6.0 Hz (Fig. 3.15a, b). The tremor intensity increased slightly during hand extension, accompanied by peaks in the EMG spectra at 6.8–7.1 Hz, depending

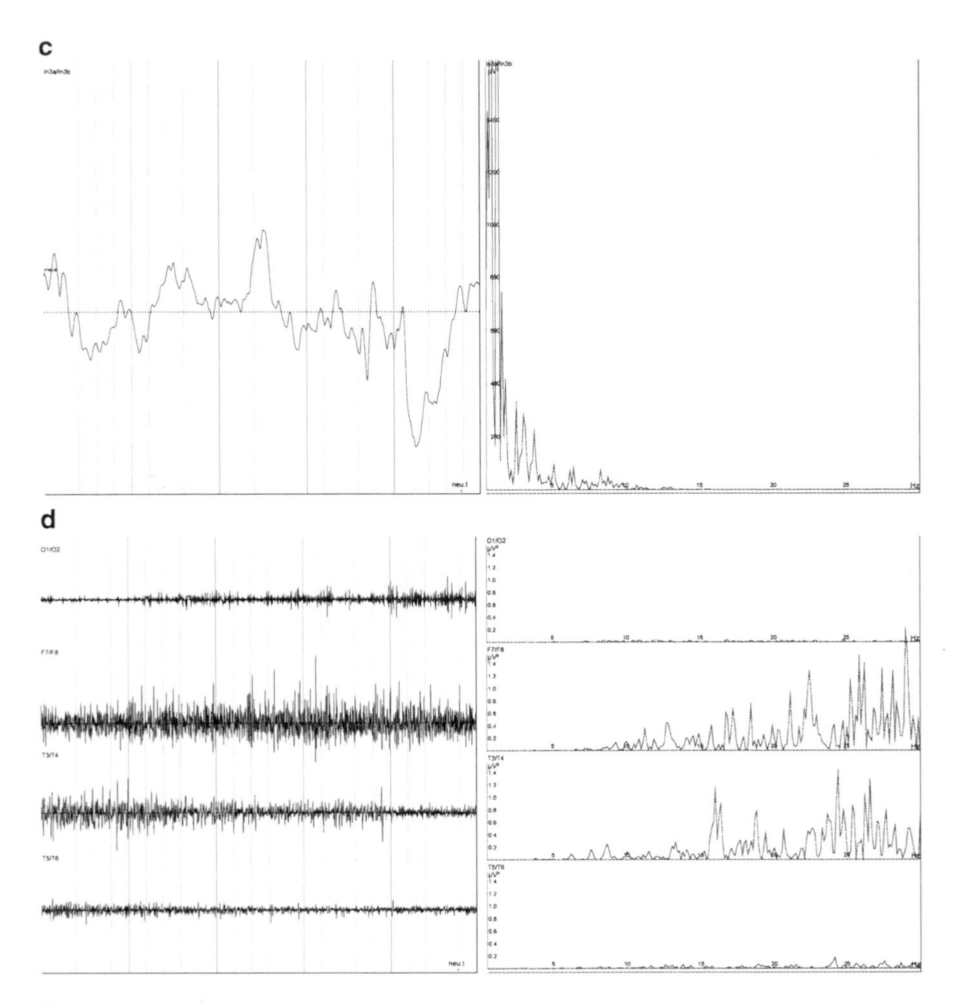

Fig. 3.14 (continued)

on the exact segment selected from the data. Harmonics were visible at two and three times the base frequency. During the top-nose test, the tremor was present in especially the lower arm muscles at approximately 7 Hz (Fig. 3.15c, d). In the EMG alternating contractions were visible in the agonist and antagonist muscles (Fig. 3.15e). During standing and lying down, the tremor decreased but did not disappear. The base frequency of the tremor did not change as a result of contralateral distraction, entrainment or loading.

Summary:

- The tremor was maximal during posture, but did not disappear completely during rest or during intentional movements.

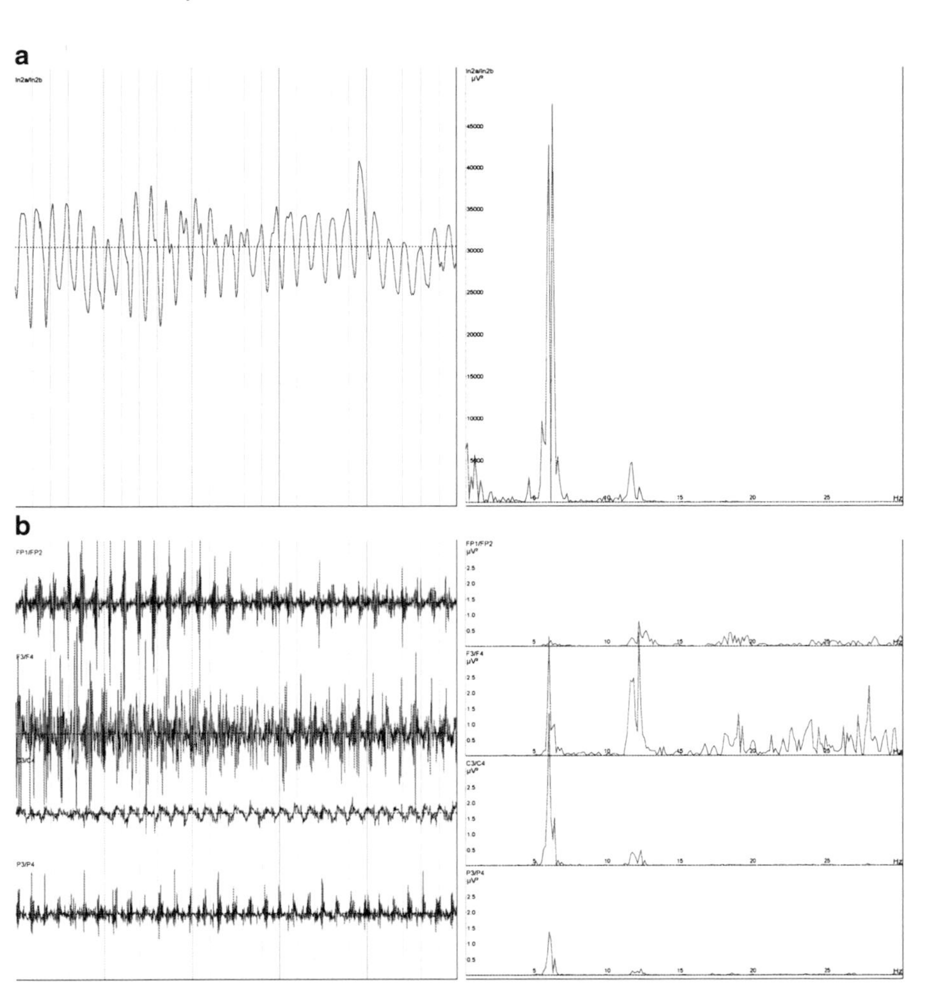

Fig. 3.15 Results of polymyography in patient 3. *Left*: Accelerometer or EMG traces (10 s), *right*: FFT results of selected data. Accelerometry from left and right hand. EMG from (*top* to *bottom*) left flexor carpi radialis, extensor digitorum communis, biceps brachii, and triceps and the same muscles on the right. Spectrum is displayed from 0 to 20 Hz. *Vertical axes* have been adapted individually for each figure. (**a**) Accelerometer results during rest. A clear peak can be observed at 6 Hz. (**b**) EMG results (same time interval as **a**) during rest. (**c**) Accelerometer results during top-nose test. A clear spectral peak can be seen at 7.1 Hz. (**d**) EMG results (same time interval as **c**) during top-nose test. A spectral peak at 7.0 Hz can be observed in particularly the lower arm extensor muscle. (**e**) Alternating bursts in upper arm extensors (first channel) and flexors (second channel) during top-nose test

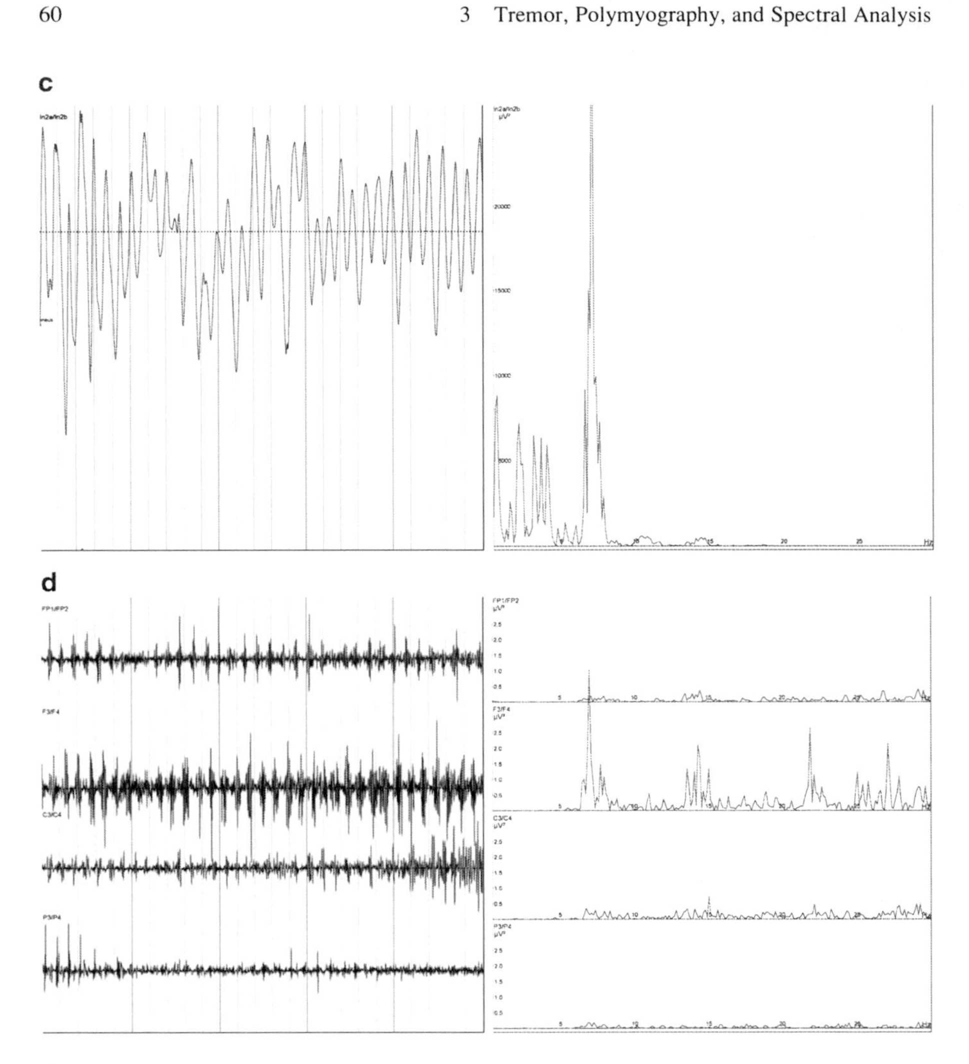

Fig. 3.15 (continued)

- The frequency of the tremor varied between 6.0 and 7.1 Hz.
- The EMG showed alternating contractions in agonist and antagonist muscles.
- The tremor was not influenced by distraction, entrainment, or weights.

Altogether, when looking at the typical distinguishing characteristics of the different types of tremor in Table 3.1, the results of the polymyography and accelerometer recordings are well consistent with essential tremor. There is no component of (enhanced) physiological tremor, since there is no change in tremor frequency on loading and the agonist–antagonist burst pattern is not synchronous. Since this patient did not respond to any medications for essential tremor, the

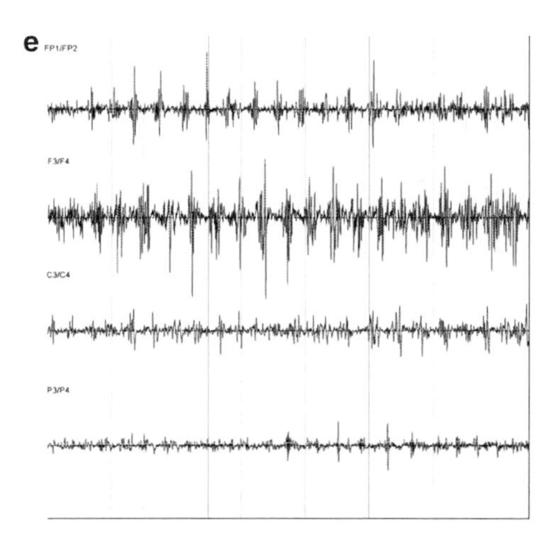

Fig. 3.15 (continued)

possibilities for an intracranial operation were now further explored based on the conclusions of the clinical examinations and the polymyography and more extensive clinical work-up.

In all patients in this chapter, polymyography helped to determine the diagnosis. In two cases, polymyography mostly served to confirm the working diagnosis of the neurologist. However, in the case of Patient 1, who was already considered for an intracranial operation to treat what was initially diagnosed as an essential tremor, the tremor turned out to be a psychogenic tremor.

3.5 Other Applications of Spectral Analysis in Neurology

Since most physiological signals are rhythmical, spectral analysis is generally helpful to objectively describe the frequency content of these signals. The most important example for applications of spectral analysis in clinical neurophysiology, apart from EMG, is of course the electroencephalogram or EEG (see Chap. 4 for a more extensive explanation of the EEG). In an EEG brain electrical activity is recorded by electrodes that are placed on the scalp. Spectral analysis is used to analyze the various rhythms in the EEG. An important and well-known rhythm is the alpha rhythm, which was already described by Hans Berger in the late 1920s. Because it was the first EEG rhythm to be described, Berger termed it the alpha rhythm. This rhythm has a frequency approximately between 8 and 13 Hz, can typically be clearly observed in the raw EEG, specifically on occipital electrodes (Fig. 3.16), and increases in amplitude when the eyes are closed or when a subject relaxes.

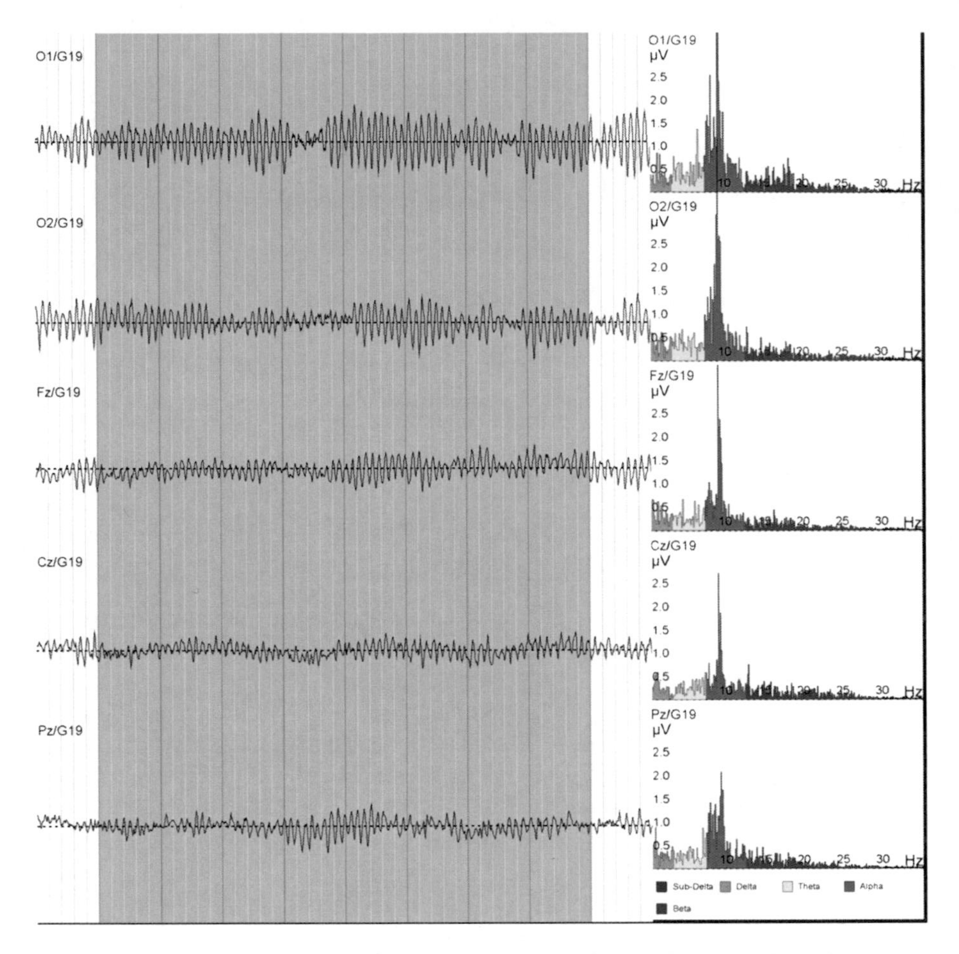

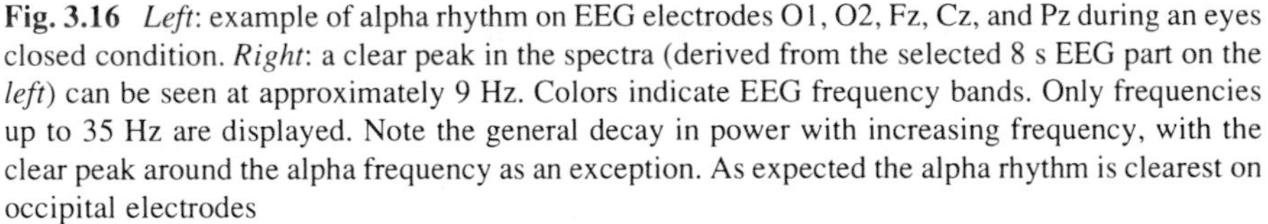

Fig. 3.16 *Left*: example of alpha rhythm on EEG electrodes O1, O2, Fz, Cz, and Pz during an eyes closed condition. *Right*: a clear peak in the spectra (derived from the selected 8 s EEG part on the *left*) can be seen at approximately 9 Hz. Colors indicate EEG frequency bands. Only frequencies up to 35 Hz are displayed. Note the general decay in power with increasing frequency, with the clear peak around the alpha frequency as an exception. As expected the alpha rhythm is clearest on occipital electrodes

Alpha is also suppressed during mental work, such as performing a calculation. The alpha frequency is highly individual. Mu rhythm is related to alpha rhythm and can be found over the sensorimotor cortices: this rhythm is suppressed when performing a motor action, such as making a fist. The clinically important EEG rhythms are given in Table 3.2; a more detailed description of these rhythms can be found in many clinical EEG handbooks (e.g., Niedermeyer and Lopes da Silva 2005).

The presence and distribution of EEG rhythms highly depends on maturation of the brain, i.e., the EEG of babies younger than 1 year is completely different from the EEG in adults. Also, on the other end of the scale, the EEG changes in the elderly. Whereas EEG frequencies generally increase in young children, they

Table 3.2 Clinically relevant EEG rhythms. Note that the frequency boundaries of the various rhythms may slightly differ between sources

Name	Symbol	Frequency range (Hz)	Presence/relevance
Delta	δ	0–4	Drowsiness/sleep
Theta	θ	4–8	Infancy/childhood, drowsiness/sleep
Alpha	α	8–13	Adults: eyes closed, physical relaxation, mental inactivity
Beta	β	13–30	As "fast" alpha, also frontally, medication-induced, mixes with mu
Gamma	γ	>30	*Pharmaco-EEG*, neurocognitive research

Table 3.3 Sleep stages and their distinguishing spectral and other features

Sleep stage	Description	Spectral features	Other features
W	Awake	Alpha and beta	–
N1	Drowsiness	3–7 Hz slow waves	Alpha dropout, vertex waves
N2	Light sleep	2–7 Hz slow waves	Sleep spindles, vertex waves, K-complexes
N3	Slow wave sleep	<2 Hz slow waves	K-complexes, some sleep spindles
REM	REM	Faster frequencies	Rapid eye movements

decrease in the elderly. Further, the general shape of an EEG spectrum follows a $1/f$ curve (see Fig. 3.16): slow frequencies have more power than high frequencies. This is partly due to the filtering effect of the skull that acts as a low-pass filter, thereby suppressing high frequencies. When recording EEG directly from the cortex (electrocorticography), e.g., during epilepsy surgery, higher frequencies are indeed observed.

Whereas clinical EEG evaluation usually proceeds by visual assessment, with a little help from spectral analysis to, e.g., determine the spectral alpha-peak, a visual evaluation of sleep EEG recordings is simply not tenable. A sleep recording or polysomnogram may contain up to 12 h of EEG, EMG, and electro-oculogram (*EOG*) signals, which results in simply too much data to page through and assess visually. The most important goal of assessing a polysomnogram is to determine the occurrence and duration of the different sleep stages. This is normally done according to criteria for visual assessment that were initially developed by Rechtschaffen and Kales (R&K) in 1968 and revised in 2007 by the AASM (American Academy of Sleep Medicine). The main division that is made in sleep staging is between REM (Rapid Eye Movement) and non-REM sleep. Non-REM sleep is then further subdivided in stages N1–N3, where stage N3 is the deepest sleep stage. REM sleep is distinguished by the typical eye movements. The other stages are divided based on particular EEG phenomena such as *K-complexes* and *sleep spindles* and spectral content: generally speaking frequencies become lower and amplitudes higher when the sleep stage becomes deeper. A rough overview of sleep stages according to the AASM classification and their distinguishing features is given in Table 3.3.

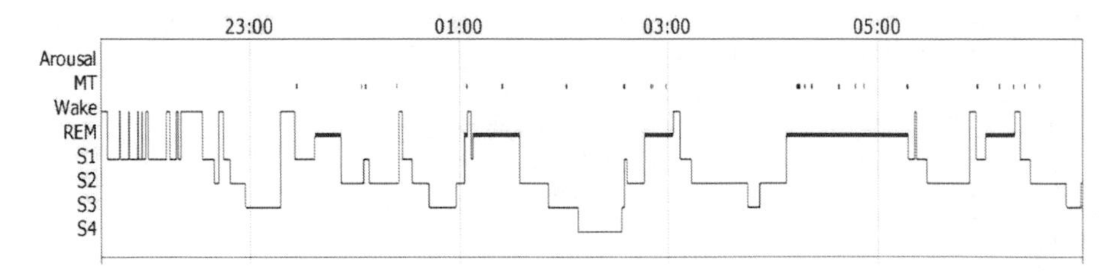

Fig. 3.17 Example of automated sleep staging of an overnight polysomnographic recording. Note that the K&S classification is still used here (N1 = S1, N2 = S2, N3 = S3 + S4)

The advantage of these definitions is that they can be captured in software programs that automatically determine sleep stages (Fig. 3.17) based on the relative contribution of the different EEG frequency bands to the signal and the presence of features such as sleep spindles and K-complexes, which leaves the clinical neurophysiologist "only" with the job to assess quality of the result of such a program and to write a report.

3.6 Answers to Questions

Answer 3.1
Suppose that a healthy human heart beats at 60–120 beats per minute, depending on its state (relaxed, excited, after exercise, with fever). This amounts to 1–2 beats per second and thus a frequency of 1–2 Hz.

Answer 3.2
The theoretical minimum frequency to sample a signal that contains frequencies up to 40 Hz is 80 Hz. However, to adequately represent the ECG signal in the time domain, with its sharp peaks, it is better to use a higher sampling rate, such as, e.g., 200 Hz.

Answer 3.3
The maximum frequency is 2,000 Hz. This is the Nyquist frequency that is equal to half the sampling frequency. The spectral resolution is determined by the length of the signal and is in this case 1/5 = 0.2 Hz.

Answer 3.4
A signal that is sampled at 100 Hz contains 100 data points per second. To avoid zero-padding, the length of the selection should be a power of 2, e.g., 128, 256, or 1,024. In these examples the selection should thus be 128/100 = 1.28 s, or 256/100 = 2.56 s, or 1,024/100 = 10.24 s.

Glossary

Accelerometer Device to measure acceleration, available for 1–3 dimensions.

Amplifier Device to amplify (enhance) the amplitude of small electrical signals, so that they can be recorded.

Amplitude Height of a signal.

(Ant)agonist Muscles that counteract each other: if the biceps is the agonist, then the triceps is the antagonist.

(Anti)symmetric Consisting of two parts that are the same in size and shape. A symmetric function $f(x)$, such as $\cos(x)$, is the same when mirrored in the y-axis: $f(x) = f(-x)$. For an antisymmetric function $g(x)$, such as $\sin(x)$, the following holds: $g(x) = -g(-x)$.

Diadochokinesis Fast execution of opposite movements, e.g., wrist supination–pronation.

EOG Electro-oculogram. Recording of the electrical activity accompanying eye movements by placing electrodes around the eyes.

Essential tremor The most common movement disorder. Clinically, essential tremor presents with action tremor (postural and kinetic) with tremor frequency in the range of 4–12 Hz. The arms are primarily affected, but neck and head, trunk and legs can also be involved.

Extension Opposite of flexion: stretching a joint.

Flexion Bending a joint using flexor muscles.

Intention component Tremor occurring when a limb is approaching a target.

K-complex EEG waveform occurring during sleep, consisting of a peak higher than 100 μV and lasting longer than 0.5 s.

(Non)periodic A motion or function is periodic when it is repeating itself, over and over again. Formally: for a periodic function $f(x)$ there is a T such that $f(x + T) = f(x)$.

Orthostatic tremor A rare tremor characterized by subjective sensation of loss of balance while standing. Clinically, a visible or palpable tremor in the trunk and lower extremities may be observed.

Pharmaco-EEG Electrophysiological brain research in preclinical and clinical pharmacology. By considering EEG spectral properties before and after medication, the effect of that medication on the brain can be assessed. Often used in clinical trials of new medications on healthy volunteers and patients.

Pronation Hand held horizontally with the palm down.

Psychogenic Resulting from a psychological disorder, also nonorganic.

Sleep spindles Burst of EEG activity occurring during sleep stage II, consisting of 12–16 Hz waves lasting for 0.5–1.5 s.

Spectrum Here: a representation of a signal in terms of its frequency components.

Supination Hand held horizontally with the palm up.

Supine Position lying down on the back.
Vertex waves EEG wave during sleep occurring on central(-parietal) electrodes, shaped as a peak–wave–peak complex.
Wavelet analysis A form of time-frequency analysis that allows to localize activity at a certain frequency in time.

References

Online Sources of Information

http://www.scholarpedia.org/article/Tremor. Extensive overview with many references on the classification of tremors and the pathophysiology of the tremors in Table 3.1
http://www.sengpielaudio.com/calculator-period.htm. Simple, but accessible online calculator to transform frequency to period and vice versa
http://en.wikipedia.org/wiki/Fourier_series. Overview of the mathematics of Fourier series
http://en.wikipedia.org/wiki/Frequency_spectrum. Examples of frequency spectra of different signals, some from acoustics
http://en.wikipedia.org/wiki/Aliasing. Examples of aliasing

Books

Findley LJ and Koller WC (1994) Handbook on tremor disorders. Informa Health Care (Chapter 6 in particular). Also available from http://books.google.com
Lyons RG (2004) Understanding digital signal processing, 2nd edn. Upper Saddle River, NJ, Prentice Hall
Niedermeyer E, Lopes Da Silva F (2005) Electroencephalography: basic principles, clinical applications, and related fields, 5th edn. Lippincott Williams and Wilkins, Philadelphia, PA, Also available from http://books.google.com
Rechtschaffen A, Kales A (eds) (1968) A manual of standardized terminology, techniques, and scoring system for sleep stages of human subjects. US Department of Health, Education, and Welfare Public Health Service – NIH/NIND, Bethesda, MD

Chapter 4
Epilepsia, Electroencephalography, Filtering, and Feature Extraction

After reading his chapter you should know:

- What the EEG can and cannot measure
- Why filtering can help distinguish between brain and other activity in the EEG
- What types of filters there are
- The basic mathematical concepts involved in filtering (additional material)
- How filters in the time and frequency domain are related
- How filters should and shouldn't be used
- How EEG features can help differentiate between different types of epilepsies
- How EEG can be used in other neurological applications

4.1 Patient Cases

Patient 1

A 91-year-old patient is brought into the ER. Earlier that day he suffered from severe headache and he had to vomit. His help reports that she saw limb contractions. He did not bite his tongue, nor did he suffer from *urinary incontinence*. He now suffers from weakness on the left side of his body, which is fortunately improving. He reported to have been confused and have a headache yesterday night, as well. He was admitted to the hospital, with a preliminary diagnosis of an *ischemic* cerebrovascular accident (CVA, also known as *stroke*) accompanied by an epileptic seizure.

Patient 2

Since a couple of months, a male 9-year-old patient is suffering from periods of absent-mindedness. During these periods, that occur once to several times a day and

N. Maurits, *From Neurology to Methodology and Back: An Introduction to Clinical Neuroengineering*, DOI 10.1007/978-1-4614-1132-1_4,

last approximately 5 s, he stops talking and his eyes turn away. After such a period, he often has forgotten what he was doing. Throughout these periods he cannot be contacted; his facial color is normal and he does not show limb or smacking movements. When the patient is asked to hyperventilate by sighing for 1 min by the child neurologist he visits with his parents, he suddenly stops and stares for 3 s. Although clinically, the symptoms are typical for absence epilepsy, a more definite diagnosis is needed before treatment can be started.

4.2 Electroencephalography: Measuring the Brain in Action

Seizures that are accompanied by limb movements, as described in patient 1, can occur as a result of epilepsy, but other causes such as syncope (fainting) are always considered by clinicians. Sometimes, no underlying organic disorder can be identified and the seizures may be considered psychogenic. On the other hand, although most people would probably associate epilepsy with the impressive so-called grand-mal (*tonic-clonic*) seizures in which a patient may drop to the floor and vigorously contract his limb muscles in a pronounced rhythm, sometimes loosing control over his bladder and biting his own tongue, epilepsy may also present itself with seizures as in the example of patient 2, which are more subtle and may seem much more benign. Since this book is not intended to be a primer on epilepsy, it suffices to say here, that many variants exist, both in children and adults. What all epileptic seizures have in common is that they are the result of aberrant brain activity.

It is not completely understood how epilepsy develops, although several genes have now been associated with forms of generalized epilepsy and specific epilepsy syndromes in children. These genes were found to code for proteins that are essential building blocks of *ion channels* in neurons. Part of the story of epilepsy may thus be that (temporarily) dysfunctioning neuronal ion channels induce abnormal brain rhythms, which in turn lead to abnormal behavior (seizures).

So what methodology can be used to assess the functioning of neurons in the brain, noninvasively? Before answering that question it is important to understand the physiology of brain neurons to some extent. We here focus on pyramidal neurons that are situated in the outer layers of the *cortex* and are of utmost importance for human behavior. Pyramidal neurons are nerve cells with a triangular-shaped cell body (hence the name; see Fig. 4.1).

Like all neurons, they communicate with each other using electrochemical processes: within the neuron its activity is determined by electrical processes, whereas communication between neurons is achieved chemically through *synapses* and *neurotransmitters*. Each neuron has one outgoing branch (the axon that originates from the cell body and branches to connect to other neurons) and by synapsing onto each other neurons can form large networks.

All neurons can be excited electrically. During rest, they maintain a stable voltage difference across their membrane by active (ion pumps in the membrane) and passive (ion channels open to *diffusion* of ions due to concentration differences)

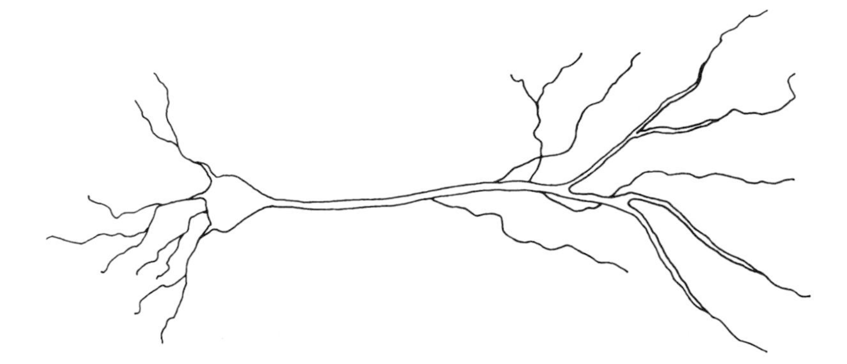

Fig. 4.1 Schematic of a pyramidal neuron with (from *left* to *right*) the cell body (soma) with its dendrites, the axon, and its branches

processes. This results in a steady-state potential of approximately −70 mV across the membrane. The potential is negative on the inside with respect to the outside, due to an uneven distribution of mainly sodium, potassium, and chloride ions. When a neuron is excited by another neuron by means of an *inhibitory* postsynaptic potential (IPSP; inhibitory synaptic contact on cell body), its permeability for potassium and chloride ions increases (negative ions go in, positive ions go out), resulting in an increased (more negative) membrane potential. As a consequence, a current of decreasing intensity starts to flow along the membrane, thereby hyperpolarizing the neuron and making it less likely to fire, i.e., to send out an action potential along its axon. On the other hand, when an *excitatory* postsynaptic potential excites the neuron (EPSP; excitatory synaptic contact on dendrite), the neuron's permeability increases nonselectively, thereby locally decreasing the membrane potential (making it less negative). This results in a locally depolarizing current along the membrane, making firing more probable. An action potential occurs when the summation of all EPSPs and IPSPs results in a depolarization of the membrane beyond a certain threshold. In that case an all-or-nothing response occurs; in the axon, close to the cell body, the membrane permeability to sodium and potassium ions increases momentarily, leading to a sudden collapse and even reversal of the membrane potential. Subsequently, the original membrane potential is quickly restored. This electrical change is the action potential, which always has the same amplitude and lasts only about 1 ms. The local current depolarizes the adjacent membrane, thereby leading to the same sequence of events in neighboring areas and allowing the action potential to propagate along the axon. The action potential, being very brief, does not penetrate far into the extracellular space. Postsynaptic potentials, however, last for over 100 ms and change the membrane potential by several millivolts. Furthermore, pyramidal neurons have dendrites penetrating through several cortical layers, thereby guiding currents generated in deep layers to more superficial areas. They are closely packed, in a parallel orientation and receive similar input in large groups. One afferent axon may excite thousands of pyramidal neurons at once, resulting in similarly directed and timed activation. Together, these properties allow part of the summated pyramidal neuronal postsynaptic electrical activity to travel through all the tissues covering the cortex, to reach the scalp. As a result different parts of the scalp will reach different potential levels, which can be recorded by placing *electrodes* on two different positions on the scalp and deriving the potential difference. This type of recording is called an electroencephalogram (literally: "writing of electrical brain activity") or EEG.

Hence, the method of choice to look at the brain in action is the EEG. One of its strong points is that it allows to observe changes in brain potentials with an unchallenged high temporal *resolution*. Since some (inter-*ictal*) epileptic phenomena are very brief (in the order of milliseconds), there is only one other noninvasive technique available that can be used to assess and diagnose epilepsy: magnetoencephalography (or MEG). While EEG records electrical brain activity, MEG measures its magnetic counterpart. Importantly, the laws of *electromagnetism* imply that any electrical current in a wire induces a magnetic field around it. Similarly, electrical currents in cortical neurons generate magnetic fields that can be detected with MEG. Unfortunately, compared to EEG, MEG is expensive (because of its technology that involves supercooling and heavy magnetic shielding) and static (it cannot be moved around to record at, e.g., the bedside). Other brain imaging techniques, such as functional magnetic resonance imaging (fMRI) and positron emission tomography (PET), are very good at (spatially) localizing brain activity, but not so much at pinpointing and characterizing brain activity in time. For clinical practice, EEG is thus the main tool to assess brain activity. In Sect. 3.4 the EEG and its rhythms were already briefly introduced. Here, we consider in more detail how EEG recordings can help diagnose epilepsy.

As mentioned before, an EEG recording basically consists of placing electrodes on the scalp and deriving potential differences between pairs of electrodes. Since the potential difference is very small (in the order of microvolts), *amplifiers* are needed to visualize the EEG signal. Nowadays, these amplified potentials are typically stored digitally on computers using dedicated software. To be able to compare recordings with each other, within as well as between patients, it is essential that electrodes are placed on unambiguously defined positions. The international 10–20 system of electrode placement provides such a system, in which the whole scalp is covered. It uses two distances (between *nasion* and *inion* and between left and right *preauricular* points) to generate a system of lines that intersect at intervals of 10 and 20% of these distances (see Fig. 4.2).

A standard set of electrodes consists of 21 electrodes which can be attached in the right positions one by one using measuring tape and glue or paste. More recently, elastic caps in different sizes have been introduced that allow positioning of all electrodes at once. These caps may also contain 64, 128, or even 256 electrodes. To ensure good contact (reflected in low *impedance*) between electrode and skin, conducting pastes that contain ions in solution (e.g., salt pastes) are used in between. Before applying this paste, the skin is sometimes lightly abraded with scrub paste, to remove excess skin oils that would otherwise decrease electrical conductivity. A typical routine EEG recording will last for 20–30 min and will include recording during rest while awake and if possible, a limited recording of

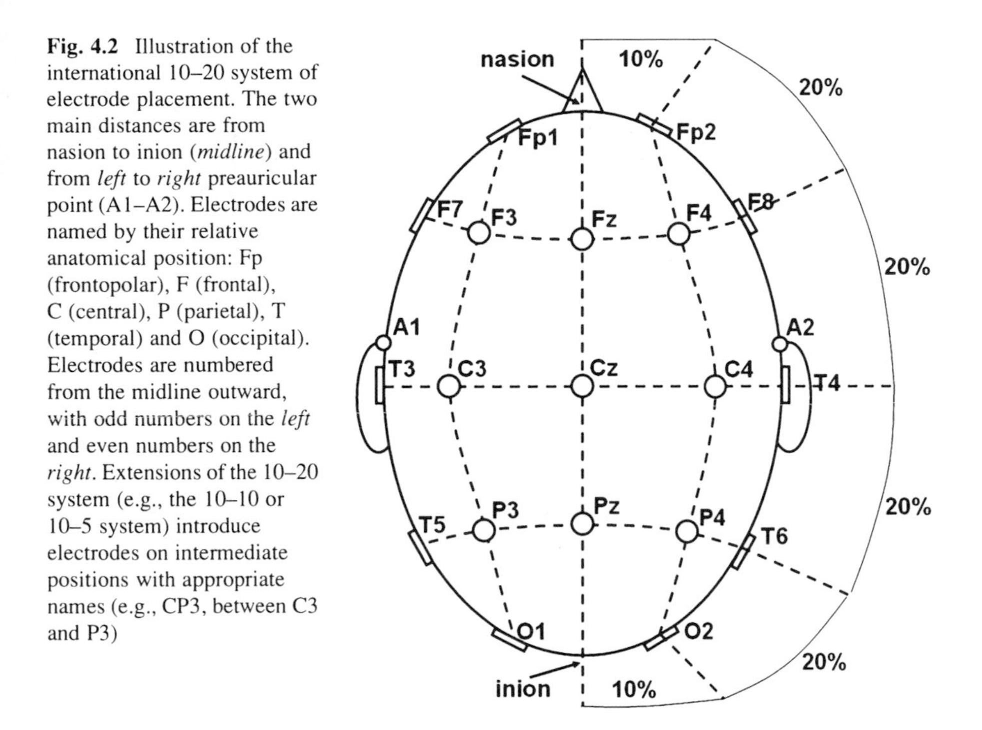

Fig. 4.2 Illustration of the international 10–20 system of electrode placement. The two main distances are from nasion to inion (*midline*) and from *left* to *right* preauricular point (A1–A2). Electrodes are named by their relative anatomical position: Fp (frontopolar), F (frontal), C (central), P (parietal), T (temporal) and O (occipital). Electrodes are numbered from the midline outward, with odd numbers on the *left* and even numbers on the *right*. Extensions of the 10–20 system (e.g., the 10–10 or 10–5 system) introduce electrodes on intermediate positions with appropriate names (e.g., CP3, between C3 and P3)

(light) sleep. A typical clinical EEG report will contain information on the presence of (ab)normal rhythms, their *reactivity*, and their symmetry over the two hemispheres, in terms of amplitude and location. For example, while awake, the patient will be asked to close and open his eyes to observe alpha reactivity. In healthy subjects the alpha rhythm (8–13 Hz, see Sect. 3.4 and Fig. 4.3) should disappear when the eyes are opened.

Despite the sometimes striking phenomenology of epilepsy, the disease is not always easy to diagnose using EEG. During epileptic seizures the EEG is always (strongly) abnormal, but since seizures typically occur infrequently, most EEGs are actually recorded between seizures. In some patients, epileptic phenomena or activity can be observed inter-ictally, but in many, the EEG is completely normal between seizures. An approach that sometimes helps is to stimulate the brain such that the threshold to develop an epileptic seizure is decreased. This can be done by sleep depriving the patient before recording, or, for certain types of (reflexive) epilepsies, by stimulating the brain visually by flashing light at increasing frequencies (Fig. 4.4) or by suddenly making a loud sound.

Finally, hyperventilation normally induces generalized slow waves, but may increase the presence of some types of epileptic phenomena (e.g., so-called 3 Hz spike-and-wave discharges) in patients. An EEG will be assessed as being abnormal when it contains epileptic activity, abnormal slow waves, abnormal amplitudes (in terms of asymmetries or generalized amplitude changes) or abnormal reactivity, frequency, or distribution. Since judging clinical EEGs takes considerable practice

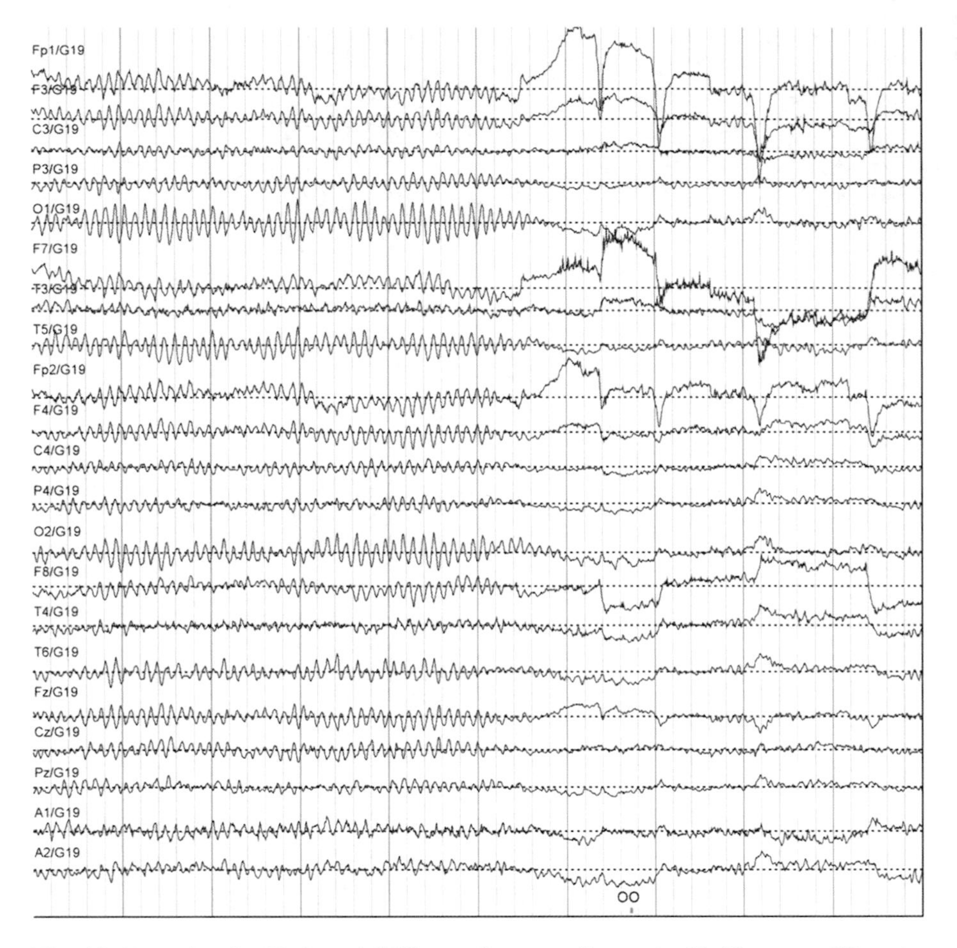

Fig. 4.3 Example of a 21-channel EEG recording according to the 10–20 system. When the patient is asked to open his eyes (OO, middle), there is adequate brain reactivity (decreased alpha rhythm). When opening the eyes, blink artifacts are clearly visible on *derivations* that contain frontal and frontopolar electrodes

and whole books have been written on the topic, here only two illustrations of typical EEG features of epilepsy are given (Fig. 4.5).

4.3 Filtering

EEG is not very selective in measuring electrical activity. This implies that besides measuring brain activity, an EEG recording may also contain electrical activity from other sources. This type of EEG activity is called an artifact since it is unwanted for when assessing the EEG. Artifacts can be of biological or technical nature and

a

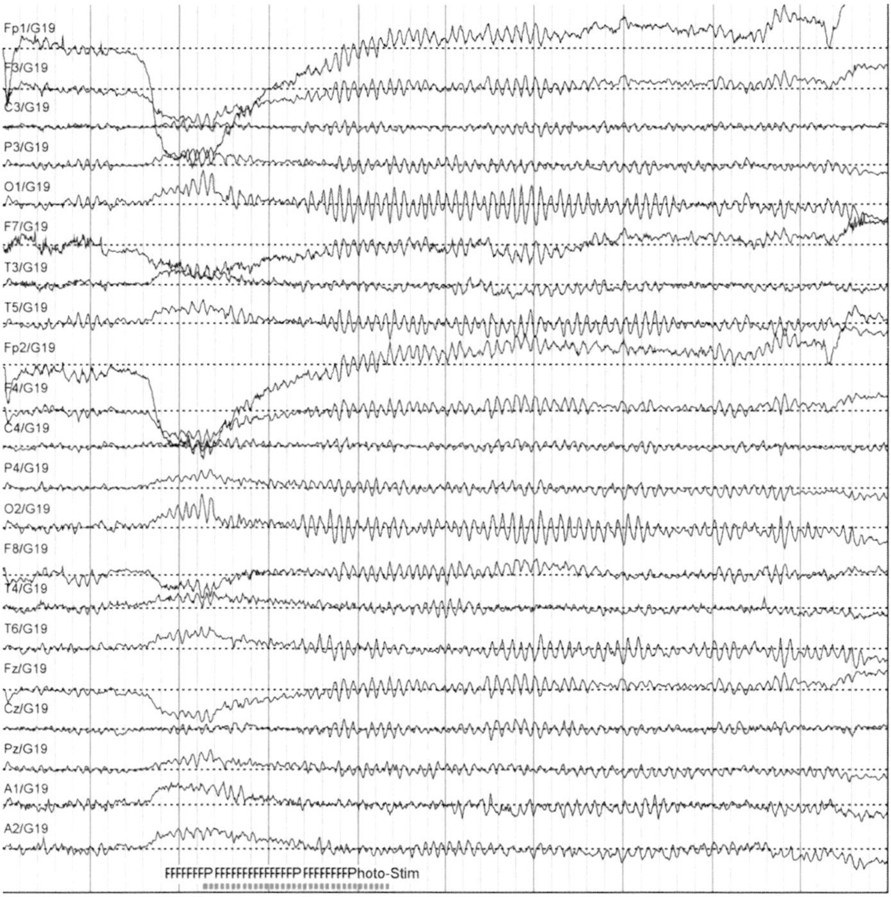

Fp1/G19
F3/G19
C3/G19
P3/G19
O1/G19
F7/G19
T3/G19
T5/G19
Fp2/G19
F4/G19
C4/G19
P4/G19
O2/G19
F8/G19
T4/G19
T6/G19
Fz/G19
Cz/G19
Pz/G19
A1/G19
A2/G19
FFFFFFFP FFFFFFFFFFFFFFP FFFFFFFFPhoto-Stim

Fig. 4.4 (**a**) Normal and (**b**) abnormal reaction to light flash stimulation (indicated with Photo-Stim at the bottom of the figures); eyes closed

can – with some experience – often be recognized as not originating from the brain. What helps identify artifacts is the fact that brain potentials usually originate from particular brain areas and from there propagate through all tissues (gray and white matter, *meninges*, cerebrospinal fluid, skull, and scalp) to reach the electrodes (volume conduction). This implies that an electrical brain potential usually occurs on several adjacent electrodes simultaneously, with decreasing amplitude when the distance to the most active brain areas increases. Thus, large amplitude signals on isolated electrodes and joint activity on multiple, but nonadjacent electrodes are highly unlikely to be the result of brain activity.

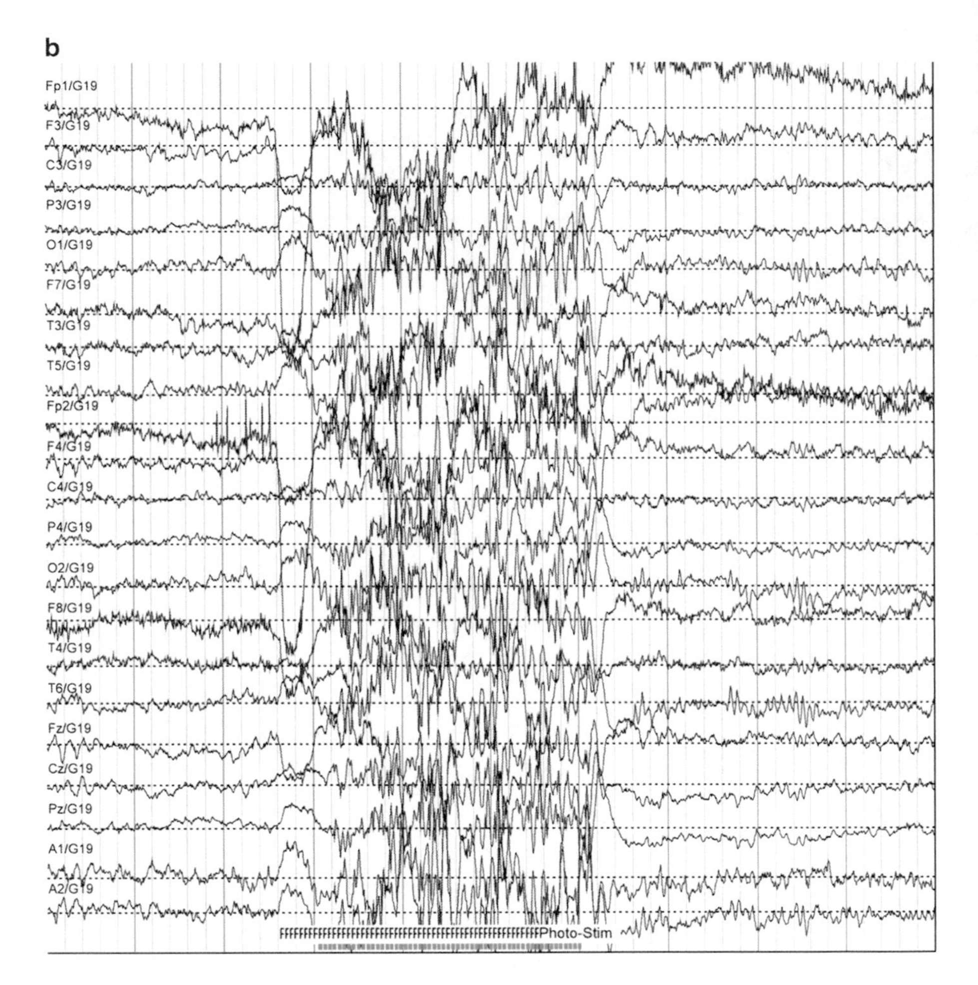

Fig. 4.4 (continued)

4.3.1 *Biological Artifacts*

Biological artifacts originate from the patient. The most common biological artifacts are blink artifacts, eye movements, muscle artifacts, movement artifacts, and electrocardiographic and pulse wave artifacts. Since the retina of the eye is electrically charged (positive at the front and negative at the back), blinks and other eye movements lead to changes in electrical potentials on electrodes close to the eyes (such as the frontopolar Fp electrodes) and even on more posterior electrodes (see Fig. 4.6). Fortunately, when eye movements are recorded separately by attaching additional electrodes under the eye and to the outer *canthi*, the EEG can be corrected for these artifacts (Sect. 7.3.4.2).

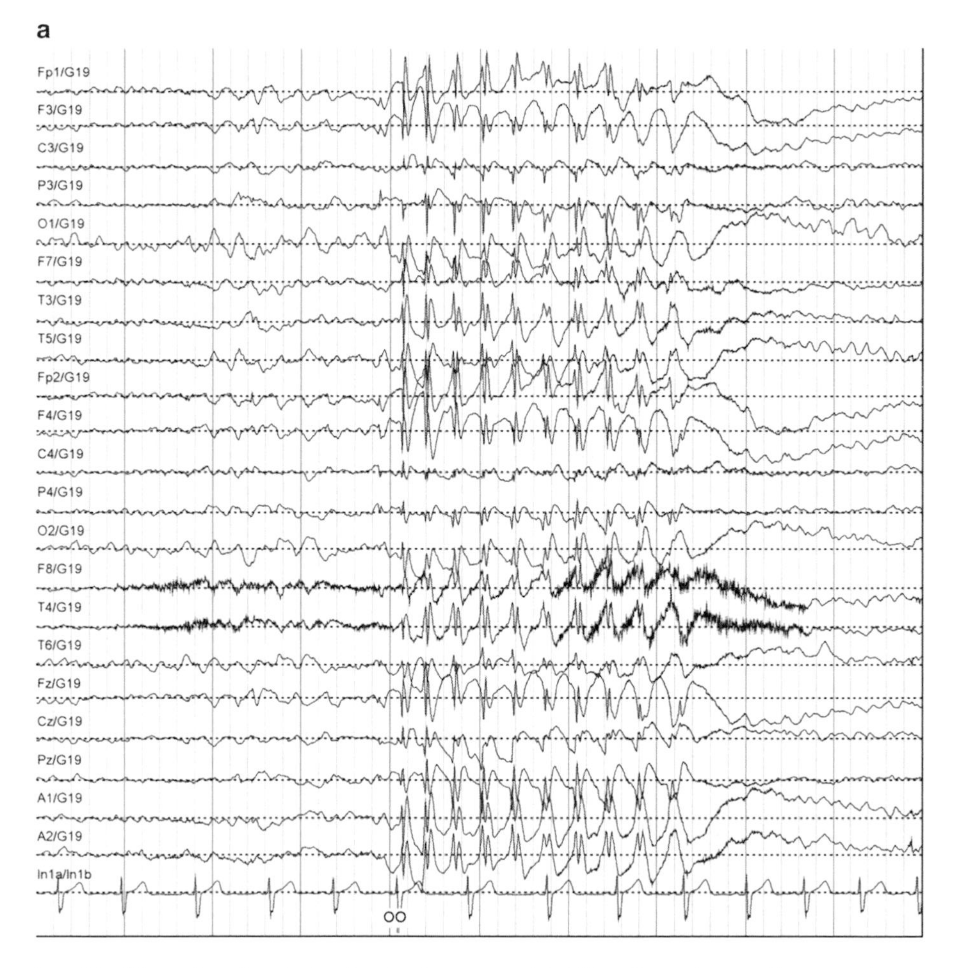

Fig. 4.5 Illustration of typical epileptic phenomena. (**a**) 3 Hz spike-and-wave discharges during absence epilepsy (note that the vertical scale has been adapted for a clear view of the spike–wave complexes) and (**b**) onset of tonic-clonic seizure (at 6 s) followed by repetitive spikes during the tonic phase (at 7–8 s) and generalized spike-and-wave discharges during the clonic phase of a generalized seizure (at 9–10 s). *Vertical lines* are 1 s apart

Muscle artifacts develop because electrodes make no distinction between recording electrical activity due to neuronal activation and electrical activity due to muscle contraction (Fig. 4.7). When a patient moves, clenches his jaw, talks or frowns, the EEG is temporarily drowned in high amplitude broad frequency signal. EMG cannot be removed and the only option is to try and prevent it (Sect. 7.3.4.2).

Movement artifacts often co-occur with muscle artifacts (that are the result of movement in most cases) and typically consist of slow, large amplitude waves. Although movement artifacts can sometimes be removed as described in Sect. 4.3.2, their frequency band may overlap with phenomena of interest in the

b

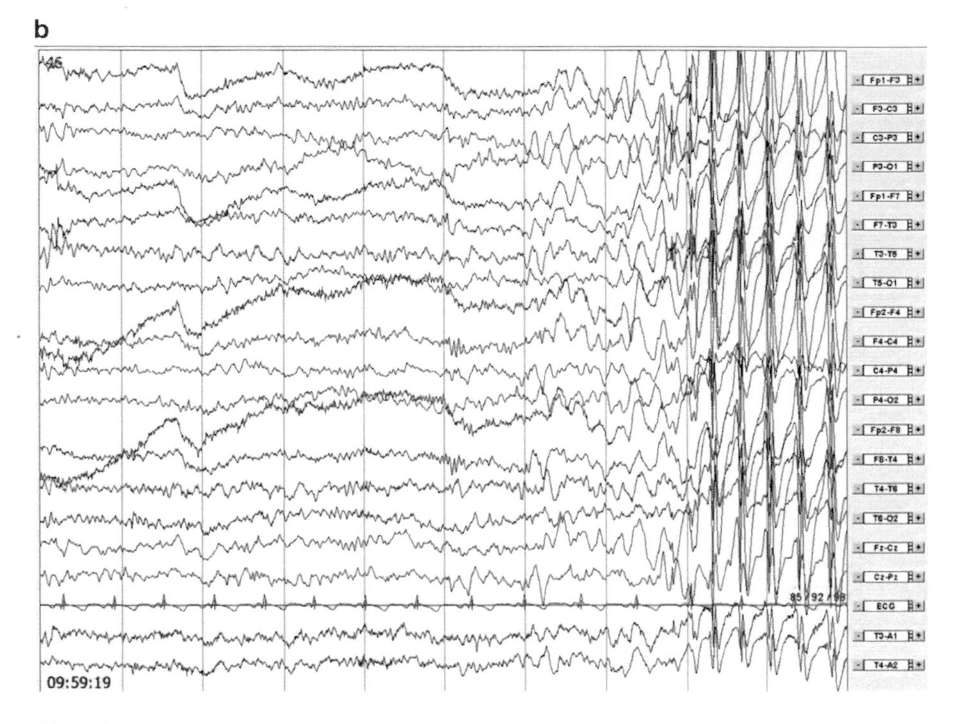

Fig. 4.5 (continued)

EEG and care must be taken not to "throw the baby out with the bathwater." As is the case for muscle artifacts, they are best prevented.

Finally, electrocardiographic artifacts arise in the EEG when electrodes are placed far enough apart to pick up the electrical activity of the heart. Pulse wave artifacts arise when an artery is pulsating directly under the electrode, thereby moving the electrode. Both types of artifacts can often be circumvented by replacing the electrode to a slightly different position.

4.3.2 Technical Artifacts

The most common technical artifact is due to electrical interference from power lines and other technical equipment close to the EEG amplifier. In Europe its frequency is 50 Hz, but it can be 60 Hz in other countries. With modern technical advances in EEG amplifiers, this type of artifact has become easier to deal with, to such an extent that it is now even possible to record in the electrically "dirty" environment of an intensive care unit (ICU). However, this so-called "mains" artifact may still occur on isolated or all electrodes. In case the artifact is present on isolated electrodes, it is likely due to high impedance, e.g., because the conducting paste was not yet applied, or because

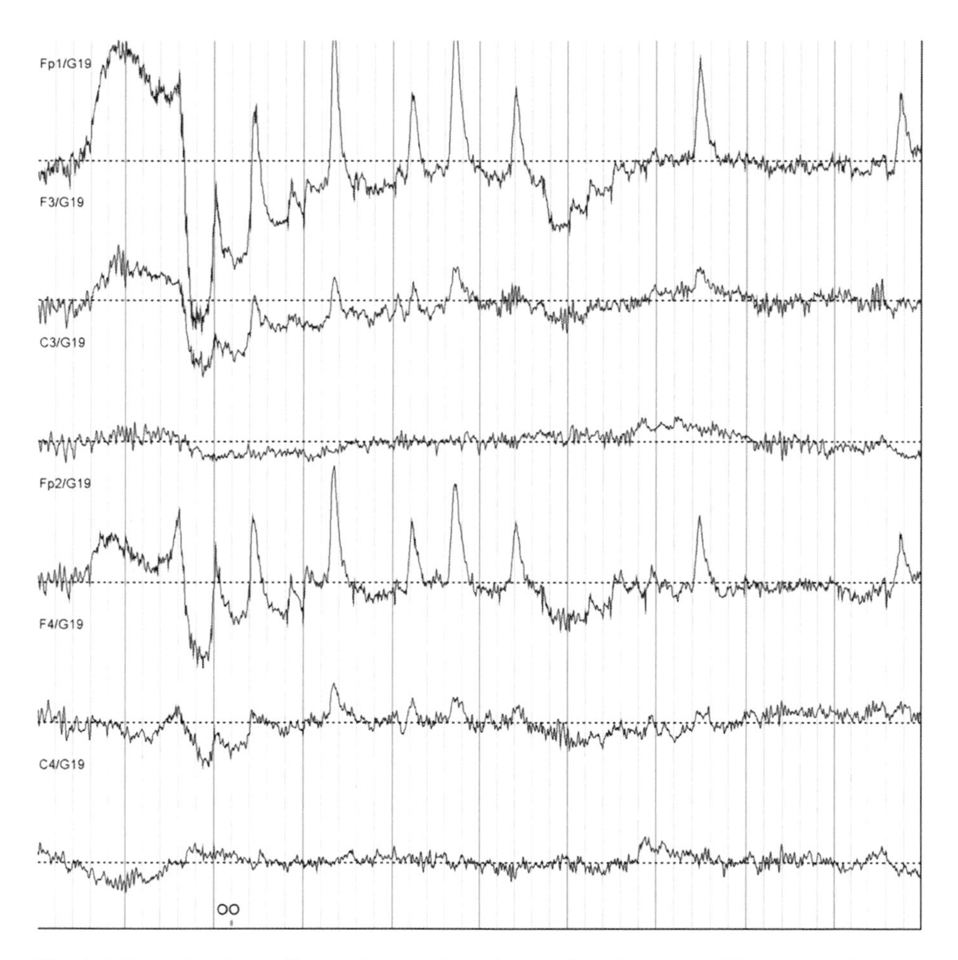

Fig. 4.6 Example of eye blink artifacts in channels Fp1 (first channel) and Fp2 (fourth channel). Note that eye blinks also affect channels that lie more posteriorly, such as F3 (second channel) and F4 (fifth channel), but that it does not influence even more posterior channels such as C3 and C4 (third and sixth channels)

the electrode wire is broken. If mains artifact is found on all electrodes, it may be because the reference electrode has a lot of mains artifact (thereby influencing all electrodes) or the surroundings are heavily contaminated by electrical interference (e.g., when recording close to an MR scanner or in the ICU). An example of mains artifact on isolated channels is given in Fig. 4.8.

Low amplitude mains artifact may be difficult to distinguish from EMG artifact to the naked (inexperienced) eye, but viewing the power spectrum easily allows to decide on the type of artifact. EMG has a very widespread spectrum whereas the mains artifact spectrum contains single peaks at 50/60 Hz and its harmonics, see Fig. 4.9. Just as biological artifacts, technical artifacts are best prevented, or dealt with at the beginning of the recording (see Sect. 7.3.4.2).

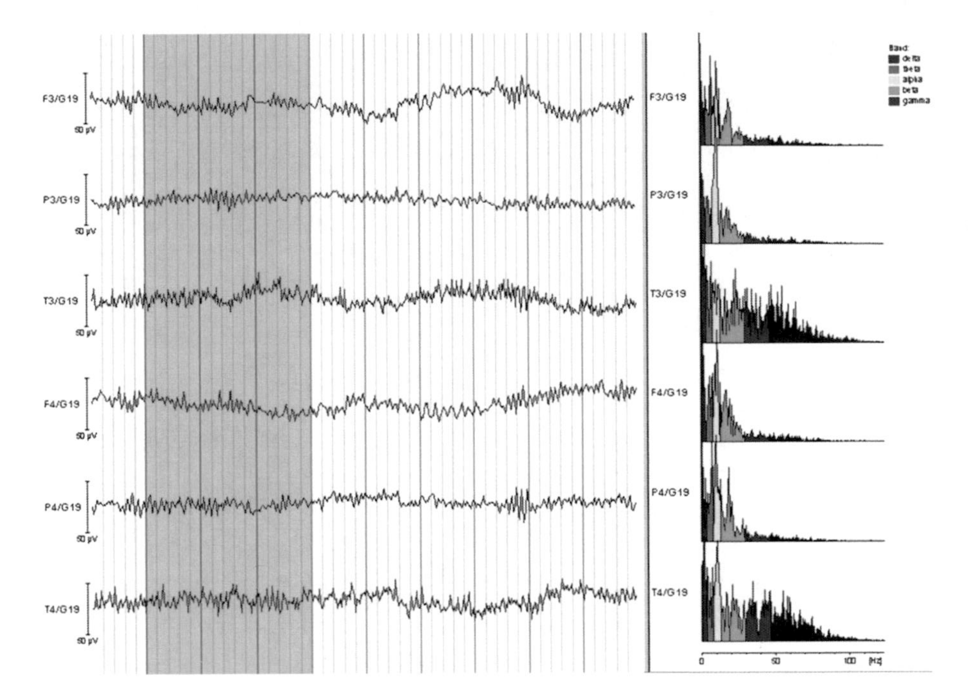

Fig. 4.7 Example of muscle artifact on EEG channels T3 (third channel) and – to a lesser extent – T4 (sixth channel), probably due to jaw clenching. The spectra on the right show the typical shape of an EMG spectrum on T3 and T4: maximum power around 30–50 Hz with a slow decrease until approximately 250 Hz. The channels without muscle artifact show a typical 1/frequency EEG spectrum

Question 4.1 Which of the channels in Fig. 4.9 suffers from mains artifact and which from EMG artifact and why?

However, when artifacts cannot be circumvented, they can often be removed by filters. A filter (in the understanding of this chapter) allows to selectively remove part of a signal, thereby reducing or enhancing certain aspects of the signal. The trick of filtering is thus to choose the filter properties such that unwanted signals are removed while the signal of interest still remains.

4.3.3 Filtering in the Frequency Domain

In this chapter only digital filters will be discussed; i.e., filters that act directly on digital signals. The functioning of a filter is most easily understood by considering

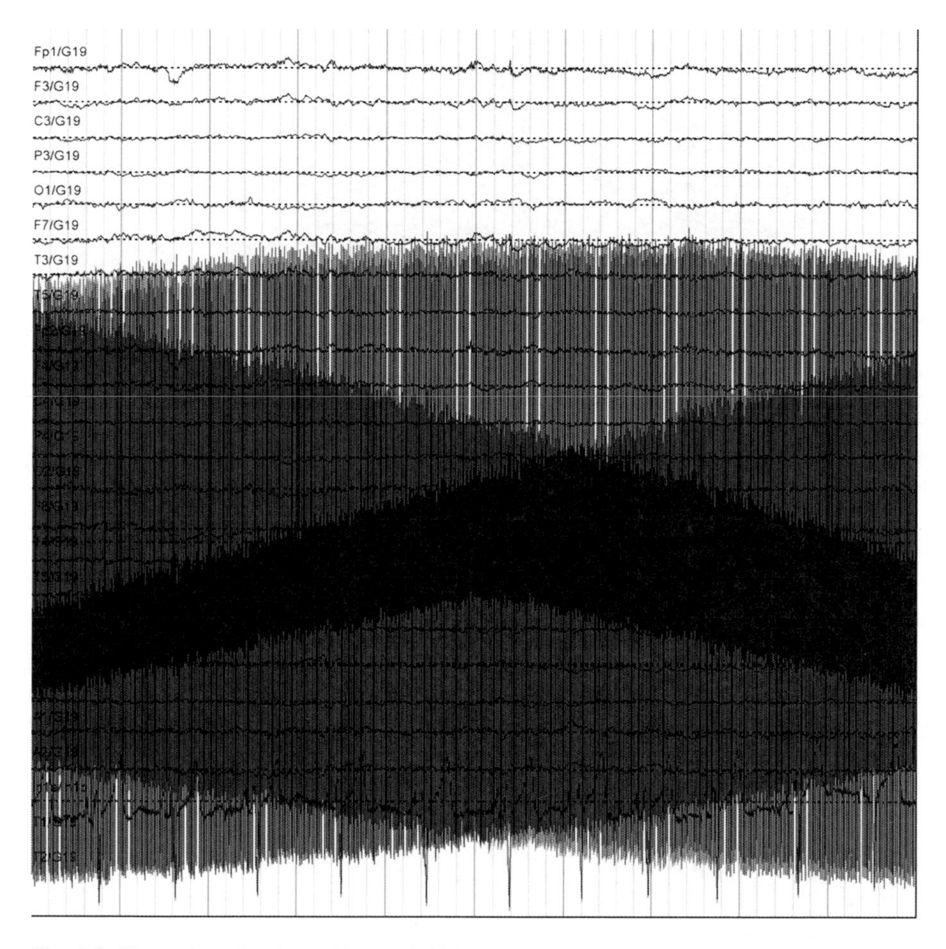

Fig. 4.8 Illustration of mains artifact with high amplitude on two isolated channels

the filtering process in the frequency domain, since a filter should remove exactly those frequencies that are unwanted. This could theoretically (mathematically) be accomplished by Fourier transforming the signal from the time to the frequency domain, setting all coefficients of the transformed signal belonging to unwanted frequencies to zero and inverse Fourier transforming the adapted signal in the frequency domain. This way of understanding the filtering process is illustrated in Fig. 4.10.

As mentioned before, filters should be chosen such that the unwanted frequencies in the signal are removed, whereas the frequencies of interest in the signal remain. This implies that a filter can only be adequately chosen when the spectral fingerprints of both the signal of interest and the artifact are known. In addition, a filter can only be chosen optimally when the spectra of signal of interest and artifact do not overlap; in other cases the filter can only remove part of the artifact.

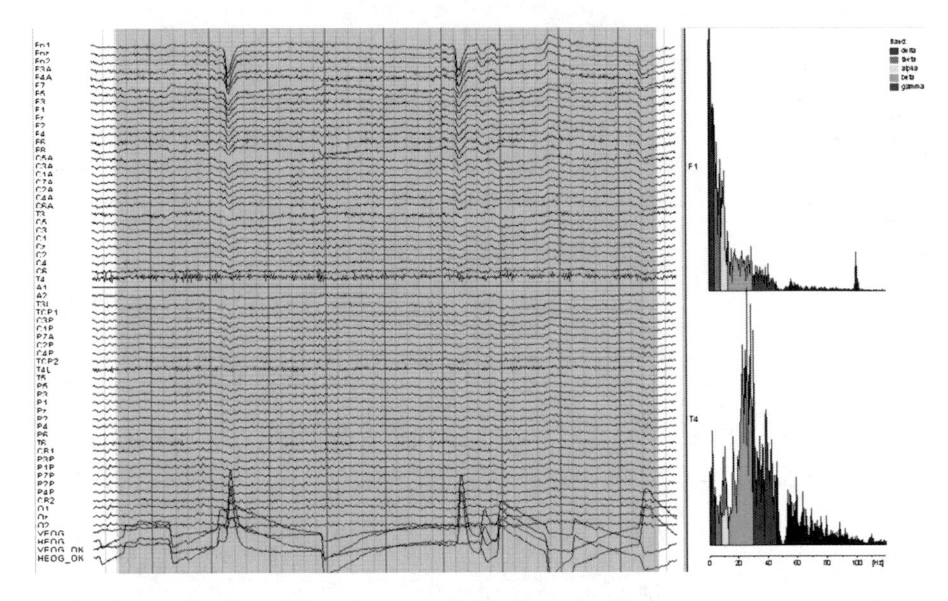

Fig. 4.9 Sample of 128-channel EEG data with artifacts. Only 64 channels are shown. *Left*: raw data, *right*: spectral data of channels F1 (ninth channel) and T4 (thirtieth channel) between 0 and 125 Hz. Note that this EEG was recorded using a 50-Hz notch filter

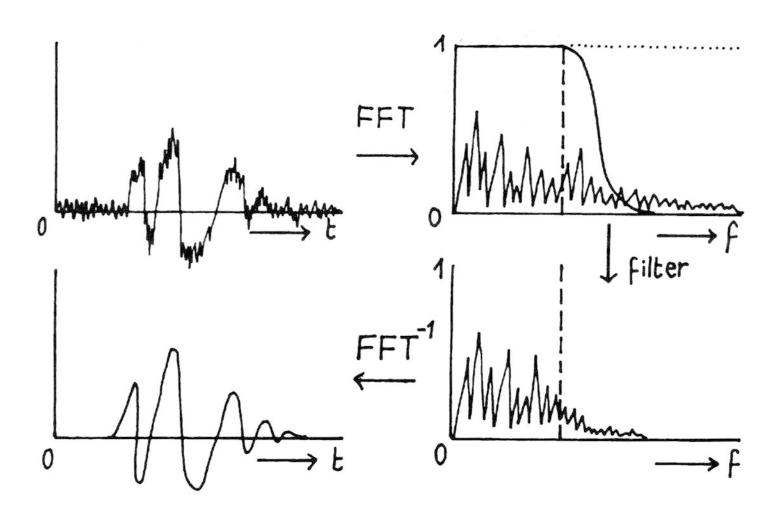

Fig. 4.10 Principle of filtering via the frequency domain. In a first step, the signal is transformed to the frequency domain using FFT. Then, in a second step, unwanted frequencies are set to zero by multiplication with the transfer function (in this case for a low-pass filter) and finally, the signal is transformed back to the time domain using inverse FFT (FFT^{-1}). *f* frequency; *t* time

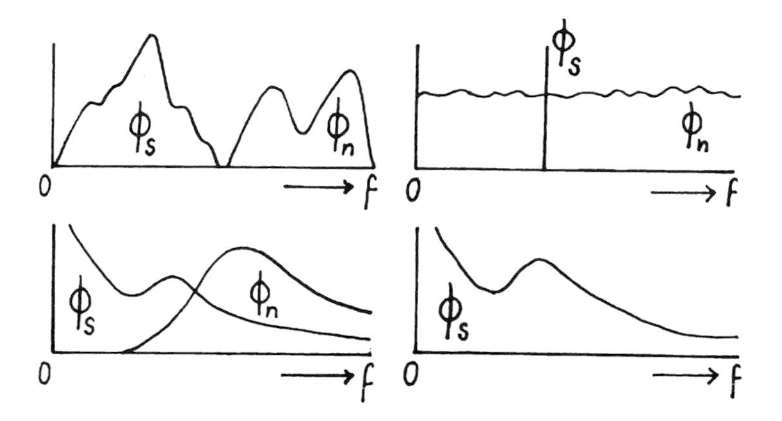

Fig. 4.11 Top left: the spectra of the signal (ϕ_s) and the noise (ϕ_n) are disjoint. *Bottom left* and *top right*: the spectra of signal and noise overlap. *Bottom right*: noise is absent

Question 4.2 Can artifact successfully be removed from the signal in the examples in Fig. 4.11?

A filter is defined in the frequency domain by its transfer function which consists of two parts: the frequency response function and the phase response function. The reason for this duality is that the Fourier transformed signal is complex valued for each frequency. This implies that each of these values has both an amplitude and a phase. Thus, a filter can only be completely known by defining what it does to both the amplitude *and* the phase at each frequency (see Box 4.1). If a filter changes the phase of the Fourier transformed signal at certain frequencies, this implies that peaks may change latency in the time domain; an unwanted phenomenon in most cases. Many digital filters however are "zero-phase shift" filters, indicating that phase is not changed by the filter. In these cases, the phase (shift) of the transfer function is constantly zero and the filter is otherwise completely defined by its amplitude response function or, when expressed as a fraction between 0 and 1, gain.

Box 4.1 Update on Complex Numbers

Complex numbers were invented to be able to define the roots of any polynomial equation including, e.g., $x^2 + 3 = 0$. They are numbers of the form $z = a + bi$, where a is the real part and b the imaginary part. The number i is defined as $i^2 = -1$.

A complex number can also be thought of as a point (a, b) in a Cartesian coordinate system in which the x-axis is the real axis and the y-axis the imaginary axis. This is called the complex plane. This allows to define a complex number $z = a + bi$ equivalently as a pair of numbers (r, φ) where

(continued)

Box 4.1 (continued)

$$r = |z| = \sqrt{a^2 + b^2} \quad (1)$$

is the magnitude, modulus, or absolute value and

$$\varphi = \arg(z) = \pm \arctan\left(\frac{b}{a}\right) \quad (2)$$

is the argument or phase.

This alternative representation of complex numbers is called the polar form and the original value can easily be recovered as $(a, b) = (r \sin(\varphi), r \cos(\varphi))$. Alternative expressions for z are the trigonometric form

$$z = r[\cos(\varphi) + i \sin(\varphi)] \quad (3)$$

or the exponential form

$$z = re^{i\varphi}. \quad (4)$$

The complex conjugate z^* of a complex number z is defined as

$$z^* = a - bi. \quad (5)$$

4.3.4 Types of Filters

Filters can be categorized in different ways, but an important distinction between filters is made by the specific effect they have on different frequencies in the signal: low cut-off (or high-pass), high cut-off (or low-pass), band-pass, or band-stop (see Fig. 4.12).

These four qualifications of filters make a distinction between which frequencies in the signal are preserved and which are attenuated. For example, when the goal is to remove low frequencies (e.g., slow signal changes due to movement in an EEG) from the signal, a low cut-off filter can be used. On the other hand, when high frequencies need to be removed (e.g., 50 Hz mains artifact in an EEG), a high cut-off filter can be used. In the latter example, since mains noise is limited to a particular frequency, it is also possible to opt for a band-stop filter, removing only the frequencies just around the artifact frequency. Finally, a band-pass filter is simply a combination of a high-pass and a low-pass filter, which is often used for EEG recordings to attenuate both the lowest and the highest frequencies, i.e., to remove both slow- and fast-changing unwanted signals.

Another application of a low-pass filter is to avoid aliasing (see Sect. 3.3.2): when all frequencies above the Nyquist frequency (sampling frequency/2) are cut

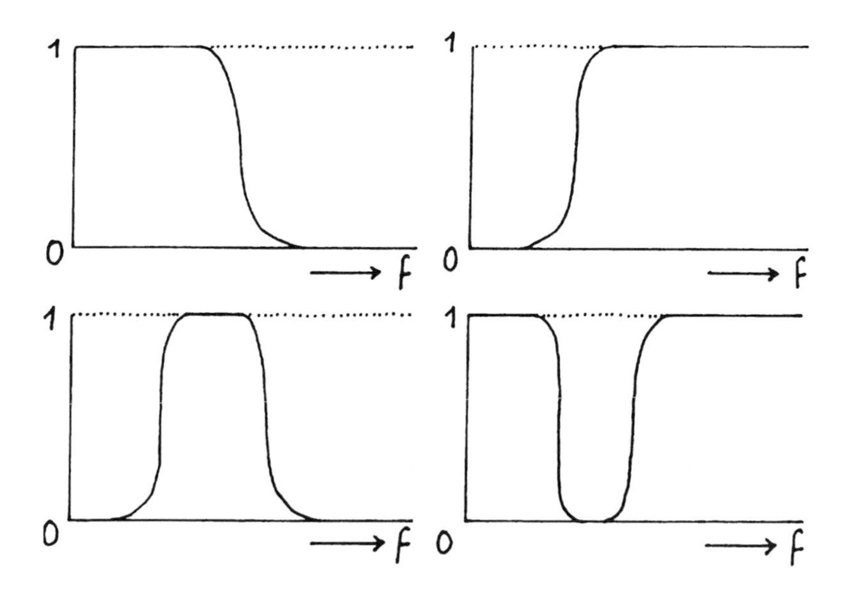

Fig. 4.12 Illustration of the four types of filters by their transfer functions. *Top left*: low-pass filter, *top right*: high-pass filter, *bottom left*: band-pass filter, *bottom right*: band-stop filter. *f* frequency

off before sampling, the artifactual frequencies due to aliasing will no longer be present. Note however, since filters are not perfectly "sharp," in practice a factor of 3 with respect to the sampling frequency is taken into account.

In practice, zero-phase shift filters of a specific type (Butterworth, Chebychev, elliptic) are fully defined by their cut-off frequency (or frequencies in case of a

> *Question 4.3* How should an anti-alias filter be chosen when sampling at 250 Hz?

band-pass filter) and order. The cut-off frequency is the frequency at which 50% of the amplitude is preserved. In case of a low cut-off filter, this means that for the lowest frequencies the amplitude is set to zero (stop band), whereas for the highest frequencies the amplitude is completely preserved (pass band). In a frequency band around the cut-off frequency, the preservation of the amplitude slowly rises from 0 to 100%. The steepness of a filter is defined by its order; the higher the order, the steeper the filter. A first order filter will reduce the signal amplitude by half every time the frequency doubles (i.e., when the frequency goes up an octave). A second order filter will reduce the signal amplitude by a factor of 4, every time the frequency doubles, etc. Instead of the order of the filter, the steepness or roll-off of the filter can also be defined in dB/octave (see Box 4.2). A reduction in signal

Box 4.2 Update on Decibels

Although the decibel is mostly known as a measure of sound intensity, more generally it is a logarithmic unit of measurement that compares the size of a physical quantity relative to a reference level. A decibel is a tenth of a bel.

When the quantity under consideration measures power or intensity, decibel is defined as

$$10^{10} \log \frac{p_1}{p_2}, \tag{1}$$

where p_1 and p_2 must have the same dimension and p_2 is the reference power.

When the quantity under consideration measures amplitude, which is typically considered as the square root of power, decibel is defined as

$$10^{10} \log \frac{{a_1}^2}{{a_2}^2} = 10^{10} \log \left(\frac{a_1}{a_2}\right)^2 = 20^{10} \log \frac{a_1}{a_2}, \tag{2}$$

where a_2 is now the reference amplitude.

amplitude of a half implies that power is reduced by $-20 \log½ = 6$ dB. Hence a first order filter has a roll-off of 6 dB/octave and similarly, a second order filter has a roll-off of 12 dB/octave.

Although Chebychev and elliptic filters are steeper than Butterworth filters, which might seem more optimal for filtering purposes, they suffer from ripples in the pass band (and the stop band in case of the elliptic filter). This means that the gain is not exactly 1 or 0, but fluctuates around this number. In many practical applications, a Butterworth filter is therefore used as zero-phase shift filter.

4.3.5 Filtering in the Time Domain: Smoothing

So far, only filtering in the frequency domain has been discussed, mainly because it is easier to understand the basic concepts. However, there are some problems with filtering, especially for the sharpest filters. Suppose, we want to filter out one particular frequency, say at 1 Hz. The effect of a perfect filter would be to subtract an infinitely long 1 Hz sine wave from the signal in the time domain, even for finite length signals. This means that in effect artifactual 1 Hz sines are introduced to the signal in the time domain. Generally speaking, the extent of a filter in the frequency domain is inversely related to its extent in the time domain and vice versa. This is

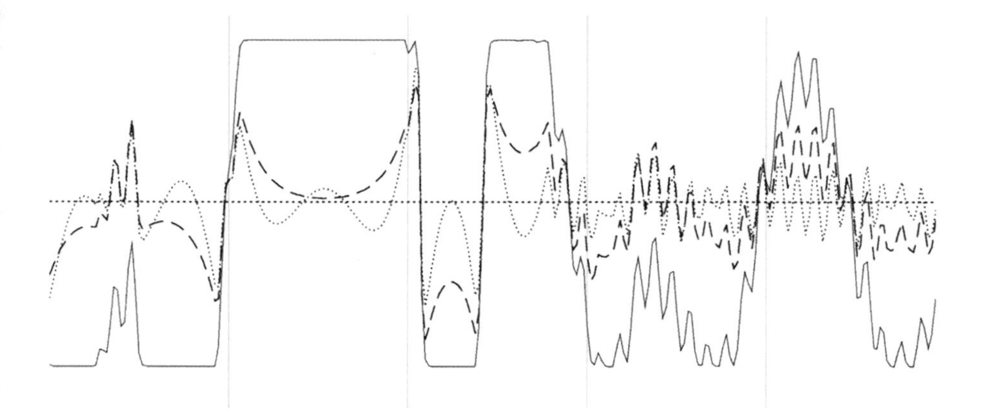

Fig. 4.13 Example of the effect of an increase in sharpness of the filter on the signal in the time domain. *Drawn line*: original signal, *dashed line*: signal filtered with a first order butterworth filter (6 dB/oct), *dotted line*: signal filtered with a third order Butterworth filter (24 dB/oct). The effect is particularly large when the original signal increases sharply

one of the reasons why very sharp filters are not used in practice: they introduce artifactual ripples in the signal (see Fig. 4.13).

Because of the inverse relationship between the extent of filters in the time and frequency domain, filtering in the time domain does not solve the problem, but some filters are (more) easily applied in the time domain. The most applied filter in the time domain is a smoothing filter that acts as a low-pass filter. For each data point in the original signal, a smoothing filter calculates a weighted sum of the surrounding values to find the filtered value. A smoothing filter is typically symmetrical and can take 3, 5, 7, or many more data points into account. The simplest approach is to average the surrounding values, but this has the disadvantage that some of the temporal resolution is lost. Instead, weights can be used that are largest for the middle value, drop off to zero toward the beginning and end of the filter, and together add up to 1. The last requirement ensures that the total amount of information in the signal is preserved. For a filter with three weights, these could be ¼, ½, and ¼ (a Hanning smoothing window). This type of filtering can of course only be performed on digital signals since the future of analog signals is not known.

4.3.6 Relationship Between Filtering in the Frequency and Time Domain

The weights of a smoothing filter indicate how every data point surrounding the data point of interest is contributing to the filtered value. Vice versa one could consider how a particular unfiltered data point influences all surrounding filtered values by reversing the weights of the smoothing function in time. This is called the impulse response function (IRF). For a symmetric smoothing filter like the Hanning filter described above, the IRF is simply the same as the smoothing filter (see also

Box 4.3). The IRF thus yields the response of a filter to an impulse (a delta function, which is infinity at time zero and 0 everywhere else, with an integral of 1). The filtered signal could thus be obtained by multiplying the IRF with the unfiltered value at each point and summing all these weighted IRFs. Mathematics (Box 4.3) can be used to show that the IRF is related to the transfer function (the description of the filter in frequency space) by the Fourier transform.

Once this is known, the relationship between the extent of a filter in the frequency domain and the extent of a filter in the time domain can be explored by

Box 4.3 Mathematics of Filters

The smoothed value S_i^f of a sampled signal S at position i can be mathematically described as:

$$S_i^f = \sum_{j=-n}^{n} w_j S_{i+j}. \tag{1}$$

Here, a symmetric filter with $2n + 1$ weights is used. This can be rewritten to:

$$S_i^f = \sum_{j=-n}^{n} w_j S_{i+j} \overset{j=-k}{=} \sum_{k=-n}^{n} w_{-k} S_{i-k} = \sum_{k=-n}^{n} \mathrm{IRF}_k S_{i-k}, \tag{2}$$

where IRF_k is the kth element of the IRF as described above.

This is equivalent to:

$$S^f = \mathrm{IRF}^* S, \tag{3}$$

where * denotes convolution. Thus, the filtered signal is equal to the convolution of the IRF with the unfiltered signal.

Applying Fourier transform to (3) and using that the Fourier transform of a convolution of two signals is the product of the Fourier transformed signals, results in:

$$F(S^f) = F(\mathrm{IRF} * S) = F(\mathrm{IRF}) \cdot F(S). \tag{4}$$

Hence, in the frequency domain the filtered signal can be obtained by multiplying the unfiltered signal with the Fourier transform of the IRF, which must thus be the transfer function. Hence the IRF and the transfer function are related through Fourier transform.

applying Fourier transforms to some examples. An interesting example is the ideal filter in the frequency domain: a step function (one for all passing frequencies and zero for all frequencies that are stopped). When this function is Fourier transformed a sinc function ($\sin(x)/x$) with infinite extent results. One way to make a distinction between filters in the time domain based on their extent is the division between finite (FIR) and infinite impulse response (IIR) filters. Moving average filters are examples of FIR filters, whereas the ideal low- or high-pass filter just mentioned provides an example of an IIR filter.

4.3.7 Dos and Don'ts of Filtering

Although filters are generally very useful to remove artifacts (Sect. 7.3.4.2), there are some practical rules regarding when and how to use filters. The most important aspect of filters to keep in mind is that very steep filters (approximating ideal filters) introduce artifactual waves (ripples) in the signal. An example was already given in Fig. 4.13. Thus, a shallower roll-off is preferred. Vice versa, since the cut-off frequency only indicates the frequency at which 50% of the amplitude is preserved, one should realize that filtering does not imply that all frequencies in the pass-band are passed and all other frequencies are stopped. This explains the phenomenon illustrated in Fig. 4.14.

The best practice is to avoid artifacts as much as possible. This can be achieved by optimizing electrode impedance, replacing and repositioning electrodes when necessary, avoiding movements and muscle contraction (by having the subject relax as much as possible and seating him comfortably), instructing the subject to blink as little as possible (or provide for blinking pauses), and by removing sources of mains noise. Details about each of these approaches are described in Sect. 7.3.4.2. When filtering is needed, it should be limited and only slow roll-offs should be used. Typical EEG filters would use a pass-band of 0.01–70 Hz (at 12 dB/oct), when sampling at least at 200–250 Hz. Notch filters are best not used at all, except in very noisy environments, such as an ICU.

4.4 Feature Extraction

In general, feature extraction can be used to focus on phenomena of interest in a signal, by *dimension reduction*. In that sense it has a similar use as EEG filters: it can be used to discern signals of interest from noise. This way of processing the EEG data is especially useful, when the amount of data in relation to the relevant information is large. Although mathematical methods of dimensionality reduction such as *PCA* and ICA can also be considered as feature extraction methods, the focus is here on feature extraction applications and methods that are particularly relevant for EEG.

Fig. 4.14 Spectrum of an EEG signal, (**a**) before and (**b**) after filtering with a high-pass filter (cut-off 10 Hz, 12 dB/oct) and a low-pass filter (cut-off 8 Hz, 12 dB/oct) which should remove all frequencies in case of ideal, sharp filters. Spectral power remains because of the slow roll-off of both filters, as is illustrated in the transfer functions: (**c**) 10 Hz high-pass at 12 dB/oct, (**d**) 8 Hz low-pass at 12 dB/oct, and (**e**) the two filters combined. The shape of the filtered spectrum clearly reflects the shape of the transfer function of the combined filters

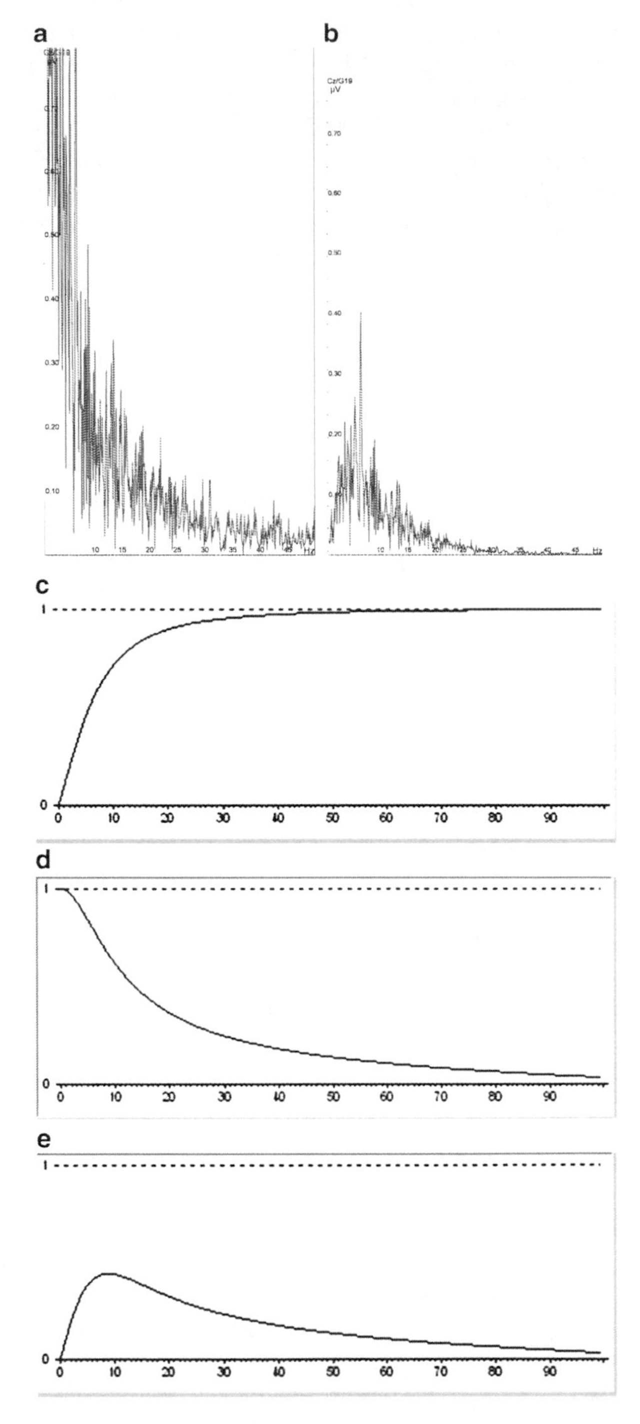

4.4.1 Template Matching and Wavelet Analysis

One method to identify specific features in the EEG is template matching which entails finding a smaller part in the entire EEG signal which matches a predefined template signal. This is done by sliding the template along the EEG signal, calculating the local cross-correlation (i.e., the sum of the products of all values of the template with the temporally matching values of the signal) and finding the largest cross-correlation over all temporal shifts. Mathematically, the cross-correlation is defined as

$$(t * s)(n) = \sum_{k=-\infty}^{\infty} t^*(k)s(n+k), \tag{4.1}$$

for a template signal t and an EEG signal s, both sampled discretely in time. Here, t^* is the complex conjugate of t (see Box 4.1), which is equal to t itself for real numbers as would be expected from an EEG template. Typically t would only be nonzero for a limited number of sample points. If the template (closely) matches the local EEG signal, large template values get multiplied by large EEG signal values, resulting in a large value for the cross-correlation. Calculating the cross-correlation can be a lengthy procedure, but a mathematical theorem (see also Box 4.3) allows to calculate the cross-correlation (or convolution) as a simple product in Fourier space. In Sect. 5.6.1 an example of template matching in neuromonitoring is discussed.

Wavelet analysis is basically a specific case of template matching in which multiple templates (wavelets; derived from a so-called mother wavelet by scaling (stretching in time) and translation) are all matched to the signal. A wavelet is a wave-like oscillation with an amplitude that starts at zero, becomes nonzero for a brief period of time, and decreases to zero again. Different wavelets in one "family" look similar, but have different constituent frequencies. In contrast to Fourier analysis, wavelet analysis allows to locally determine the frequency content of a signal. Thus, if parts of the EEG signal with a specific frequency signature need to be identified – such as sleep spindles (see Sect. 3.5) – wavelet analysis may be helpful.

4.4.2 EEG Spike Detection

An example in which the amount of interesting data in the EEG can be much smaller than the total amount of data is the case of epileptic spikes. Epileptic spikes are a hallmark of epilepsy and therefore very important for diagnosis, particularly because they can occur inter-ictally. Additionally, they can be used for localizing an epileptic focus in the preparatory phase before epileptic surgery (see also Sect. 8.3). In principle spikes are identified by visual evaluation of the entire dataset, which can be an enormous amount of work in case of, e.g., overnight recordings during

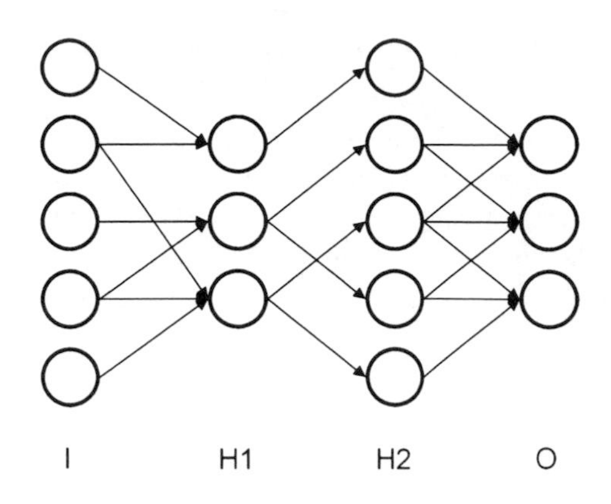

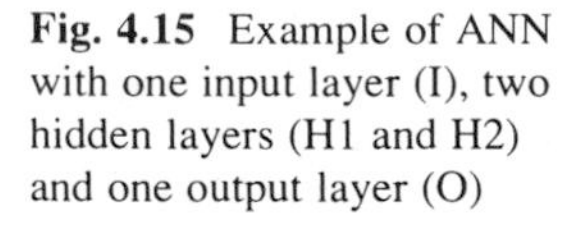
Fig. 4.15 Example of ANN with one input layer (I), two hidden layers (H1 and H2) and one output layer (O)

sleep. Automatic spike detection would thus be very advantageous. Even when the amount of false positives is large, a considerable reduction of the amount of data that needs to be reviewed and assessed manually can be achieved. For automatic detection, it is first and foremost important to decide on what makes a spike a spike. Usually, spikes are defined as very sharp waveforms with a duration of 20–70 ms. If the waveform is (a little less) sharp, and a little longer in duration (70–200 ms), it is called a sharp wave. To detect spikes an EEG segment can be divided into waves and their constituent halfwaves (line segments connecting two consecutive local extrema) after which durations, amplitudes, and sharpness (calculated as the second derivative) at the top are determined. If these values exceed certain thresholds, a peak may be detected. Template matching (see Sect. 4.4.1) can also be used to find epileptic spikes based on a template of the spike morphology.

However, the above definition of a spike is not very specific, in that it also includes, e.g., eye blinks and artifacts. Hence, the context in which the spikes appear should also be taken into account; i.e., a spike is not just a spike because of its specific morphology, but also because of what happens at the same time in other channels and what happens just before and after the occurrence of the spike. Thus, using only local wave morphology is usually not enough to identify epileptic spikes. Gotman and Wang (1991) have therefore made spike detection dependent on the "state" of the patient, which can vary from active wakefulness to slow wave EEG. Alternative newer methods for spike detection employ artificial neural networks (ANNs) or wavelet analysis (see Sect. 4.4.1).

ANNs are mathematical models/tools whose structure is based on biological neural networks. They consist of interconnected groups of artificial neurons or nodes, often configured in three or more layers (input layer, one or more hidden layers, and output layer, see Fig. 4.15). Depending on the (weighted) sum of the inputs, a neuron will fire, i.e., a node will send input to its output nodes. An important application of ANNs is classification. In order to perform such a task, the ANN first needs to be trained on a training data set, containing examples of inputs that belong to the different classes that need to be distinguished. As a result of learning (processing the training data set and finding the network that matches each input from the training data set to its output) the number of nodes, their interconnectivity, and the weights may be adapted. In case of spike detection, the examples in the training data set would need to contain both spikes and nonspikes.

4.5 EEG Findings in Individual Patients

In both patients 1 and 2, EEG was recorded for approximately half an hour using 21 electrodes in a standard 10–20 setting. The typical procedures used to assess EEG reactivity (eyes opening and closing) and to evoke epileptic activity (light flashing, making a loud noise) that were explained in Sect. 4.2 were applied.

Patient 1

After this patient had been admitted to hospital, he was examined extensively. Both on a computed tomography (CT) image and on magnetic resonance imaging (MRI), edema was observed fronto-temporally in the right hemisphere, causing swelling of the brain and possibly epileptic activity. Although the edema was surrounding a white matter lesion, it was not exactly clear what had caused this lesion. Since left limb shocks were continuing, an EEG was made to exclude *status epilepticus*. EEG activity was found to be generally normal in the left hemisphere: the background EEG was sufficiently fast and some reactivity and *differentiation* were present. In contrast, periodic lateralized epileptic discharges (PLEDs) were observed continuously in the right hemisphere (Fig. 4.16).

Approximately after 10 min of EEG recording, the patient suddenly opened his eyes, turned his head to the right, then to the back, and to the left. He flexed his left elbow and then made grasping movements with his right hand and flexed and extended his left elbow in a tonic-clonic manner. In addition his heart rate increased and the frequency of the PLEDs increased to 2 Hz after which rhythmic delta activity was observed for almost 2 min. After this seizure, the EEG was temporarily flat (Fig. 4.17).

Even though patient 1 suffers from a generalized seizure during the EEG recording, there is no status epilepticus. The patient received antiepileptic drugs, resulting in disappearance of the seizures and a general improvement of his condition. After 6 days he was discharged from the hospital.

Patient 2

In general, sufficiently fast EEG activity, with proper differentiation and reactivity, was observed in this young patient. However, very frequently *paroxysmal* activity occurred, in which isolated peak–wave complexes (Fig. 4.18) and series of peak–wave complexes occurring at a frequency of 3 Hz (Fig. 4.19) were found in mainly right frontocentral areas, spreading to neighboring areas.

At the same time left midtemporal peak–wave complexes were observed. The 3 Hz peak–wave complexes occurred both spontaneously, as well as during

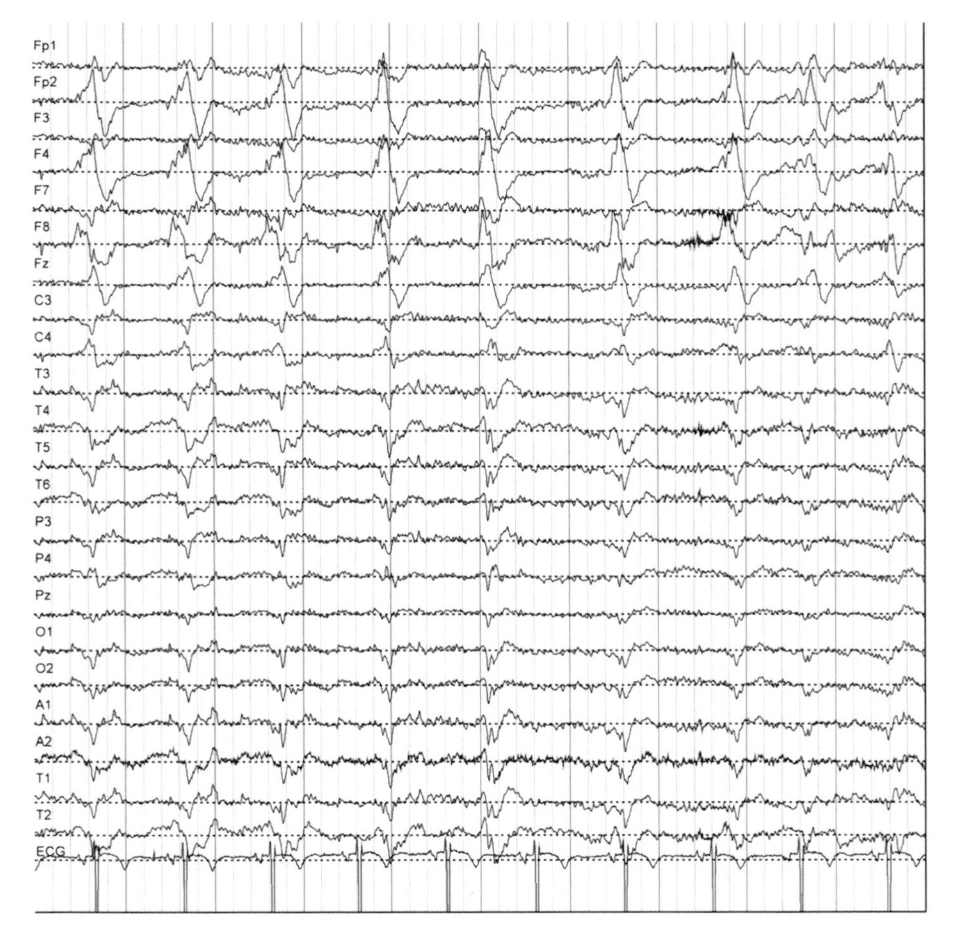

Fig. 4.16 Continuous PLEDs in the right hemisphere (all even channels) in patient 1. The PLEDs are particularly present in frontocentral areas and sometimes spread to the left hemisphere. The bottom channel is the ECG

hyperventilation. During these periods, the patient opened and closed his eyes, stopped hyperventilating and continued later, without noticing the absence himself. After a series of 3 Hz peak–wave complexes, the EEG quickly returned to normal. Thus, the EEG was normally built up, with aspects of absence epilepsy.

The EEG findings confirmed the clinical diagnosis of absence epilepsy. Absence epilepsy is most often diagnosed during childhood and one of the idiopathic generalized epilepsies, meaning that there is no other cause of epilepsy (such as a tumor) and that the entire brain is affected by it (reflected in the loss of consciousness). It can often be treated with antiepileptic medication, which was also the first approach in this patient.

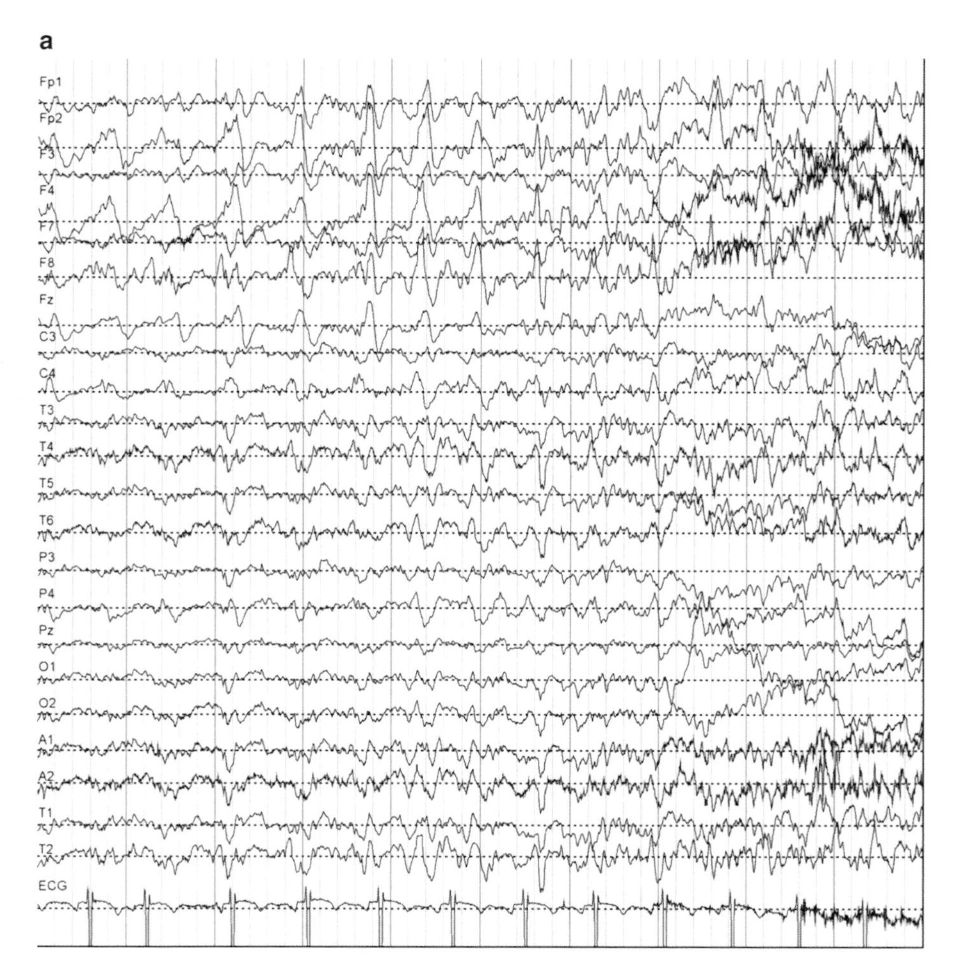

Fig. 4.17 EEG in patient 1 during his epileptic seizure. (**a**) Start of the seizure; heart rate and PLED frequency increase, (**b**) seizure with 4 Hz delta activity, (**c**) ample muscle artifact is observed during the seizure because of the tonic-clonic limb movements, (**d**) flat post-ictal EEG

4.6 Other Applications of EEG Recordings in Neurology

Since the development of other neuroimaging techniques such as CT (using X-rays) and MRI (using magnetic fields), the clinical use of regular EEG recordings has been mostly limited to the diagnosis of epilepsy. However, new technical and scientific developments have renewed interest in the EEG as a technique that can be used to assess brain function cheaply, noninvasively, and under various circumstances. In Sect. 3.5 the use of EEG to diagnose sleep disorders is illustrated. Here, we focus on applications of the EEG in circumstances where it is important to

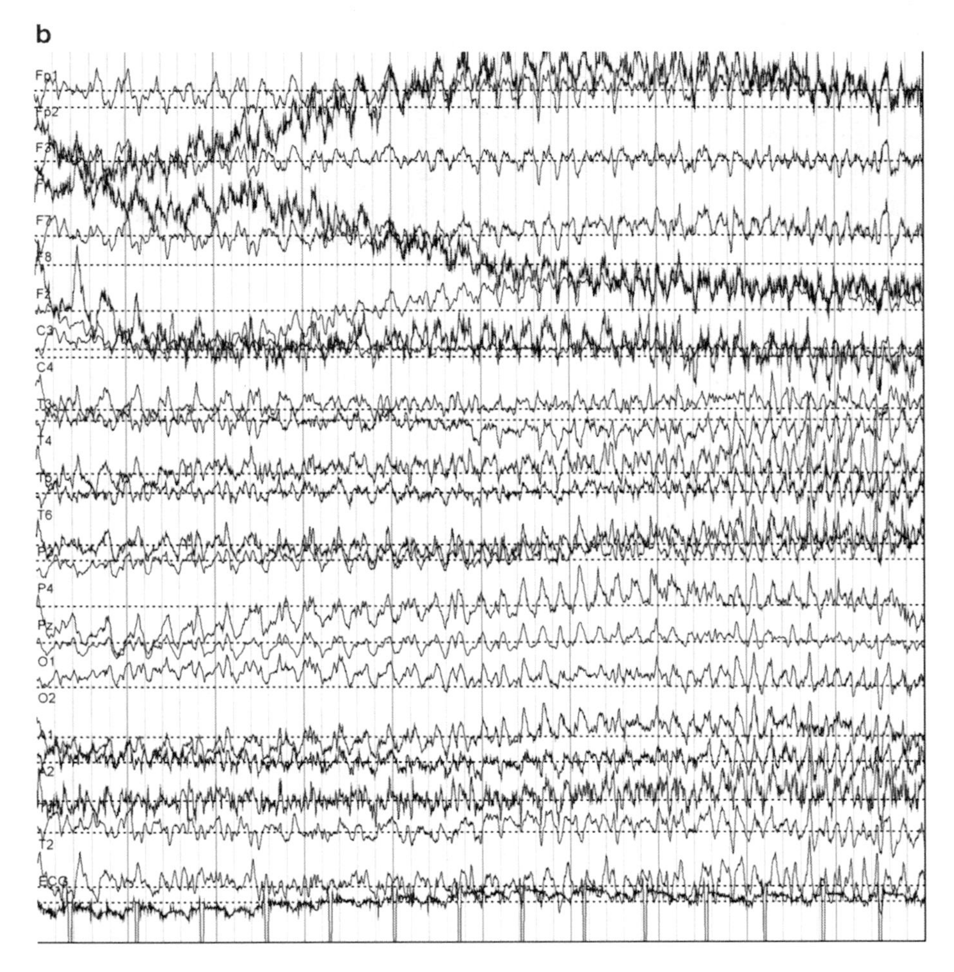

Fig. 4.17 (continued)

assess brain function when other clinical assessments are limited. What two of these applications have in common is that digital signal analysis techniques are essential, allowing the assessment of the EEG by nonexperts in those specific circumstances.

4.6.1 Monitoring of Carotid Artery Endarterectomy

The purpose of carotid endarterectomy is to remove *atherosclerotic plaque* from the carotid artery. At some stage during this procedure, the artery is clamped to temporarily stop the blood flow and allow the surgeon a clear view of the plaque. This is a very risky stage in the surgery for the brain when *collateral arteries* are

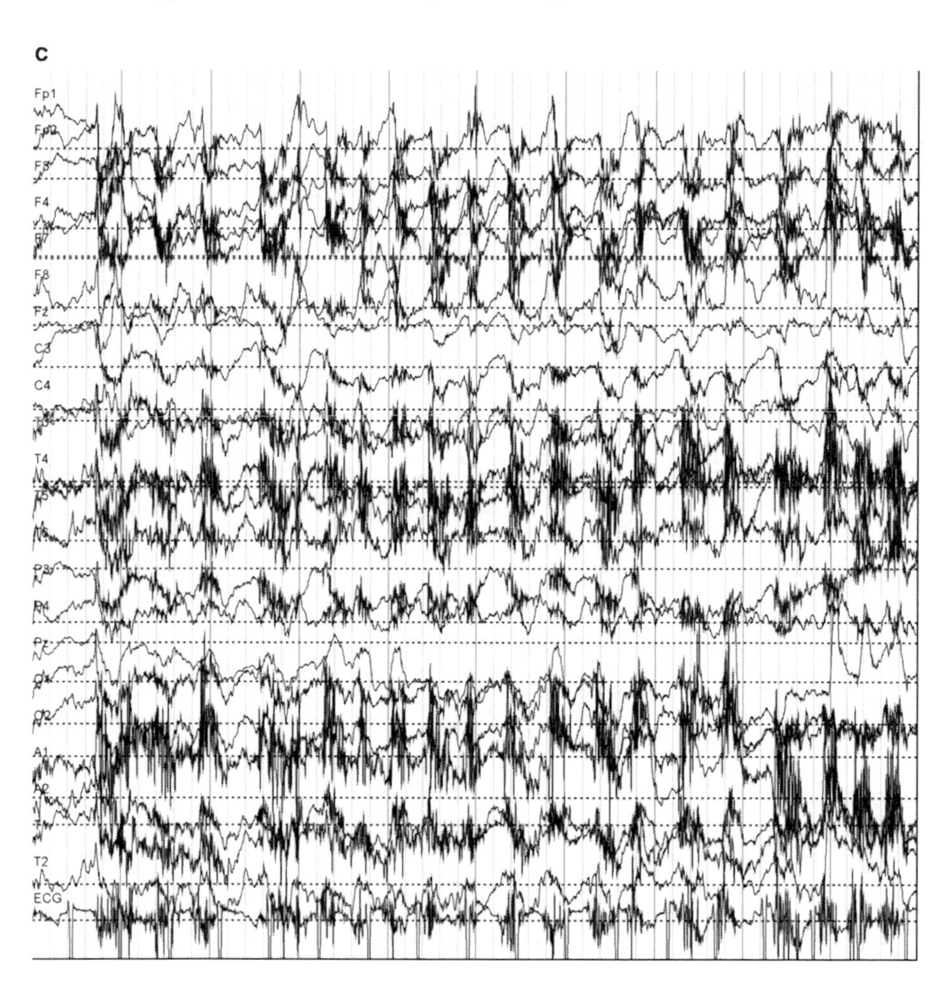

Fig. 4.17 (continued)

not sufficiently developed to compensate for the hypoperfusion on the clamped side. In those cases part of the brain may suffer from *hypoxia* or *ischemia* which can potentially lead to permanent brain injury. Since EEG is sensitive to both hypoxia and ischemia, and allows a continuous, noninvasive, objective, and rapid assessment of brain function, it is routinely used to monitor oxygen supply to the brain during carotid endarterectomy. Mild *hypoperfusion* is not visible in the EEG, but it is usually well tolerated. However, a further reduction of the cerebral blood flow leads to slowing and/or a decrease in amplitude of the EEG, and a flattening of the EEG signal is related to severe hypoperfusion. Thus, the EEG already detects neuronal dysfunction, when it can still be reversed. An easy and quantitative way to analyze such changes in the EEG is spectral analysis

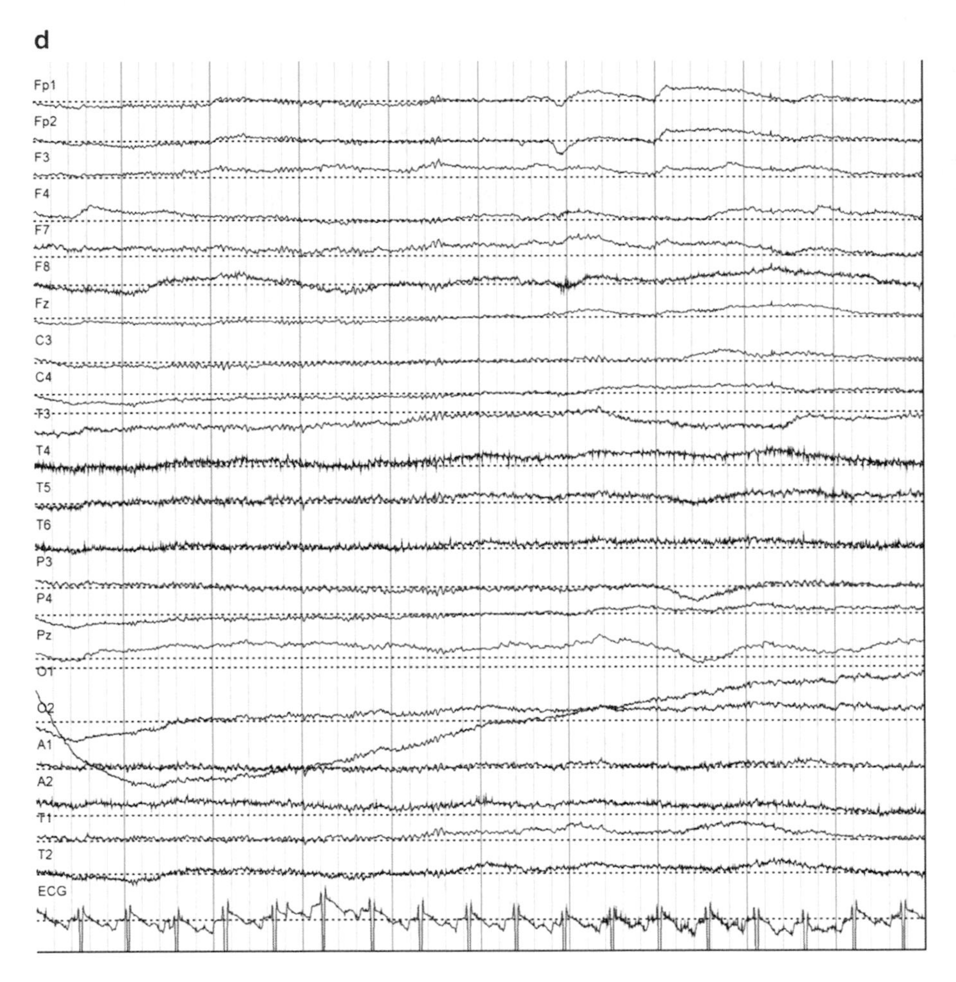

Fig. 4.17 (continued)

(see Sect. 3.3). Since EEG also has some disadvantages in the operating room, being sensitive to artifacts, noise, and anesthesia, transcranial duplex ultrasound (see Sect. 10.5.1) is often employed as an additional technique.

4.6.2 Monitoring of Status Epilepticus in Intensive Care Units

Of all patients who are admitted to the ICU, a large part suffers from nonconvulsive status epilepticus (NCSE) related to, e.g., acute brain injury, intracranial hemorrhage, or intoxications. Because there is no motor activity in NCSE, it cannot be reliably diagnosed by clinical examination only. In addition, when (convulsive)

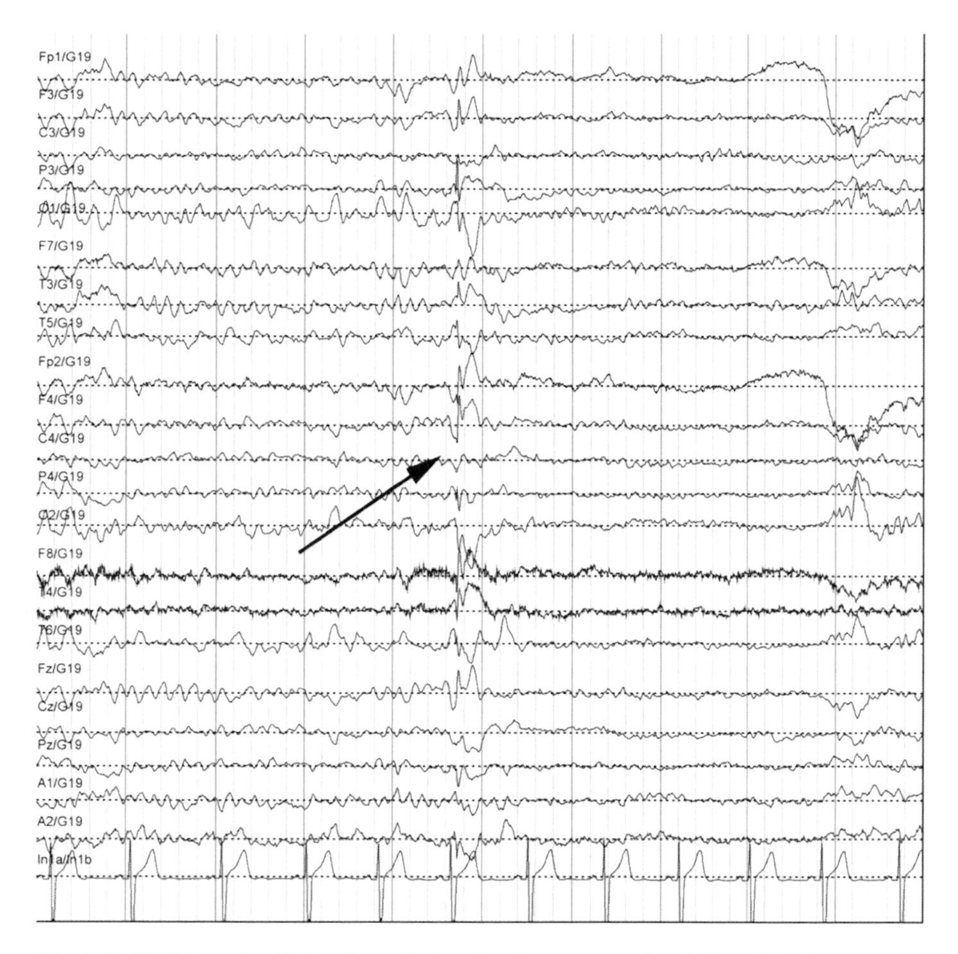

Fig. 4.18 EEG in patient 2 showing an isolated peak–wave complex, indicated by the *arrow*

status epilepticus is treated, the motor activity may disappear while seizure activity still persists. In both cases, only continuous EEG recordings can reveal whether seizure activity is (still) present and whether it is controlled. Furthermore, continuous EEG recordings can capture sporadic, spontaneous epileptic activity, where routine EEG only provides a short 30 min impression of brain activity and can miss a seizure. A disadvantage of continuous EEG monitoring is that large quantities of data are generated which need to be assessed at a regular basis to provide adequate treatment as soon as it is required. One approach is to train ICU personnel to recognize possible pathophysiological events and have them warn a neurologist who can further assess the EEG. To avoid long delays in assessment, it is also possible to remotely assess the EEG using network technology. Finally, several automatic quantitative EEG analysis methods can help to detect abnormalities. One example is to display the percentage of alpha activity in the EEG per 30 s window, as *convulsive* activity is often accompanied by a relative increase in alpha activity.

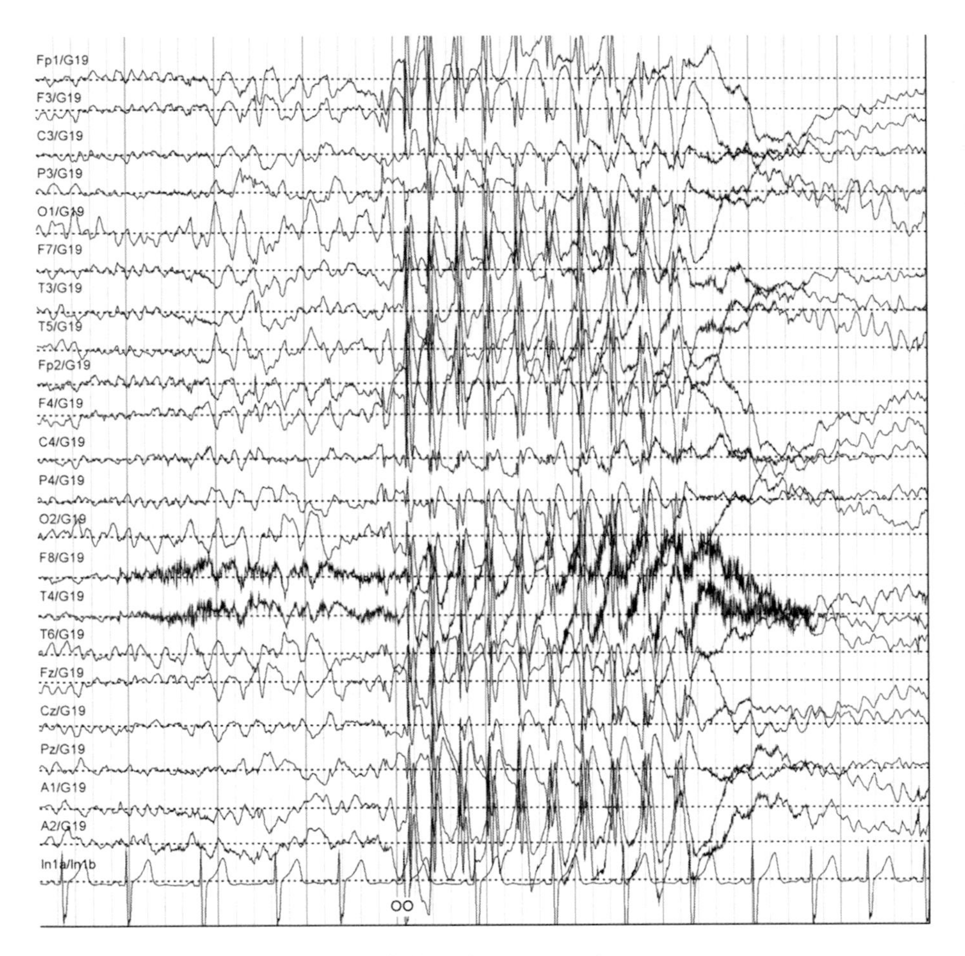

Fig. 4.19 EEG in patient 2 showing 3 Hz peak–wave complexes

4.6.3 Confirmation of Brain Death

In some countries the EEG plays an important role in diagnosing brain death. In those cases one of the criteria to confirm brain death after the clinical tests have been completed is an isoelectric EEG (see Fig. 4.20).

The precise definition of an isoelectric EEG differs, but it does *not* include burst-suppression EEGs (alternation of flat lines (less than 2 μV EEG amplitude) with burst-like activity, occurring, e.g., in acute intoxications), low-voltage slow EEGs consisting of low frequency waves (occurring, e.g., in comatose patients), or flat lines over limited areas of the brain only. The guidelines for demonstrating brain death have particular requirements about the number and position of the electrodes and the duration of the recording. What makes this endeavor particularly difficult are the artifacts due to the presence of instruments in the ICU (such as ECG monitors, respirators, and warming blankets) or artifacts of biological origin (cardiovascular, muscular, or respiratory).

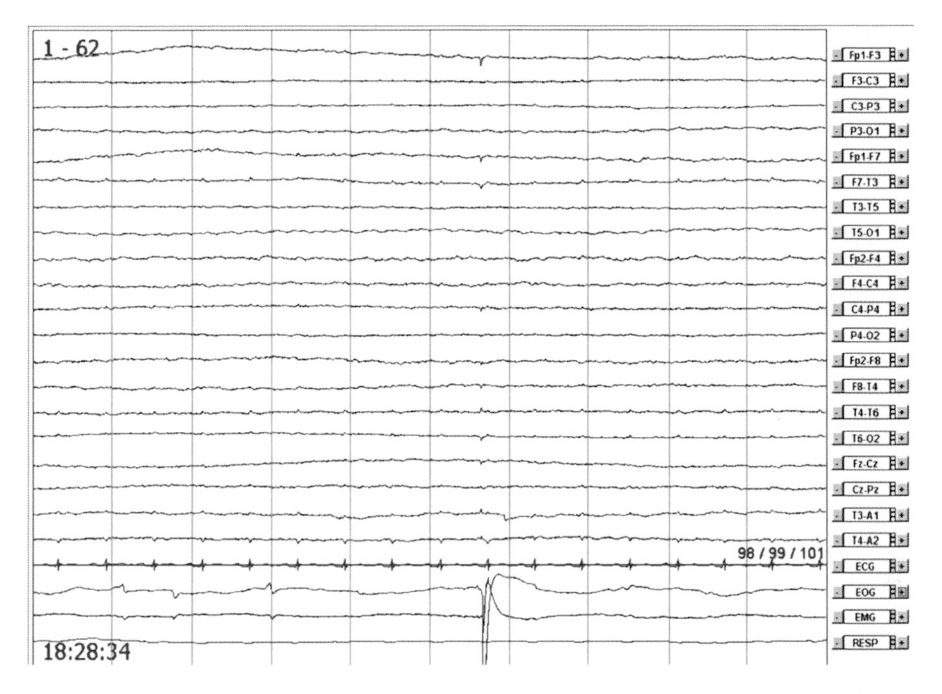

Fig. 4.20 An isoelectric EEG (21 EEG channels in 20 derivations, ECG, EOG, EMG, and respiration). Note that the only activity in the EEG temporally correlates with either heart beats or movements

4.7 Answers to Questions

Answer 4.1

The spectra show that there is mains noise on channel F1 (peak at 100 Hz, the first harmonic of 50 Hz) and EMG noise on channel T4 (broad spectrum peaking around 20–30 Hz, typical of EMG).

Answer 4.2

For the top left example, the spectra of signal and noise do not overlap, allowing to define a filter that selectively removes the noise. In the top right example, the signal spectrum contains a single frequency, whereas the noise is white (containing all frequencies). In that case, the most optimal filter would select only the frequency of the signal (or a very narrow band around it). In the bottom left example, the signal and noise spectra overlap to a large extent. This could be an example of EEG (φ_s) and EMG (φ_n) spectra. Here, an optimal filter can only be defined such that the noise is removed as much as possible but not without removing some of the signal as well. Finally, in the bottom right (EEG) example, filtering is unnecessary because noise is absent.

Answer 4.3
When sampling at 250 Hz, the Nyquist frequency is 125 Hz. Although theoretically, a low-pass filter with a cut-off frequency of 125 Hz would suffice to avoid aliasing if filters were perfect, in practice a low-pass filter with a cut-off frequency of 70 or 80 Hz would be used.

Glossary

Amplifier A device that increases the amplitude of a signal.
Atherosclerotic plaque Thickening of an artery wall due to build-up of fatty materials such as cholesterol.
Axon Long extension of a neuron that conducts electrical activity away from its cell body.
Canthus Corner of the eye where the upper and lower eyelids meet.
Clonic The second phase of a tonic-clonic seizure in which the body makes rhythmic jerking movements.
Collateral artery An artery which establishes a parallel circulation as in a by-pass.
Convulsive Here: rapid alternating muscle contraction and relaxation as in an epileptic seizure.
Cortex As in cerebral cortex: the outer layer of the cerebrum or forebrain.
Dendrite Branched extensions of a neuron that conduct the electrochemical stimulation as received from other neurons to the cell body.
Derivation The weighted sum of several (usually subtraction of two) EEG channels.
Differentiation Here: distribution of EEG rhythms over the scalp. As an example, differentiation is normal when alpha is maximal in occipital areas.
Diffusion The motion of particles due to a concentration gradient.
Dimension reduction Here: simplification of a signal by reducing the number of variables influencing it.
Electrodes Here: small metal (iron, tin, gold) disks used to make contact with the body to record electrical signals from it.
Electromagnetism One of the fundamental forces of nature describing the interaction between electrically charged particles.
Excitatory Increasing the chance of future occurrence of an action potential.
Hypoperfusion Decreased blood flow.
Hypoxia State of oxygen deprivation.
ICA Independent component analysis: a mathematical technique to blindly separate sources of activity in a compound signal such that all components are statistically independent.
Ictal During a seizure.
Impedance Resistance to time-varying electrical current.

Inhibitory Decreasing the chance of future occurrence of an action potential.
Inion The protuberance of the occipital bone at the lower posterior part of the skull.
Ion channel Protein that forms a pore in a membrane, allowing the transport or diffusion of ions.
Ischemia Insufficient blood supply, typically due to vascular obstruction.
Meninges System of membranes covering and protecting the central nervous system.
Nasion The intersection of the frontal and two nasal bones of the skull; the middle of the nose bridge.
Neurotransmitter Chemical substances in the body that help to relay signals from neurons to other cells.
Paroxysm Sudden fit or outburst.
PCA Principal component analysis: a mathematical technique to derive several uncorrelated variables from a possible correlated dataset such that the first component explains a maximum amount of variance.
Preauricular point An indentation in front of the ear, just above the jawbone.
Reactivity Here: response of EEG activity to certain stimuli, e.g., reactivity of the alpha rhythm is normal when it disappears upon opening the eyes.
Resolution The capability to differentiate different scales of a measure.
Stroke Loss of brain function due to an obstruction in blood supply.
Synapse Contact between the axon of one neuron and a dendrite or cell body of another.
Tonic The first phase of a tonic-clonic seizure in which the body becomes completely rigid.
Urinary incontinence Involuntary leaking of urine.

References

Online Sources of Information

http://en.wikipedia.org/wiki/Epilepsy. Overview of types of epilepsy, etiology, treatment and epidemiology
http://en.wikipedia.org/wiki/Eeg. Extensive overview of what EEG measures, how it is used and interpreted
http://en.wikipedia.org/wiki/Filter_(signal_processing). Overview of concepts related to filtering of (digital) signals
http://en.wikipedia.org/wiki/Digital_filter. More technical explanation on how digital filters are actually designed and implemented
http://en.wikipedia.org/wiki/Artificial_neural_network. Basics of ANNs
http://en.wikipedia.org/wiki/Wavelet_analysis. Basics of wavelet analysis, including the mathematics

Books

Ebersole JS, Pedley TA (2003) Current practice of clinical electroencephalography, 3rd edn. Lippincott, Williams and Wilkins, Philadelphia (Chapters 22 and 25 in particular)

Fisch BJ (1999) Fisch and Spehlmann's EEG primer. Basic principles of digital and analog EEG, 3rd edn. Elsevier, Amsterdam

Luck SJ (2005) An introduction to the event-related potential technique. MIT Press, Cambridge (Chapter 5 in particular)

Niedermeyer E, Lopes Da Silva F (2005) Electroencephalography: basic principles, clinical applications, and related fields, 5th edn. Lippincott Williams and Wilkins, Philadelphia. Also on http://books.google.com

Papers

Gotman J, Wang LY (1991) State-dependent spike detection: concepts and preliminary results. Electroencephalogr Clin Neurophysiol 79:11–19

Wilson SB, Emerson R (2002) Spike detection: a review and comparison of algorithms. Clin Neurophysiol 113:1873–1881

Chapter 5
Multiple Sclerosis, Evoked Potentials, and Enhancing Signal-to-Noise Ratio

After reading this chapter you should know:

- Why evoked potentials can be used to assess functioning of sensory pathways
- How signal-to-noise ratio (SNR) can be improved for evoked potentials
- How reliability of evoked potentials can be assessed
- How a cortical evoked potential is calculated from the raw EEG signal
- What are the clinically most used evoked potentials
- What factors determine stimulation frequency and number of stimulations for an evoked potential

5.1 Patient Cases

Patient 1

This 16-year-old female patient suffers from double vision (diplopia) and sensory disturbances in the right side of her face. In addition, she could not walk for about 15 min after getting out of bed one morning. She has remaining numbness and cramps in her legs, since. Neurological testing shows that her right trigeminal nerve, the main sensory nerve of the face, is indeed not functioning well. Two years ago, she suffered from an inflammation of her left optical nerve (optic neuritis). At that time, an MRI showed three white matter lesions indicative for *demyelination*. She was treated with anti-inflammatory drugs successfully. A new MRI now shows a lesion in the white matter in the midbrain.

N. Maurits, *From Neurology to Methodology and Back: An Introduction to Clinical Neuroengineering*, DOI 10.1007/978-1-4614-1132-1_5,

Patient 2

Since a few months, this male 38-year-old patient suffers from a tingling feeling in his right arm, right leg, and right side of his tongue. In addition, he has double vision and diminished vision in his right eye. All these complaints suddenly develop, last for a few minutes, and then disappear again. These episodes now take place up to several times a day. In between these attacks, he still suffers from decreased sensation in the right side of his body. Earlier investigations explored the spinal fluid for *oligoclonal bands* (which can be indicative of a disease of the central nervous system when their number is increased) and an MRI for white matter lesions, thinking that the patient may suffer from multiple sclerosis (MS), a demyelinating disease of the central nervous system. The number of oligoclonal bands was normal and although some white matter lesions were seen on the MRI, they were more consistent with a disease of the small vessels in the brain than with MS. The neurologist now thinks that the patient may suffer from inflammation of the small blood vessels in the brain, MS, or another demyelinating disease.

5.2 Evoked Potentials: Assessing Sensory Pathway Functioning

Both patients 1 and 2 suffer from complaints (double vision, sensory and, in patient 1, motor disturbances) that could originate from damage in different locations in the central nervous system. Since both have suffered from these problems on multiple occasions, they might have MS. MS is a disease in which the fatty myelin sheath around axons in both the brain and the spinal cord may be damaged, resulting from an attack by the patient's own immune system on the myelin cells. Since damage can occur anywhere in the white matter of the central nervous system, the symptoms in MS can vary widely, from sensory, motor and visual problems, via bladder and speech difficulties to pain and fatigue. In MS, symptoms often occur in attacks and symptoms may go away in between. One of the first attacks in MS often involves the optic nerve. If MS is the cause of the visual problems in patients 1 and 2, i.e., if the myelin sheath insulating nerve fibers of the optic pathway is damaged, one would expect signal conduction to be deteriorated resulting in altered processing of visual stimuli in the brain. One way to assess the functioning of the visual pathway is to perform a visual evoked potential (VEP) recording.

An evoked potential (EP) is a potential change recorded from a peripheral nerve, muscle, or the brain resulting from ("evoked by") stimulation of a sense organ or part of the nervous system.

5.3 Obtaining Evoked Potentials

5.3.1 *Enhancing Signal-to-Noise Ratio*

Most brain EPs cannot be seen in routine EEG recordings, because their amplitude is so low ($<20\ \mu V$) that they drown in the background activity. The motor EP is an exception because it yields a strong EMG response from an otherwise relaxed muscle (see Sect. 11.3). The difference between the brain EP and the background activity is that only the first is tightly coupled in time to the stimulus onset, as long as the stimulation conditions are strictly maintained. The time-locked properties of an EP allow to enhance its SNR by averaging (see Box 5.1). This procedure was first described by Dawson (1954).

Not all types of noise are removed by averaging, such as noise time-locked to the stimulus which can be a stimulus artifact or part of the EP belonging to the previous stimulus. Mains noise can also remain in the average, particularly when a constant interstimulus interval (ISI) is used.

Question 5.1 How can a constant ISI cause mains noise to remain in an EP?

Furthermore, the average can be strongly negatively influenced by large amplitudes in single trials, e.g., due to patient movement. Typically, individual trials in which the amplitude exceeds a predefined level are therefore discarded. When the evoked potential itself differs from trial to trial, the resulting average will also be distorted, usually resulting in broader peaks. Sometimes, it is known beforehand that the EP will vary over time, such as during intraoperative monitoring. Under these circumstances other approaches need to be taken to facilitate the recording of EPs (see Sect. 5.6.1).

Another method to enhance the SNR is filtering (see Chap. 4). By filtering, only the frequencies that contain physiologically relevant information are maintained. Usually a band-pass filter is employed (see Sect. 4.3.2). The precise filter properties depend on the EP. For example, for a brainstem auditory EP (BAEP), very high frequencies are relevant, because the first components of this EP already occur after a few milliseconds (see Sect. 5.4.3). Therefore, a band-pass filter for BAEPs would pass frequencies from, e.g., 30 to 3,000 Hz. On the other hand, for the (cortical) auditory EP (AEP) which contains later components, only frequencies from, e.g., 1 to 70 Hz would be passed. Note that the application of filters can also have some disadvantages (see Sect. 4.3.5), which causes some to calculate EPs using only a high-pass filter to just filter out slowly varying artifacts (movement, slow drifts due to sweating).

5.3.2 Stimulation Protocols

Since an EP can only be assessed when the SNR is high enough, we need several hundreds to thousands of repetitions of the stimulation (see Sect. 5.3.1). To finish this procedure as quickly as possible, one would be tempted to increase the stimulation frequency. But the stimulated part of the nervous system needs sufficient time to process the stimulus completely and return to its initial status (this time

Box 5.1 Improving the SNR by Averaging

The SNR is usually defined as the ratio between the average signal power P_{signal} and the average noise power P_{noise}:

$$\text{SNR} = \frac{P_{\text{signal}}}{P_{\text{noise}}}. \tag{1}$$

Both powers must be measured in the same units and over the same bandwidth (range of frequencies). In representation (1) the SNR of single trial EPs typically varies between 1:1 and 1:100.

To obtain an interpretable EP the SNR must generally be increased to 10:1. As the SNR will improve in proportion to the square root of the number of averaged trials, typically 500 to a few thousand trials are needed. Thus, if a single EP trial has an SNR of 1:4, an SNR of 10:1 can be obtained by averaging over $(4 \times 10)^2 = 1{,}600$ trials.

The signal+noise model for evoked EEG activity assumes that the activity e_i in the ith trial is equal to the sum of the evoked potential s and noise n_i which differs from trial to trial:

$$e_i(t) = s(t) + n_i(t). \tag{2}$$

When the noise is *stationary*, uncorrelated to the stimulus, and of zero mean, averaging of (2) over N trials yields:

$$\frac{1}{N}\sum_{i=1}^{N} e_i(t) = \frac{1}{N}\sum_{i=1}^{N}(s(t) + n_i(t)) = s(t) + \frac{1}{N}\sum_{i=1}^{N} n_i(t) = s(t),$$

when N is large enough in relation to the SNR.

is called the refractory period), before the next stimulus is given. Higher stimulation rates may otherwise result in lower component amplitudes. Very early components generally recover quickly whereas later components need more time. This allows BAEPs to be acquired at a rate of ~10–20 Hz, whereas AEPs cannot be acquired at a rate higher than ~2 Hz. In addition, EP amplitude may also decrease when stimulation is highly regular, due to *habituation*. To further prevent habituation, random ISIs are also employed, particularly for later EPs. When a random ISI is used, the distance between the stimuli varies in an unsystematic manner around a mean value, e.g., 2 ± 0.5 s. A random ISI also has the advantage that artifacts that are *phase-locked* to the stimulus are prevented.

Question 5.2 Does the higher stimulation frequency of BAEPs compared to AEPs actually allow quicker acquisition? Remember Sect. 5.3.1.

An exception to the stimulation frequency rule are the steady-state evoked potentials (SSEPs). SSEPs are used to investigate how the brain processes sensory stimulation at a specific frequency. The most commonly investigated SSEP employs visual stimulation at high frequencies (flicker stimulation from 5 up to 75 Hz). When the retina is stimulated at these high frequencies, the brain generates electrical activity at the same (or multiples of the) frequency of the visual stimulus. The stroboscopic flashing employed in a routine EEG recording results in a SSVEP: a long sequence of sinusoidal waves in the EEG at the frequency of stimulation. This allows testing the brain's response to photic driving, which is of particular interest in epilepsy (Sect. 4.2). Other clinical neurological applications of SSEPs are scarce, however.

5.3.3 Assessing Reliability

Since the EP can vary depending on the amount of noise that is present or the cooperation of the subject, it is important to first assess the reliability of an EP. The first step to assess reliability of an EP is to determine its reproducibility. This is routinely done by acquiring two sets of averages under identical circumstances and overlaying them. A disadvantage of this method is that during a prolonged recording session, unwanted changes can occur (in the environment or in the subject) that result in different averages. To circumvent this, an odd-even average can be made, in which all odd and even trials are averaged separately.

To further assess reliability, it is important to determine the noise level in the recording and with that, an estimate for the SNR. Averaging the signal in an interval before stimulus presentation as well as after stimulus presentation, allows to estimate the noise level before the stimulus. The lower the amplitude in the pre-stimulus interval, the lower the noise level. In Sect. 5.3.5, it is discussed how the pre-stimulus noise level is taken into account when interpreting EPs. For this purpose, it is important that the pre-stimulus interval is long enough (usually 10–20% of the entire interval). To assess the noise level in the recorded signals, it is also possible to make a so-called "dummy" average, in which an average is calculated without stimulation. In that case, a flat curve should result. Any remaining activity is then likely to be stimulus-related and thus of technical origin. A disadvantage of the dummy average is that it prolongs the recording time considerably. This can be avoided by calculating a "plus-minus" average, by

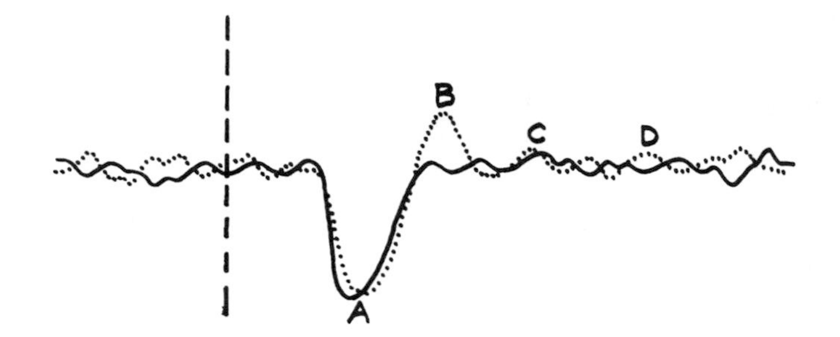

Fig. 5.1 An overlay of two EP averages (*solid* and *dotted lines*) obtained in the same person. The vertical dashed line indicates the stimulus onset

alternatively adding and subtracting trials (or by subtracting the odd and even averages). The EP and any stimulus-related artifacts should then be averaged out and only the average noise remains.

5.3.4 Quantifying and Interpreting EPs

An EP signal consists of several positive and negative potential changes, called components. A potential change is only considered to be a component when it is not an artifact, reproducible (see Sect. 5.3.3), and rises above the noise level.

> *Question 5.3* Which of the potential changes A–D in Fig. 5.1 can be considered to be a component and why?

Components are usually assessed by their latency and amplitude. Latency can be the time from stimulus presentation to the peak of the component (peak latency) or to the onset of the component (onset latency). As the onset of a component is often hard to determine due to variability in the baseline (usually the mean of the signal in the pre-stimulus interval), peak latency is mostly employed. The amplitude of a component can be measured from its peak to the peak of the previous component, to the peak of the next component or to the baseline. The advantage of the first two options is that they are independent of variations in the baseline (see Fig. 5.2).

EP latencies are influenced by age (older people in general have increased latencies) and length (longer distances to travel increase latency). EP amplitudes are also influenced by (patho)physiological variables of the nervous system such as age, demyelination (causing increased latencies and broader peaks), and *axonal degeneration*. Often amplitudes are decreased when these factors play a role, but

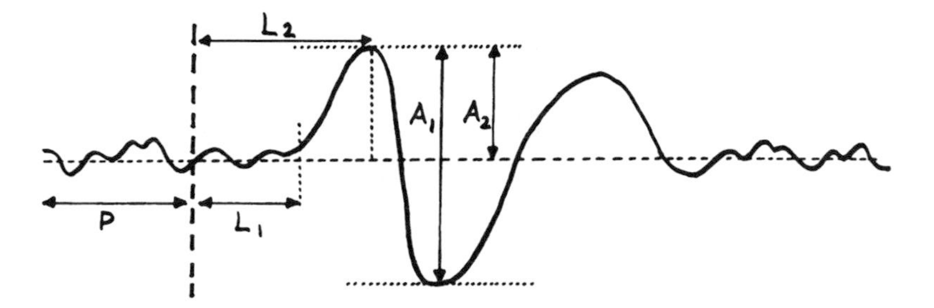

Fig. 5.2 Measures of EP latencies and amplitudes. P is the pre-stimulus interval; the mean amplitude in this interval determines the baseline. L1: onset latency, L2: peak latency, A1: peak-to-peak amplitude, A2: peak-to-baseline amplitude

there are also pathological conditions in which EP amplitudes are increased instead of decreased. One example is the "giant SEP" which can occur when cortical *inhibition* is diminished. EP amplitudes also strongly depend on the distance to the source of the activity due to volume conduction effects: the deeper the source the more attenuated a potential is by the low-pass filtering (see Sect. 4.3) effects of tissue. Hence, increased skin, subcutaneous fat layer or skull thickness, and deeper positions of nerves all decrease EP amplitudes.

EP components are usually named by their *polarity* and mean latency in a normal population (e.g., N75, P100 and N145 for the visual EP components). Since latencies depend on subject length, which varies on average between countries, names are not always consistent in the literature. Alternatively, components can just be named by their polarity and order of occurrence (e.g., N1, P2, N2 for event-related potential components (see Sect. 7.3.3)) or by their order of occurrence only (peak I to V for the BAEP).

5.3.4.1 Multichannel EPs

EPs are usually recorded with a limited number of electrodes, simply because they were developed decades ago, when assessment took place visually using oscilloscopes allowing only a limited number of channels to be displayed. In addition, computer power was restricted, so that online average calculation could be accomplished for only a few channels simultaneously, as well. The choice of the electrodes was therefore such that the main potential changes due to stimulation could be captured. As an example, to obtain the first cortical component of the somatosensory EP (Somatosensory evoked potentials (SEP); see Sect. 5.4.2) which is generated in the primary somatosensory area, it suffices to use a derivation of two electrodes, one anterior to the central sulcus (e.g., Fz) and one posterior to the central sulcus (e.g. P3/P4 or CP3/CP4). Yet, cortically generated EPs actually vary in polarity and amplitude over the head as a function of time in a complex manner and a few electrodes may not be enough to capture more subtle changes in EP *topography* due to pathology. Also,

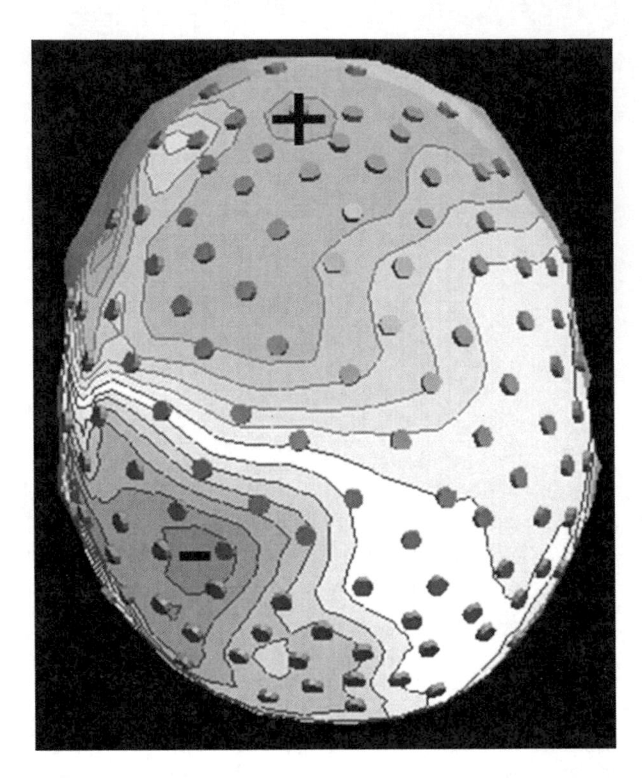

Fig. 5.3 Example of a SEP potential distribution at the peak latency of the N20 component displayed in a topographical mapping, recorded using 128 electrodes

the standard EP electrode configurations assume that the underlying anatomy (the pattern of *sulci* and *gyri*) is consistent between subjects. This is not necessarily true, particularly not when the pathology itself has changed the cortical morphology, as can be the case for tumors and neurodegenerative diseases. In the latter case, a component may still be generated fully, but have its maximum at another electrode. One way to circumvent these problems is to employ multichannel EPs, in which 32, 64, or even 128 electrodes are used, distributed over the head in a systematic extension of the 10–20 system (see Fig. 5.3 and Sect. 4.2).

Multichannel EPs allow to perform topographical analyses. Examples are instantaneous maps in which an EP distribution is displayed for one moment in time or sequential maps, consisting of several instantaneous maps regularly spaced in time. Both colors and equipotential curves can be used to represent the potential distribution (see Sect. 7.3.3.1). The number of channels needed to capture all details of the EP depends on the spatial frequency content. Similar to temporal sampling (see Sect. 3.3.2), there is a Nyquist frequency for spatial sampling: the spatial frequency of the electrode distribution must be at least twice as high as the highest spatial frequency in the EP. For all EPs discussed in this chapter an interelectrode distance of 2–2.5 cm, as in a 64 electrode configuration, suffices to obey the spatial Nyquist criterion. In clinical practice, multichannel EPs are not yet used routinely.

In our own group, we have extensively studied the additional value of multichannel EPs for clinical neurology (see van de Wassenberg et al. 2008), showing that SEP component amplitudes can indeed be estimated more accurately. Another important application of multichannel EPs is source localization (see Chap. 8).

5.3.5 Steps in Calculation of Evoked Potential Averages

Clinically, EP averages are normally calculated online by dedicated software in the EP machine. Although this calculation could be considered as a black box, parameters in this calculation can usually be adapted, and for research purposes offline EP calculations can also be useful. The main steps in calculating an EP are:

1. Filtering
   As discussed in Sect. 5.3.1, filters are used to prevent unwanted signal from influencing the EP. A large part of Chap. 4 is devoted to the details of filtering.
2. Segmentation
   The trials are cut from the raw (EEG) data from some time before stimulus onset to some time after stimulus onset. The exact length of the trial and the pre-stimulus interval are determined by the type of EP. The baseline must be long enough to give a good estimate of the noise level in the recording. The total trial length must be large enough to prevent activity from the previous trial from "leaking" into it.
3. Artifact correction
   For EP calculation, it is important that single trials with large amplitudes, do not influence the average. These trials are rejected when the amplitude exceeds a certain threshold. Usually, during EP recording, the single trials as well as the developing average are displayed, allowing the investigator to change the amplitude threshold for rejecting trials if necessary.
4. Baseline correction
   As the noise level is often estimated from the mean pre-stimulus interval (the baseline) and some peak amplitudes are measured from peak-to-baseline, the baseline value is subtracted from every trial.
5. Averaging
   Finally, all remaining trials are averaged. In an EP machine, an average is usually built-up as the recording takes place by adding the Nth trial to the average over the first N-1 trials as follows:

$$\text{average}_N = \frac{1}{N}((N-1)\text{average}_{N-1} + \text{trial}_N). \tag{5.1}$$

   This has the advantage that a recording can be stopped when the average does not change anymore to limit the recording time.

5.4 Types of Clinical Evoked Potentials

Evoked potentials have been studied in neurological patients since the 1950s and have had clinical utility since the 1970s, particularly because of their reliability and reproducibility in routine clinical practice. EPs are clinically useful because they can demonstrate abnormal sensory function when the history taking and/or clinical neurological examination is not clear, they can reveal dysfunction of a clinically normal sensory system when a demyelinating disease is suspected because of symptoms in another part of the central nervous system, they can help determine to what anatomical extent a disease has progressed and they provide an objective way to monitor a patient over time.

The long history of clinical EPs implies that there is a wealth of information available on the precise equipment needed, protocols used, and conventions employed to assess EPs (see References). Therefore, only a short overview is presented here for the three categories of stimulation that are most often used in clinical practice: visual, somatosensory, and auditory stimulation. The motor evoked potential is described in detail in Sect. 11.3.

5.4.1 Visual Evoked Potential

The VEP is the largest EP that is commonly used; it can even be observed in the routine EEG recording when evoked by the flash of a stroboscopic lamp during photic stimulation. It is also very sensitive to changes due to neurological diseases. On the other hand it is also highly influenced by changes in the type of stimulus and recording electrodes. This is one of the reasons why considerable effort has been spent to standardize VEPs (and other EPs) across and within laboratories, although recommendations still vary between authorities.

The VEP results from stimulation of the retina by light that passes through the cornea and lens. At the retina, light is transformed into electrical pulses that travel up the optic nerve to the optic chiasm, where the nerve fibers partly cross. Fibers originating from the medial halves of the retina cross (Fig. 5.4; solid lines), whereas the fibers originating from the lateral halves of the retina stay in the same hemisphere (Fig. 5.4; dashed lines). From the optic chiasm the fibers go via the optical tract to a specific part of the thalamus (lateral corpus geniculatum) that can be considered a relay station. From here wide fiber bundles go to occipital and parietal cortical brain areas where the visual information is processed.

VEPs are mostly evoked by alternating pattern stimulation (PVEP) using a chessboard pattern, in which the black squares become white and vice versa at a rate of ~1–2 Hz. Because the VEP is rather large, usually 100–200 trials are enough to obtain a reliable average. A PVEP requires patients to have a reasonably good visual acuity and to focus on a fixation point on the screen, which is not possible in less cooperative, comatose, visually disabled, or very young subjects. In these cases, flash stimulation (FVEP) with a stroboscopic lamp is also used.

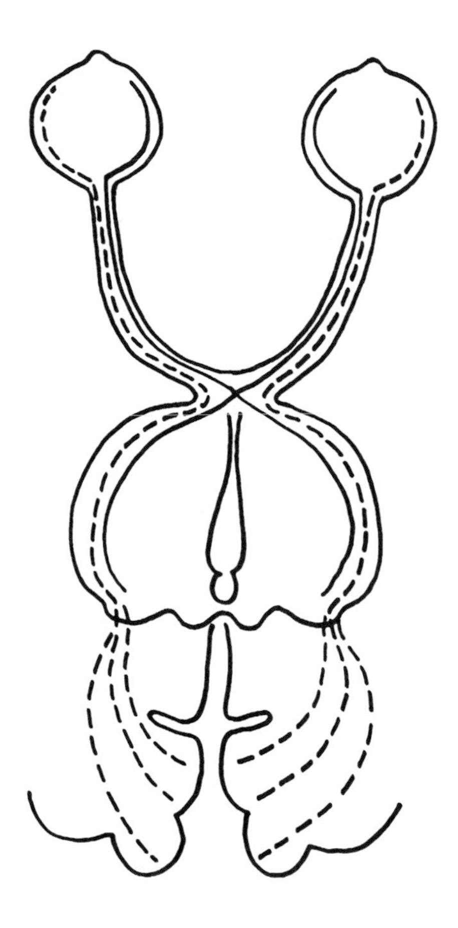

Fig. 5.4 Overview of the optical system with its main pathways. Solid lines: fibers originating from the medial retina, dashed lines: fibers originating from the lateral retina

Usually the PVEP is recorded from a few occipital electrodes (O1, O2, Oz) with Cz or Fz functioning as a reference. Both full-field stimulation and half-field stimulation are employed. Stimulation of one half of the visual field activates only one hemisphere (left retinal half-field stimulation resulting in left hemisphere stimulation and vice versa; see Fig. 5.5a), resulting in an asymmetric potential distribution on the occipital electrodes in healthy subjects. Full-field stimulation with one eye open should result in a symmetric potential distribution (see Fig. 5.5b), which is maximal at Oz. By cleverly combining and comparing full- and half-field stimulation of one or both eyes it is possible to localize abnormalities in the visual pathway to, before, or after the optic chiasm and to a hemisphere. As an example, when an asymmetry in latency is observed between left and right eye full-field stimulation, this implies a pre-chiasmatic lesion.

Full-field PVEPs normally show at least a negative-positive-negative configuration of peaks which are labeled as N75, P100, and N145, respectively (see Fig. 5.6). The P100 latency is the most important measurement for the interpretation of the PVEP, also because the other peaks are less stable and less consistently identified in healthy subjects. Lesions in the eye itself or in the optical nerve will change the morphology of

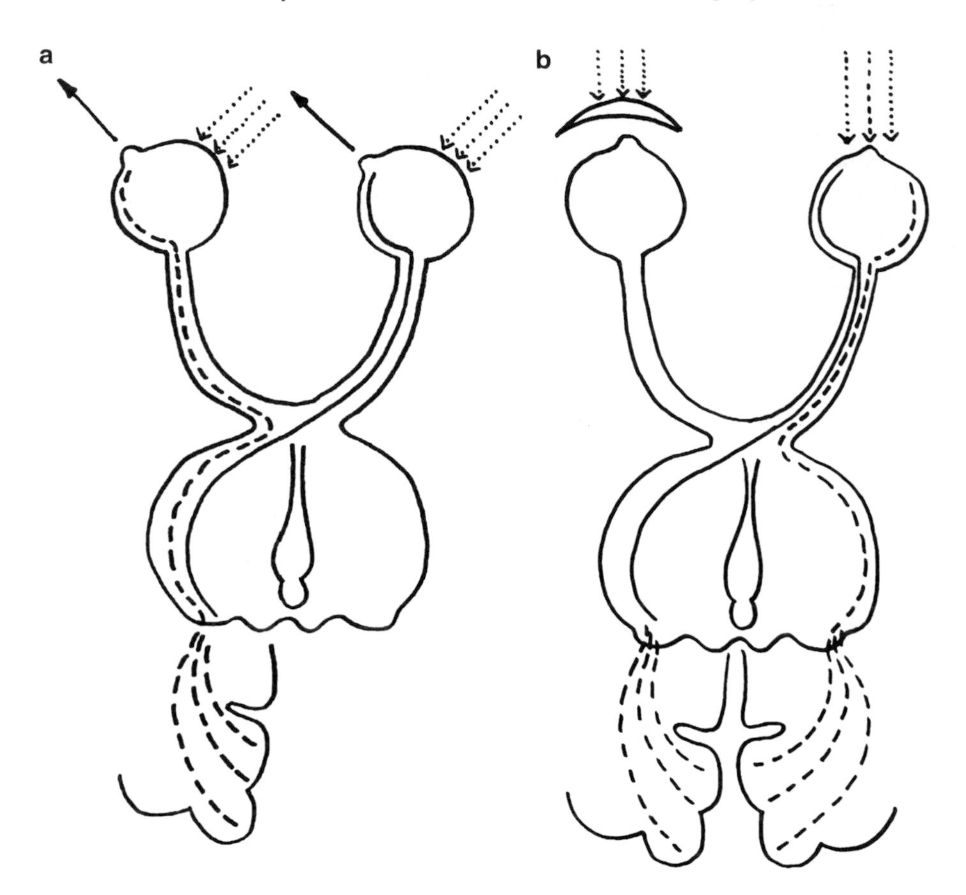

Fig. 5.5 Uni- or bilateral activation of the visual cortex as a result of (**a**) left retinal half-field stimulation of both eyes (only left hemisphere activated) and (**b**) full-field stimulation of the right eye (both hemispheres activated). *Solid arrows*: viewing direction; *dotted arrows*: incident light direction

the P100 or its latency or may even cause the P100 to be absent. Normally, the maximal amplitude of the P100 is in the midline and its latency is normally between 110 and 115 ms. In case of a strong amplitude asymmetry (>50%), the cause may be localized in or after the optic chiasm. Half-field PVEPs of both eyes are better than full-field PVEPs at evaluating visual field defects originating from after the optic chiasm.

The VEP is particularly useful in diagnosing MS and other diseases in which the myelin sheath of the optical nerve is damaged (optic neuritis). Even when there are no visual problems clinically, the PVEP is very sensitive to so-called "silent" lesions. In this sense, VEPs are just as useful as MRIs in screening for asymptomatic lesions in MS. Also, optical neuritis in the past usually results in VEP abnormalities even when clinical symptoms have ceased.

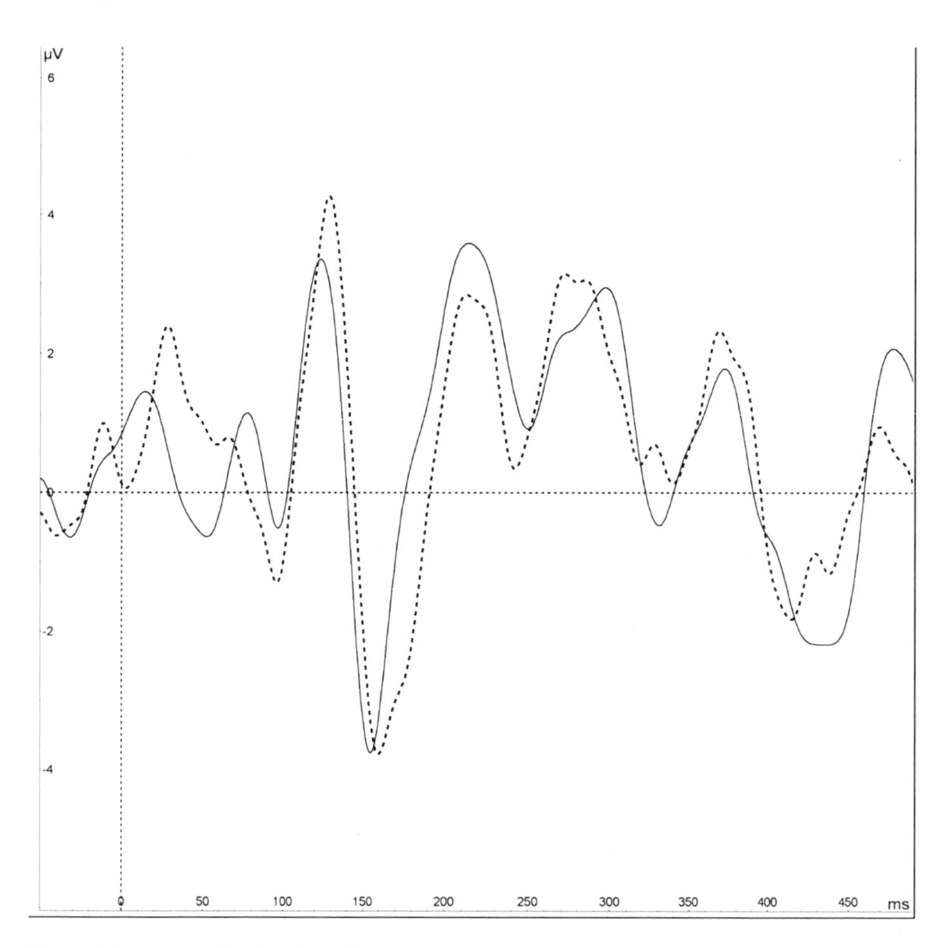

Fig. 5.6 Example of a full-field VEP in a healthy subject. The right eye was stimulated twice (*solid* and *dotted curves*). The VEP was derived from Oz-Fz. Positive up. P100 around 110 ms

5.4.2 Somatosensory Evoked Potential

A SEP is mostly obtained by electrically stimulating a peripheral (mixed) nerve or a *dermatome*; the latter is not further discussed here. The most common nerves that are stimulated are the median nerve at the wrist and the tibial nerve at the ankle. Here, only the median nerve SEP is discussed in more detail. Stimulation of the median nerve results in a *volley* that propagates along the fibers of the sensory pathway from the periphery to the brain. The volley travels through the spinal cord, via the brain stem and thalamus to the postcentral gyrus in the brain, where the primary somatosensory cortex is located. The fibers project in particular to Brodmann areas (BAs) 3b (on the wall of the central sulcus) and 1 (on the convexity of the postcentral gyrus).

A median nerve SEP is typically evoked by stimulating at a frequency of 1–3 Hz. The stimulation strength is determined such that the muscle that abducts the thumb (m. abductor pollicis brevis) just contracts: this ensures that all sensory fibers in the mixed nerve are depolarized. At a good SNR, 250–500 stimulations are usually enough for a reliable average.

Along the sensory pathway, different median nerve SEP components can be recorded. Already after approximately 9 ms, the first components (P9/N9) can be derived from anywhere on the scalp using an extracranial reference. These components reflect the volley passing from the arm to the trunk. This component is followed by several other components reflecting e.g., the volley arriving in the spinal cord (C6–C8; N13/P13) and brainstem (P14, N18). These anatomical references have been partially derived from patients with lesions of the roots C6–C8 in whom N13/P13 was absent, with high-cervical lesions in whom N13/P13 was present and P14 and N18 were absent and with lesions in the thalamus or cortex in whom all of these components were still present. The first cortical component is the N20, which results from neuronal activity in BA 3b. Alternating activity of neurons mainly in BAs, 3b and 1 subsequently leads to several cortical components: the P20, P22/P25, P27, N30, and P45 with maximal activity varying between parietal, frontal, and central positions contralateral to the stimulated side. For clinical purposes, the N20, P27, and P45 are most important (see Fig. 5.7). Typically, the cortical median nerve SEP is assessed using two derivations; CP3 and CP4 with reference to either Fz (i.e., CP3-Fz and CP4-Fz) or the contralateral earlobe (i.e., CP3-A2 and CP4-A1).

Question 5.4 What would happen to the latencies of a tibial nerve SEP compared to those of a median nerve SEP?

Since the passage of the SEP volley from one part of the somatosensory system to another generates distinct SEP components, the latency difference between two subsequent SEP components can be used to assess conduction in a certain trajectory of the somatosensory system. The absence of SEP components above a certain level indicates the position of a lesion in the somatosensory system. These properties make the SEP useful for diagnosing damage to the *plexus brachialis* or cervical roots, for diagnosing MS (through the detection of silent lesions) and other demyelinating diseases, or for diagnosing cervical *myelopathy*. Furthermore, the SEP has applications in neuromonitoring (see Sect. 5.6.1) and can help determine the prognosis of coma.

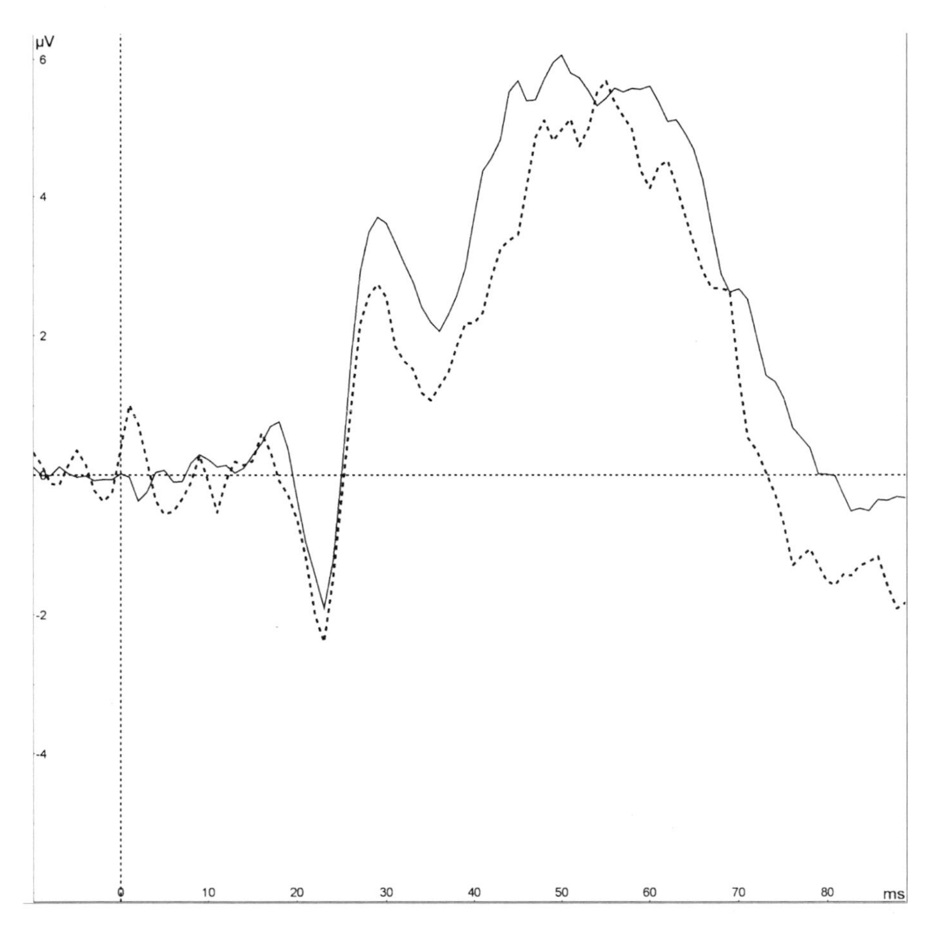

Fig. 5.7 Example of a median nerve SEP in the same healthy subject as in Fig. 5.6. The right median nerve was stimulated twice (*solid* and *dotted curves*). The SEP was derived from CP3-A2. Positive up. N20 around 23 ms, P27 around 30 ms, and P45 less clear, but around 50 ms

5.4.3 Brainstem Auditory Evoked Potential (BAEP)

A BAEP is evoked by auditory stimulation of the ear, usually by very short clicking sounds that have a frequency content from 2,500 to 4,000 Hz. There are different types of clicks: rarefaction clicks that let the eardrum vibrate towards the outer ear canal first and condensation clicks that let the eardrum vibrate in the direction of the inner ear first. Click stimulation ensures that the cochlea and with that the cochlear (or auditory) nerve are maximally stimulated. For diagnosis of hearing problems this type of stimulation does not suffice, because normal hearing can process sounds with frequencies in a much broader band. Since speech sounds are also out of the frequency range used for a BAEP (in the range of 200–3,200 Hz), a normal BAEP using clicks does not prove that a person is not speech-deaf either.

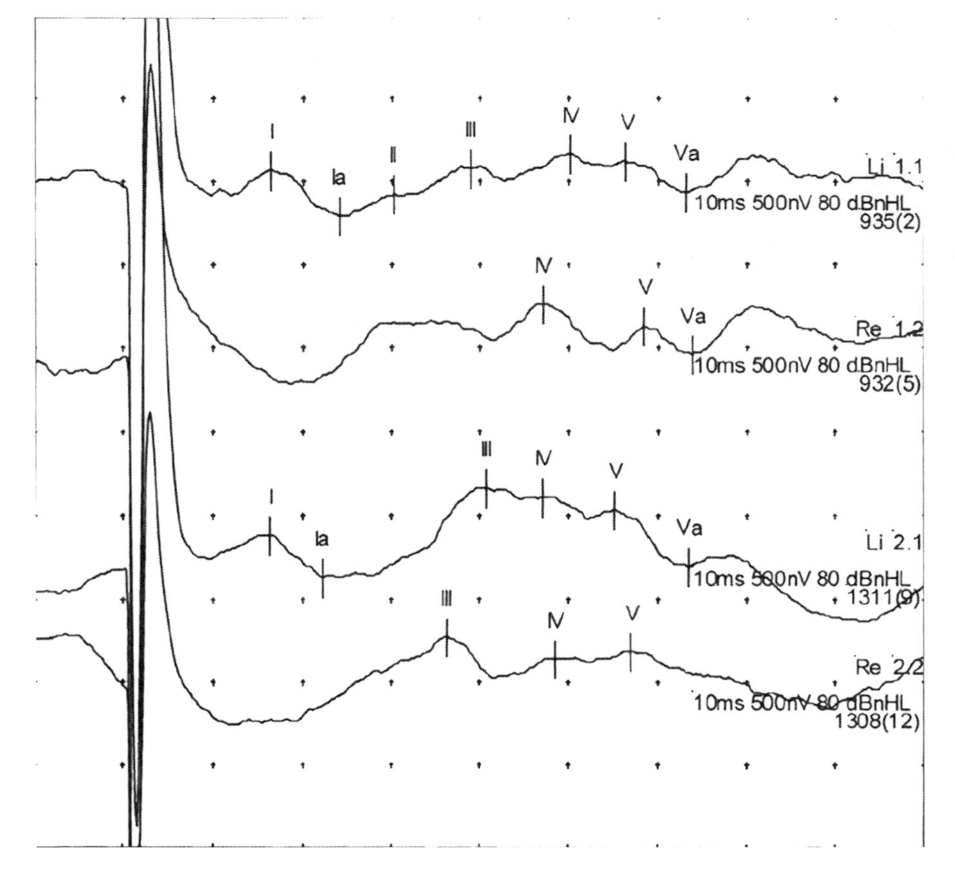

Fig. 5.8 Example of a BAEP in a healthy subject. The left ear was stimulated with rarefaction clicks at 80 dB and the BAEP was derived from the left and right hemispheres (top two graphs). The stimulation was repeated (bottom two graphs). Note that not all peaks are well reproducible. Horizontal scale: 1 ms per division

BAEPs are particularly used to assess the early part of "central hearing," i.e., the functioning of the auditory nerve, the cochlear nuclei in the brainstem, the colliculus inferior, a small brain area between the pons and the midbrain, and the auditory pathway connecting these structures. Processing of auditory stimuli in the midbrain and other subcortical areas can be assessed using middle-latency evoked potentials and cortical processing of auditory stimuli can be assessed using (long-latency) auditory evoked potentials (AEPs).

The BAEP itself is very fast: auditory processing has already progressed to the midbrain only 10 ms after stimulation. Because of the distance between brainstem/pons and the scalp, BAEPs have a very low amplitude (~1 μV) implying that at least 1,000–2,000 stimuli are needed to obtain a good SNR. The stimulation frequency is usually limited to 10–20 Hz. Typically the BAEP is derived from Cz with a reference on the ipsilateral mastoid or preauricular point. Five (positive) peaks, simply numbered I–V can be distinguished (Fig. 5.8). The latency difference between peaks I and III and peaks III and V is approximately 2 ms in healthy subjects. The latency difference between peaks I and V is called the central conduction time and is the most reliable measure to diagnose problems in the central auditory pathway up to the level of the midbrain. Age- and sex-related normal values exist for (inter)peak latencies, for left–right differences in central conduction time, and for peak V latency that can help identify abnormalities.

As it is known where the different peaks originate from approximately (I from the distal auditory nerve, II from the intracranial proximal part of the auditory nerve, III from the area between both cochlear nuclei (lower pons) and IV and V from the area just caudal of the inferior colliculus (middle to upper pons)), absence of peaks indicates the position of lesions in the auditory pathway. If peak I is missing while peaks III–V are late but still present, this usually indicates an ear dysfunction. Complete absence of all peaks can either indicate a technical problem, deafness due to an ear problem, or severe dysfunctioning of the auditory nerve. In many healthy subjects, peaks II and IV cannot be clearly distinguished, sometimes because they form a bigger complex with peaks III and V, respectively. Peak amplitudes also vary highly between subjects, giving absolute peak values with no clinical relevance.

Although different diseases can give rise to similar BAEP abnormalities, there are still a number of clinical indications for a BAEP such as tumors of the auditory nerve, pons or brainstem, and detection of silent lesions in patients thought to have MS based on another lesion with clinical symptoms and for monitoring purposes during surgery in the area involving the auditory pathway.

5.5 Evoked Potentials in Individual Patients

As mentioned in Sect. 5.4, evoked potentials can reveal the dysfunction of a clinically normal sensory system when a demyelinating disease is suspected because of symptoms in another part of the central nervous system. As MS often starts with symptoms in the visual system due to damage to the optic nerve, the VEP is easily evoked and robust, and both patients 1 and 2 suffer(ed) from visual symptoms, VEP recordings were performed in both patients according to the procedure described in Sect. 5.4.1 with additional electrodes at T5 and T6.

Patient 1

The results for the VEP evaluation in patient 1 for full-field stimulation are indicated in Table 5.1 and for half-field stimulation in Table 5.2.

The VEP responses were generally easy to evoke and well reproducible, except for the N145 during left half-field stimulation, as can be seen from Fig. 5.9 and Table 5.2.

For full-field stimulation, the latencies were normal, except for the right occipital response for right full-field stimulation. However, since more attention is paid to

Table 5.1 VEP component latency results for full-field stimulation in patient 1

	Derivation	N70 (ms)	P100 (ms)	N145 (ms)
Left eye	O1-Cz	93	112	136
	Oz-Cz	94	113	137
	O2-Cz	93	113	129
Right eye	O1-Cz	96	115	148
	Oz-Cz	96	117	149
	O2-Cz	97	119	150

Table 5.2 VEP component latency results for half-field stimulation in patient 1

	Derivation	N70 (ms)	P100 (ms)	N145 (ms)
Left field	T5-Cz	91	116	–
	O1-Cz	92	112	136
	Oz-Cz	96	111	–
Right field	Oz-Cz	94	109	151
	O2-Cz	87	112	165
	T6-Cz	87	113	165

the midline responses for full-field stimulation and the P100 latency at Oz is borderline normal, the right full-field stimulation is not considered abnormal in this patient. For half-field stimulation, latencies were normal as well (except for the left N145, of course). There was a slight ocular asymmetry as reflected in the later latencies for right eye than left eye full-field stimulation. In conclusion, nerve conduction in the visual system was normal in patient 1, despite her optic neuritis 2 years ago and her current double vision. Yet, because of the multiple old and new white matter lesions that were found on MRI, indicative of demyelination, and the recurrent neurological problems indicating damage to the central nervous system in multiple locations, she was diagnosed with relapsing-remitting MS. Although uncommon, 2–5% of MS patients experience their first symptoms before age 16 (Ness et al., 2007). Patient 1 was treated with anti-inflammatory medication (interferon-beta) to which she responded well.

Patient 2

The results for the VEP evaluation in patient 1 for full-field stimulation are indicated in Table 5.3 and for half-field stimulation in Table 5.4.

The VEP responses were easy to evoke and well reproducible as can be seen from Fig. 5.10, both for full-field and half-field stimulation.

For full-field stimulation, the latencies were normal, although borderline slow for the right eye, both ipsi- and contralaterally. For half-field stimulation, latencies were normal as well and no ocular asymmetry was observed. In summary, the VEP findings in patient 2 were normal. The MRI showed multiple white matter lesions subcortically, that were not typical for MS but could also indicate vascular problems. Furthermore, examination of the corticospinal fluid did not provide clues for MS either. The patient was therefore not diagnosed with MS and during follow-up his MRI and symptoms remained stable, also indicating that the diagnosis of MS was unlikely in this patient.

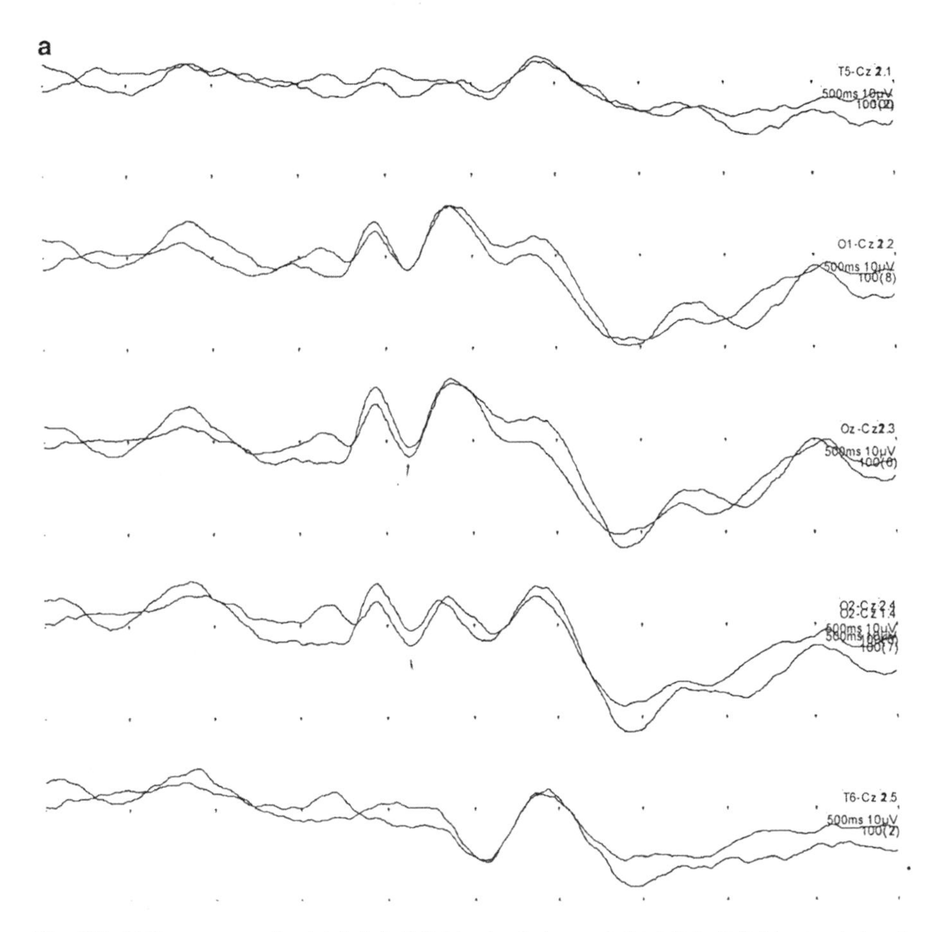

Fig. 5.9 VEP responses for (**a**) left full-field stimulation and (**b**) left half-field stimulation in patient 1. 500 ms (*horizontal axis*) and 10 μV (*vertical axis*) per division. From top to bottom: T5-Cz, O1-Cz, Oz-Cz, O2-Cz, T6-Cz

Although the VEP findings in both patients 1 and 2 were thus negative for the diagnosis of MS, they did contribute to their diagnostic process.

5.6 Other Applications of Evoked Potentials in Neurology

The applications of EPs in neurology are numerous and many have been mentioned in this chapter. In Sect. 8.4, a specific application of a multichannel median nerve SEP for preoperative function localization is discussed and details about motor evoked potentials (MEPs) are given in Sect. 11.3. In this section another two interesting applications are described.

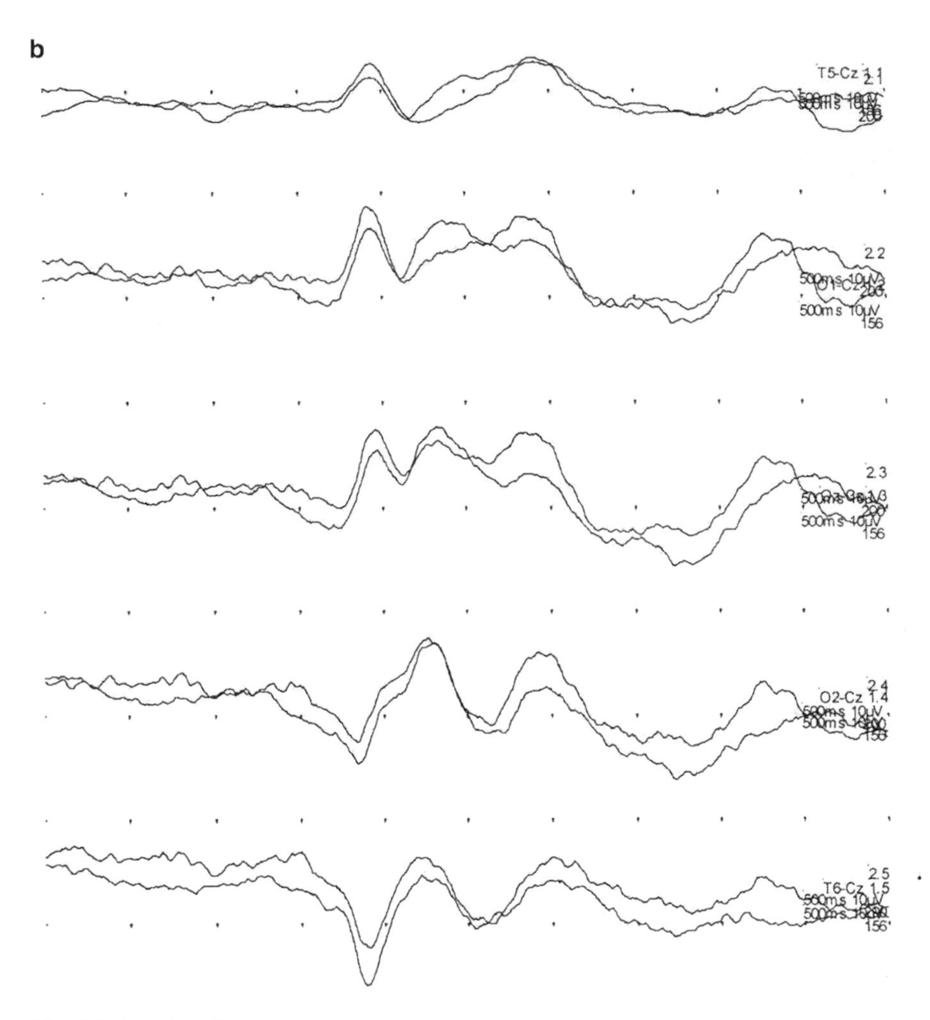

Fig. 5.9 (continued)

Table 5.3 VEP component latency results for full-field stimulation in patient 2

	Derivation	N70 (ms)	P100 (ms)	N145 (ms)
Left eye	O1-Cz	70	112	157
	Oz-Cz	83	112	161
	O2-Cz	83	113	166
Right eye	O1-Cz	91	116	167
	Oz-Cz	90	116	163
	O2-Cz	91	116	165

Table 5.4 VEP component latency results for half-field stimulation in patient 2

	Derivation	N70 (ms)	P100 (ms)	N145 (ms)
Left field	T5-Cz	81	110	174
	O1-Cz	80	106	170
	Oz-Cz	81	104	170
Right field	Oz-Cz	67	113	164
	O2-Cz	78	111	164
	T6-Cz	77	113	169

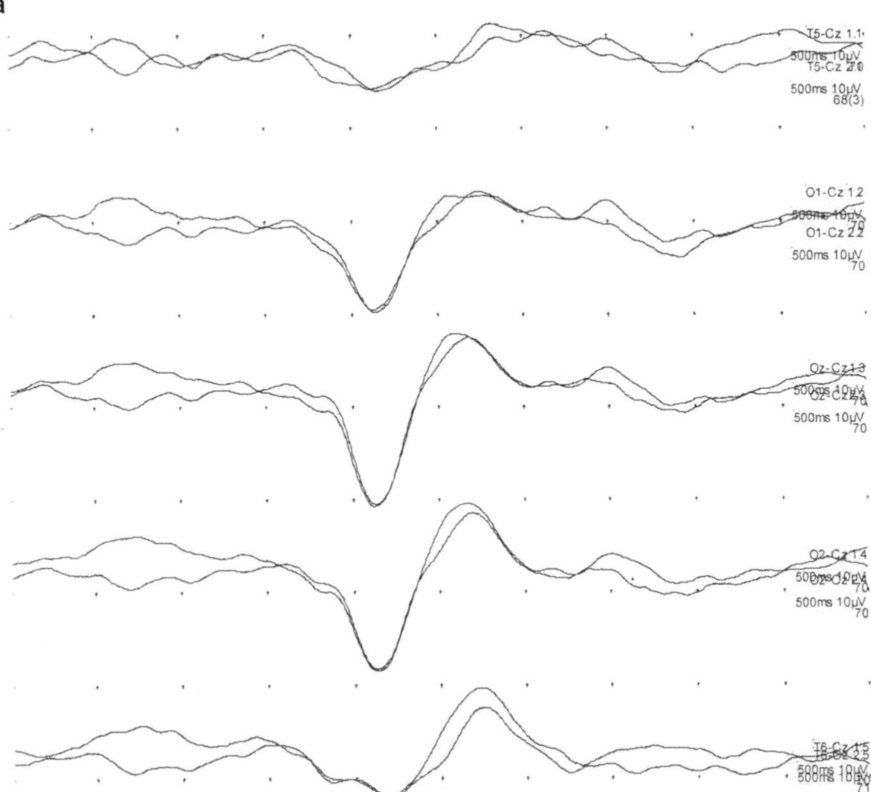

Fig. 5.10 VEP responses for (**a**) left full-field stimulation and (**b**) left half-field stimulation in patient 2. 500 ms (*horizontal axis*) and 10 µV (*vertical axis*) per division. From top to bottom: T5-Cz, O1-Cz, Oz-Cz, O2-Cz, T6-Cz

5.6.1 Intraoperative Monitoring: SEP in Scoliosis Surgery

One advantage of EPs is that they provide an almost instantaneous assessment of peripheral and central nervous system functioning, making them particularly suitable for use in situations in which this functioning can be endangered. In addition,

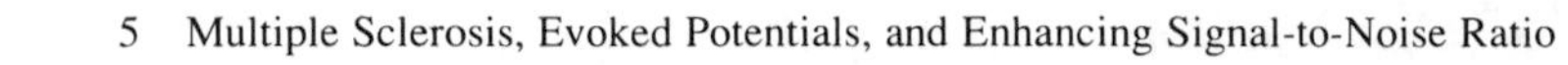

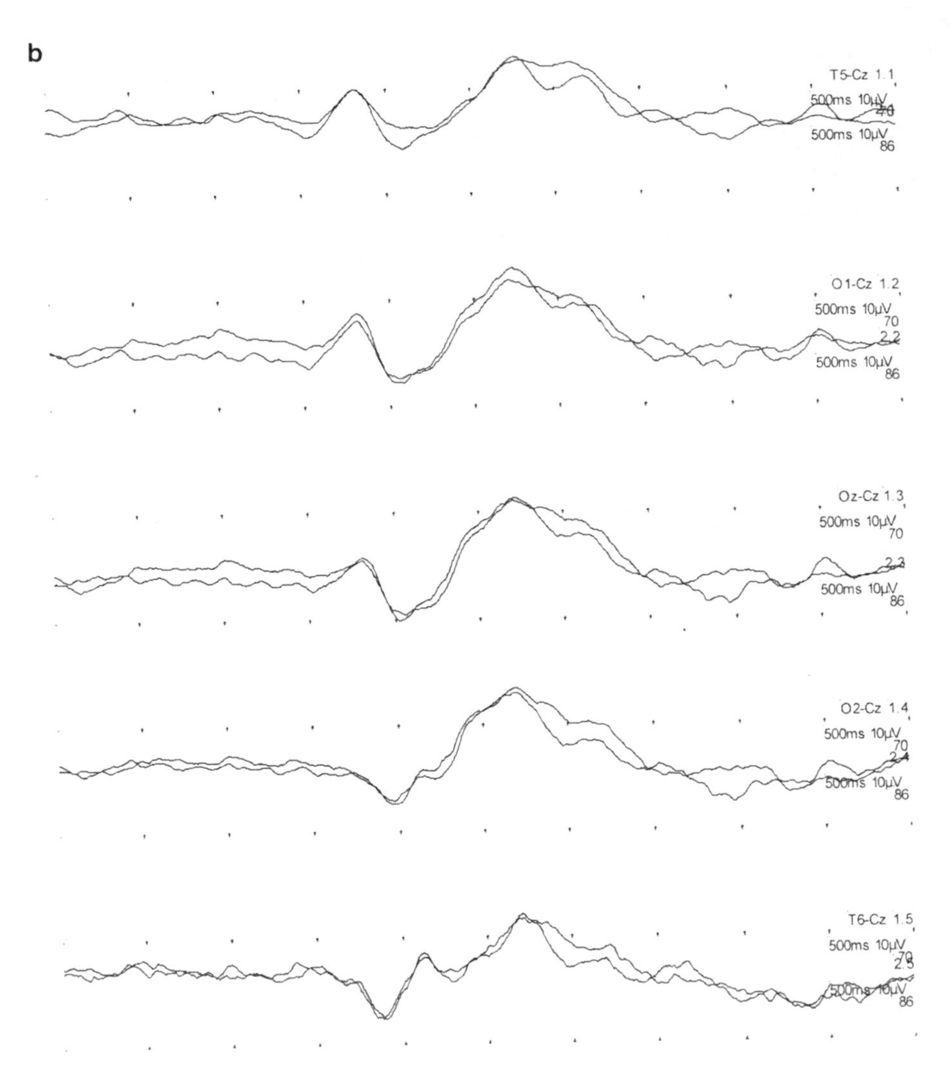

Fig. 5.10 (continued)

for many EPs cooperation of the subject is not necessary to obtain a good SNR. On the contrary, some EPs are even easier to acquire when a subject is unconscious and muscle activity cannot obscure the EP. For these reasons, some EPs are now almost routinely employed during surgical procedures which may damage the nervous system, such as spinal surgery and carotid endarterectomy.

One example is the use of SEPs to monitor spinal cord functioning during surgery to correct a *scoliotic* spine. The procedure to acquire SEPs during surgery is basically the same as during a routine recording, but there are some special features related to the difficult circumstances in the operating room. First of all, the use of anesthetics/neuromuscular blocking agents decreases/increases the amplitude of (sub)cortical SEP components and it is thus important to know what is used when. Second, it is advised to record the SEP preoperatively as well, so that absence of SEP components can be attributed to surgical intervention solely and not to technical problems related to the "hostile" operating theater environment or suboptimal electrode derivations. Finally, there is not always time to stimulate the nerve as often as would be done in a routine SEP recording. Stimulating 250–500 times at ~2 Hz takes ~2–4 min, which might simply be too long when the functioning of the spinal cord is at stake. Some special signal analysis techniques can be used to speed up SEP acquisition. One method is to use the preoperatively acquired SEP as a so-called Wiener filter. Wiener filters have a transfer function (see Sect. 4.3.3) that exactly matches the spectrum of the signal of interest, in this case the SEP.

Question 5.5 How can template matching (see Sect. 4.4.1) be used to speed up SEP acquisition?

5.6.2 BAEP to Assess Hearing Threshold

A BAEP can also be used to determine hearing threshold in patients (or very young children) who are not cooperative enough to perform other, more subjective audiometric tests. Even in animals, BAEPs can be used to assess hearing problems. In addition, a BAEP is performed when audiometric tests indicate that there may be a problem in the auditory nerve or brainstem.

When hearing threshold needs to be determined, the BAEP is first derived using a high level of stimulation e.g., 90 dBnHL (90 dB above the level at which a group of normal hearing subjects can subjectively just distinguish a tone of any frequency during silence). When peaks I, III, and V are clearly present in the BAEP the procedure is repeated at 10 or 20 dB lower stimulation intensity. The stimulation intensity is repeatedly lowered (eventually with 5–10 dB at a time) until all peaks have disappeared from the BEAP or only peak V remains. The stimulation intensity at which this peak is just visible is the hearing threshold. In most healthy adults, the hearing threshold lies between 5 and 20 dBnHL.

5.7 Answers to Questions

Answer 5.1

When the stimulus frequency is exactly a (sub)harmonic of the mains frequency (50 Hz in Europe), i.e., the ISI is a multiple of 200 ms, the mains noise has a fixed temporal relation with the stimulus, prohibiting that it will be averaged out.

Answer 5.2
Although BAEP stimulation is 5–10 times as fast as AEP stimulation, the amplitude of the BAEP is much smaller than that of the AEP (0.5–1 μV compared to 2–3 μV), implying that the number of stimulations for a BAEP needs to be at least $\sim 3^2 = \sim 10$ times higher. The net effect is that BAEP and AEP acquisition have approximately the same duration.

Answer 5.3
A is a component because it is not an artifact, reproducible, and rises above the noise level as defined by the mean pre-stimulus amplitude. B is not a component because it is not reproducible, C is not a component because it does not rise above the noise level, and D is not a component because it does not rise above the noise level nor is it reproducible.

Answer 5.4
Since the volley evoked by electrical stimulation of the tibial nerve at the ankle takes longer to travel to the spinal cord and eventually, to the cortex, latencies of tibial nerve SEPs are increased compared to median nerve SEPs. For example, the first cortical component of the tibial nerve SEP is the P39.

Answer 5.5
Similar to Wiener filtering, the preoperatively acquired SEP can be used as a template to identify the SEP in averages over limited number of trials through template matching.

Glossary

Axonal degeneration Response of a peripheral nerve to damage, leading to death and breakdown of the axon and finally of the myelin sheath.

Demyelination Damage of the fatty myelin sheath around nerve axons. The myelin sheath electrically insulates the nerve, such that nerve signal conduction can proceed efficiently.

Dermatome Area of skin that is mainly innervated by a single spinal nerve.

Gyri The ridges of the cortex.

Habituation A decrease in a response due to repeated regular stimulation.

Inhibition Suppression of activity in the nervous system.

Myelopathy Pathology of the spinal cord, in the example in the text leading to compression of the spinal cord.

Oligoclonal bands Bands that can be seen in blood serum or cerebrospinal fluid indicating the presence of antibodies.

Phase-locked Fixed in time to the same moment in the signal period.

Plexus brachialis A bundle of nerve fibers exciting from the spinal cord at level C5-T1 and running through the neck and armpit into the arm.

Polarity The sign of an EP component, i.e., positive or negative.

Scoliotic A curved spine.
Stationary The statistical properties do not change over time.
Sulci The folds of the cortex.
Topography Here: the distribution of potentials over the scalp.
Volley A simultaneous discharge of several action potentials.

References

Online Sources of Information

http://en.wikipedia.org/wiki/Evoked_potentials short overview of topics treated in this chapter, with many links to related topics

Books

Chiappa KH (1997) Evoked potentials in clinical medicine. Lippincott-Raven, Philidelphia

Ebersole JS and Pedley TA (2003) Current practice of clinical electroencephalography. Lippincott, Williams and Wilkins, Philadelphia (Chapters 27–29 and 31 in particular)

Papers

Dawson GD (1954) A summation technique for the detection of small evoked potentials. Electroencephalogr Clin Neurophysiol 6:65–84

Ness JM, Chabas D, Sadovnick AD, Pohl D, Banwell B, Weinstock-Guttman B; for the International Pediatric MS Study Group (2007) Cl7inical features of children and adolescents with multiple sclerosis. Neurology 68(16):S37–S45

Van de Wassenberg W, van der Hoeven J, Leenders K, Maurits N (2008) Multichannel recording of median nerve somatosensory evoked potentials. Neurophysiol Clin 38:9–21

Chapter 6
Cortical Myoclonus, EEG-EMG, Back-Averaging, and Coherence Analysis

After reading this chapter you should:

- Understand why jerk-locked back-averaging and corticomuscular coherence analysis can be used to investigate cortical generators of myoclonus.
- Know how EMG from multiple muscles can be employed to determine the generator of generalized myoclonus.
- Know what properties myoclonic bursts must have to allow jerk-locked back-averaging in case of (multi)focal myoclonus.
- Know the (dis)advantages of automated and manual EMG onset detection.
- Know how the significance of coherence can be determined.
- Know how the temporal relationship between two coherent signals can be determined using the phase spectrum.
- Know what nonphysiological factors can influence corticomuscular and corticocortical coherence and how this influence can be minimized.

6.1 Patient Cases

Patient 1

This 45-year-old female patient suffers from *rheumatoid arthritis* since 24 years for which she has used multiple medications, including drugs that suppress the immune system. Yet, her disease has made it necessary to replace both her hip and knee joints and she has developed a severe and progressive *polyneuropathy*. In addition, she has heart and visual problems, and suffers from *perception deafness* and obstructive sleep apnea syndrome (OSAS). These medical problems have necessitated regular hospitalization for the past few years. Now, she is hospitalized again because her general condition is deteriorating and she is not recovering well from a urinary tract infection. Because involuntary movements are observed in multiple limbs when she is admitted to hospital, a neurologist is asked to examine

N. Maurits, *From Neurology to Methodology and Back: An Introduction to Clinical Neuroengineering*, DOI 10.1007/978-1-4614-1132-1_6,

her. The movements are classified as *multifocal myoclonus* and the question arises whether the myoclonic jerks originate from epileptic activity.

Patient 2
A 25-year-old female patient is hospitalized in the department of neurology to evaluate her physical capacity and side effects of antiepileptic medication she has been using, of which the dosage has recently been increased. She has long been suffering from *cerebellar ataxia* and myoclonic movements, which are thought to be of epileptic nature. The *etiology* of her disorder is unknown and although she was born preterm and is retarded, no metabolic or genetic disorder could be established. She has been using antiepileptic medication for several years, but now complains that the *jerk*ing frequency actually increased and she now additionally suffers from muscle weakness. Before changing her antiepileptic medication, the origin of her myoclonus needs to be reevaluated.

Patient 3
This 55-year-old male suffers from a progressive tremor of the hands since 10 years. He also complains of shaking and an "unstable" feeling in his legs. Neurological examination was normal except for a distal postural tremor. He was initially diagnosed with essential tremor (ET) and treated with propranolol, which was not effective, and then with primidone, which diminished the tremor. Three years ago, he had two tonic–clonic seizures. After slowly reducing the primidone, he suffered a myoclonic status epilepticus. Treatment with other antiepileptic drugs was initiated, which reduced the tremulous and myoclonic movements and the frequency of epileptic seizures. The seizures did not result from a structural brain abnormality since MRI was normal. Since his seizures are atypical for ET, the patient was referred for further clinical neurophysiological work-up.

6.2 Simultaneous EEG-EMG Recording: Relating Brain and Muscle Activity

Patients 1, 2, and 3 all suffer from sudden, involuntary, jerky, and shock-like short-lasting movements that are called myoclonic. A usually benign form of myoclonus is sleep-onset myoclonus, in which the arms or legs briefly jerk while falling asleep. Myoclonus can be generated anywhere in the central nervous system and finding its generator helps determine diagnosis and treatment. Although clinical characteristics usually aid in distinguishing between cortical, brainstem, and spinal myoclonus, they can sometimes be misleading. Neurophysiological tools such as electroencephalography (EEG, see Chap. 4), electromyography (EMG, see Chap. 2) and evoked potentials (see Chap. 5) can be helpful. Since EMG has a high temporal resolution, it can help determine what muscles are involved in jerks, and in what order they are recruited. It also allows to discern between positive and negative myoclonus; in the first the jerk is caused by a brief muscle burst, whereas in the second the jerk is actually caused by a pause (silent period) in sustained muscular activity, leading to a compensatory movement to restore limb position (see Fig. 6.1).

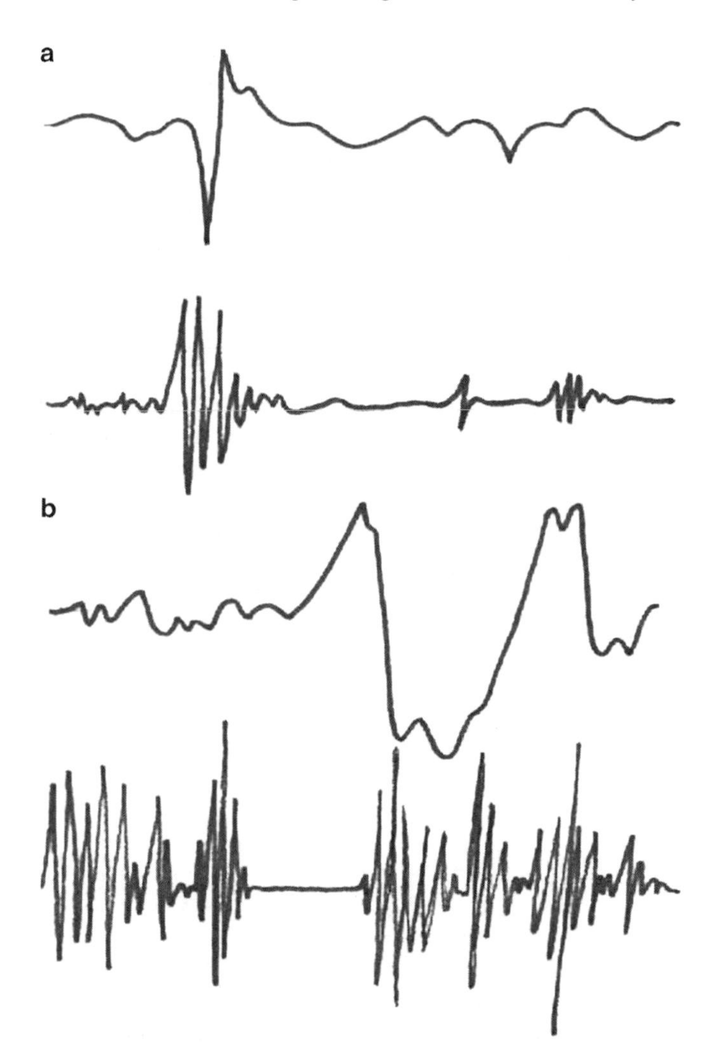

Fig. 6.1 Schematic of positive and negative myoclonus. *Top* accelerometer trace. *Bottom* EMG of involved muscle. (**a**) Positive myoclonus: a short EMG burst induces a jerk, as revealed in the accelerometer trace. (**b**) Negative myoclonus: a short pause in EMG activity during voluntary activation induces a jerk

Generalized myoclonus involves multiple muscles in a single jerk. In this case, the order in which the muscles are recruited generally reveals the origin of the myoclonus. Since recruitment of multiple muscles occurs at intervals of tens of milliseconds this process cannot be observed clinically, and EMG is recorded from these muscles to determine the recruitment order. To observe propagation of the myoclonic discharge in the spinal cord, EMG must at least be recorded from proximal and distal upper and lower limb muscles. Propagation within the brainstem can be observed by recording from at least the *masseter* muscle (its innervating nerve originates from the upper brainstem), the *orbicularis oculi* muscle (middle brainstem), and the *sternocleidomastoid* muscle (lower brainstem). In case of generalized cortical myoclonus, muscles innervated by the *cranial nerves* are recruited first, then the muscles of the upper limbs and finally, the muscles of the lower limbs. Generalized brainstem myoclonus is typically generated in the reticular formation in the middle part of the brainstem, close to the nerve innervating the sternocleidomastoid muscle (Cassim and Houdayer 2006). This muscle therefore usually is the first muscle to be recruited. The myoclonic discharge then travels up the brainstem and down the spinal cord, resulting in the masseter muscle being recruited after the orbicularis oculi muscle and upper limb muscles being recruited before lower limb muscles. Finally, generalized spinal (or propriospinal) myoclonus usually first recruits abdominal or paraspinal muscles, after which the discharge spreads up and down the spinal cord. Thus, to determine the generator of generalized myoclonus, extensive EMG recording is required.

In case of (multi)focal myoclonus, EMG can also be used to determine its origin, but since single muscles are involved, the origin cannot be determined from a recruitment order over multiple muscles. Instead, if a cortical origin is suspected, EEG can be recorded simultaneously to allow the coupling of muscle activity to (preceding) cortical activity. The two most important methods to diagnose cortical (multi)focal myoclonus are jerk-locked back-averaging and corticomuscular coherence analysis.

6.3 Back-Averaging

Already in the 1930s, Grinker and coworkers reported that there was a close temporal relationship between EMG bursts and preceding EEG spikes in a case of myoclonus (Barrett 1992). This relationship was determined by visual inspection only and was therefore rather inaccurate. In addition, it could only be observed for EEG spikes with a sufficiently high signal-to-noise ratio (see Sect. 5.3.1). Further improvement had to await technical developments and in 1975 Shibasaki and Kuroiwa described the method of jerk-locked averaging or back-averaging (Barrett 1992). This method allows to establish the precise timing between the preceding cortical activity and the EMG burst signifying the jerk-onset and additionally, the scalp distribution of the cortical activity preceding the jerk can be determined.

6.3.1 Principles of and Prerequisites for Back-Averaging

To allow jerk-locked back-averaging, both the EEG and EMG have to be recorded simultaneously. The EMG is recorded from the involved muscle(s) using pairs of electrodes a few cm apart on the belly of the muscle(s). The EEG may employ a classical 10–20 montage with 20–30 electrodes, but it can be extended to use as many as 128 channels, if the exact topography of the EEG prior to the jerk is of interest. As a minimal requirement the central electrodes C3, Cz, and C4, that overlie the bilateral central sulci, should be used. If a myoclonus has a cortical origin, it will be localized in the primary sensorimotor cortex, more laterally (C3 or C4) for upper limb muscles and more medially (Cz) for lower limb muscles. In addition, for the myoclonus to be of cortical origin, the cortical potential should precede the EMG activity by approximately 20 ms for lower arm muscles and by approximately 30 ms for lower leg muscles.

Question 6.1 Why should a cortical potential precede an EMG burst in a lower arm muscle by approximately 20 ms for the burst to be of cortical origin?

Technically, both the EEG and EMG should be sampled at a sufficiently high frequency (e.g. 1,000 Hz, to obtain a temporal resolution of 1 ms). This is most important for the EMG, since the onset of the myoclonic burst must be established with high temporal accuracy. Furthermore, it is not always possible to determine EMG onsets online, so that it should be possible to determine these onsets offline (Sect. 6.3.3).

From the patient's perspective, a jerk-locked back-average can only be determined if the frequency and consistency of the myoclonic movements are sufficiently large. If the myoclonic movements are very infrequent, recording may be very time-consuming. On the other hand, if the jerks are very frequent, other jerks may occur in the interval directly preceding a jerk, complicating analysis. As a rule of thumb, jerks should be at least 500 ms apart, and occur at least 50 times in a recording to be able to derive a jerk-locked back-average.

Question 6.2 How can the minimal duration of a recording of a patient with myoclonic jerks be determined beforehand?

6.3.2 Back-Average Calculation

The calculation of a back-average is essentially the same as for an evoked potential. However, instead of a predetermined visual, auditory, or somatosensory stimulus that the EEG can be aligned to (see Chap. 5), this point in time must be determined from the EMG. Once the EMG onsets have been determined, the EEG may be segmented around these time markers, taking an interval of 100–150 ms before the marker and 50–100 ms after the marker. Next, the segments should be inspected for ocular (blinks) or movement artifacts, so that these segments can be excluded from the average. Alternatively, automated artifact rejection methods may be used, that will exclude segments in which the amplitude exceeds a certain threshold. Finally, an average can be calculated if a sufficient number of segments remain (at least 50 as a rule of thumb). Depending on the quality of the recording, the EEG may be filtered before segmentation, e.g. by using a 1 Hz high-pass filter to remove movement artifacts or slow drifts (see Sect. 4.3.1).

6.3.3 *Tips and Tricks When Identifying EMG Onsets*

The EMG onsets that are required for back-averaging can be determined both automatically and manually. Automatic onset detection methods have the potential advantage that they can be applied online, while acquiring the data, so that a possible cortical origin of a myoclonic movement can be directly recognized. Furthermore, the result of automated methods is reproducible, which is usually not the case for manual methods due to their subjective nature. However, an obvious disadvantage of automated methods is that the onset markers may be wrongly placed, thereby distorting the average.

Both automated and manual methods can profit from preprocessing of the EMG, so that the EMG onsets can be determined more consistently and reliably. For example, the EMG may be *rectified* and *integrated* or high-pass filtered to remove movement artifacts and then rectified and integrated. An alternative to integration is *smoothing*. Note, however, that smoothing will make a sharp EMG onset less conspicuous, thereby possibly distorting the average.

> *Question 6.3* Show that a sharp increase in a signal from 0 to 1 (step function) disappears when a smoothing window of 3 points with weights 0.25, 0.5, and 0.25 is used.

6.3.3.1 Automated Methods

Automated methods usually allow to put a marker whenever the EMG amplitude exceeds a predefined threshold, either when the voltage is rising (positive direction) and the voltage threshold is reached or when the voltage is falling (negative direction) and the voltage threshold is reached. Furthermore, most methods allow to skip a certain interval (temporal delay) after a marker is set, and before the next marker is searched for. These concepts are illustrated in Fig. 6.2.

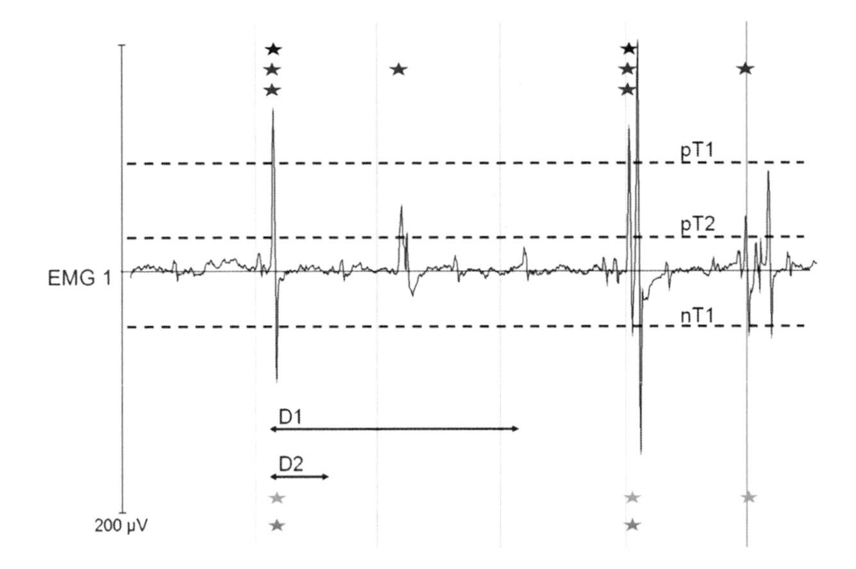

Fig. 6.2 Example of EMG data with possible choices for the threshold and direction (pT1: high positive threshold, pT2: low positive threshold, nT1: negative threshold) and temporal delay (D1: long delay of 400 ms, D2: short delay of 100 ms) to detect the onset of EMG bursts. *Stars* indicate EMG onsets for different combinations of thresholds and delays (*red*: pT2/D1, *blue*: pT2/D2, *black*: pT1/D1 and pT1/D2, *pink*: nT1/D2, *light blue*: nT1/D1)

Taking a temporal delay into account can be useful when the EMG burst is rather long but easily distinguishable from background activity. In that case, one can be sure that a second marker will not be set before the next burst occurs. However, this directly emphasizes the difficulty of automated methods; to use them blindly requires that the characteristics of the myoclonic EMG are known a priori. Therefore, a combination of automated and manual methods is typically used, thereby combining the best of both worlds. The EMG is first visually inspected to determine appropriate values for the threshold and the length of the time interval to be skipped for the next marker. Then an automated method may be used to get a first result for the detection of EMG onsets. Finally, a manual step may be taken to adapt the position of some of the markers if deemed necessary.

6.3.3.2 Manual Methods

The problem with manual detection of EMG onsets is to be consistent. As illustrated in Fig. 6.3a the initial deflections of an EMG bursts can be highly variable, even in the case of myoclonic bursts. The morphology of a burst may vary, the baseline may include a variable amount of noise or multiple bursts may coincide. In cases where the bursts are clearly delineated (Fig. 6.3b) and the baseline is well defined, it is much easier to determine the EMG onset consistently from one burst to the next.

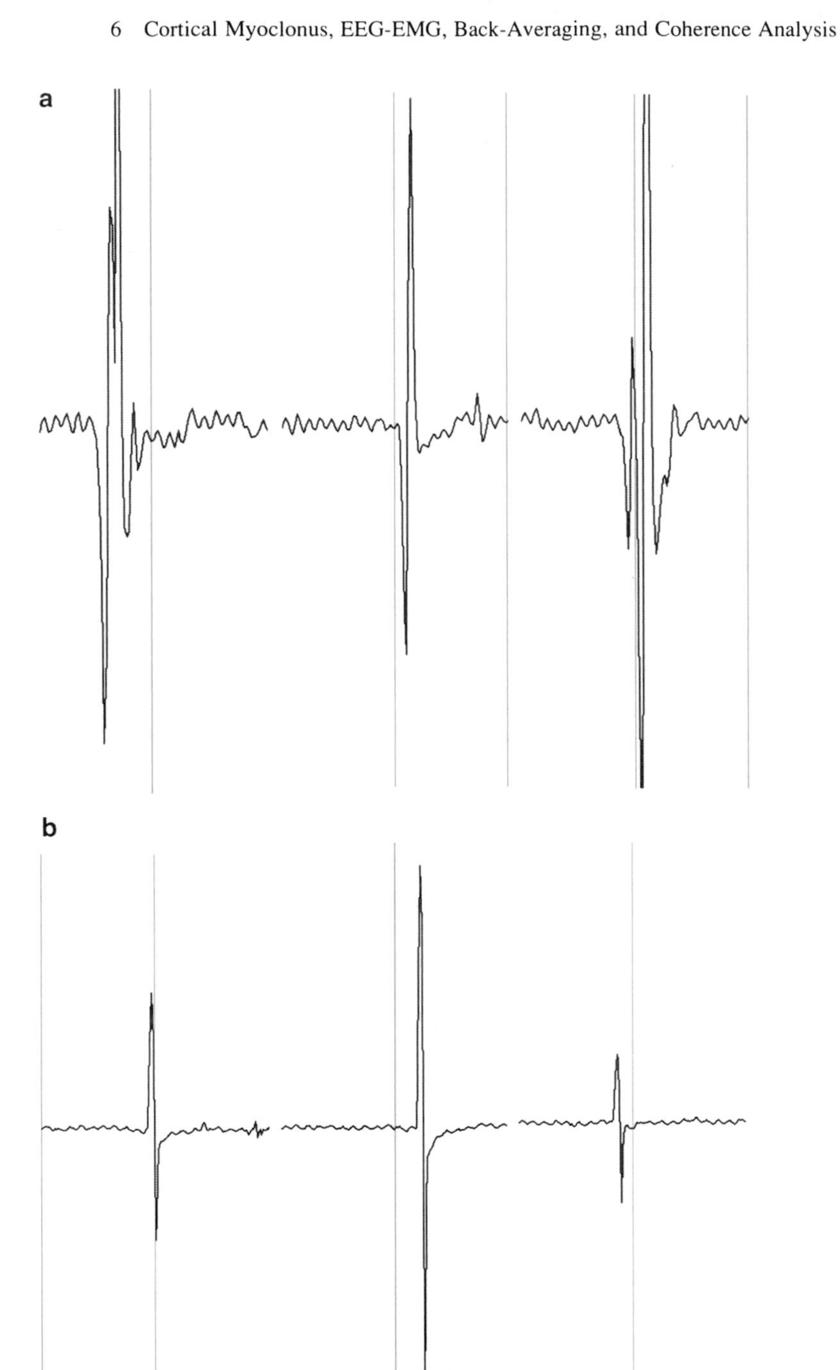

Fig. 6.3 (**a**) Three myoclonic bursts (patient 1) illustrating the variability of burst morphology. (**b**) Three myoclonic bursts (patient 2) illustrating similarity of bursts, enabling more consistent EMG onset detection, both automatically and manually. All figures show 400 ms of EMG. Vertical scales differ between (**a**) and (**b**) but are consistent within (**a**) and (**b**)

6.4 Coherence Analysis

As explained in Sect. 6.3.1, jerk-locked back-averaging is only possible when there is sufficient time between jerks. When the jerks are too frequent to be discerned (see Fig. 6.4) other methods are necessary to investigate the involvement of the cortex in generating these involuntary movements.

An alternative method that can help determine a cortical generator of myoclonus is to determine corticomuscular coherence, a measure of the relationship between cortical activity (EEG, magnetoencephalography [MEG] or local field potentials) and muscular activity as recorded by the EMG. Coherence can also be determined between two EEG signals (corticocortical coherence), between two EMG channels (musculomuscular coherence) or in general, between any two signals that have been sampled simultaneously.

In general, coherence, which describes a temporal correlation between spatially distinct parts of the nervous system, is thought to represent functional coupling between these parts (e.g. Engel et al. 1992; Varela et al. 2001). Since the motor cortex and muscles are functionally closely coupled (the motor cortex drives muscular activity), it may be expected that corticomuscular coherence analysis will provide clues on the modes of information transfer within the motor system.

Indeed, during tonic contractions, corticomuscular coherence has been found between 15 and 35 Hz using both EEG and MEG (Mima and Hallett 1999). Coherence can also be observed at lower and higher frequencies depending on muscular force or movement type, although some (healthy) people do not show corticomuscular coherence at all. If corticomuscular coherence is present it is usually localized over the primary sensorimotor cortex following more or less a somatotopic organization, i.e. upper extremity muscles are localized more laterally than lower extremity muscles. Typically, motor cortex activity has been found to precede (lead) EMG activity as expected. However, it should be noted that coherence is always the net result of several simultaneously occurring phenomena. Corticomuscular coherence can be achieved by efferent information flow (cortical motor commands to muscles), afferent information flow (proprioceptive/sensory input to the cortex from the periphery), and by a third rhythm generator such as a subcortical brain area, influencing both rhythms in the cortex and the muscle. In principle, time lag estimation from the phase spectra (see Sect. 6.4.2.2) can help solve this ambiguity.

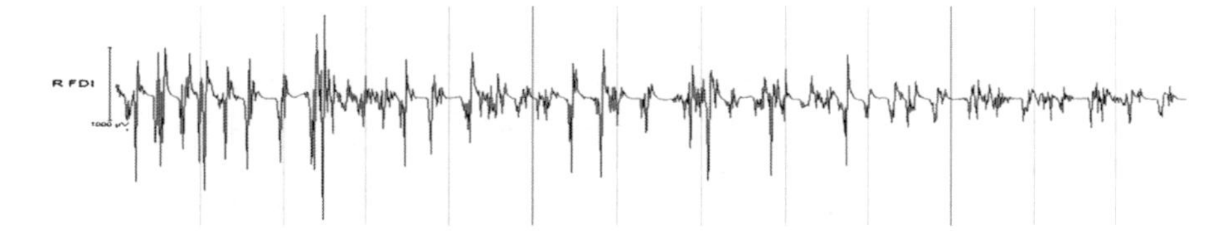

Fig. 6.4 Example of right first dorsal interosseus (FDI) EMG recording of myoclonic bursts that occur very frequently, hampering the accurate detection of EMG onsets necessary for back-averaging

6.4.1 *Coherence: Signal Correlation in the Frequency Domain*

Coherence is an extension of Pearson's correlation coefficient to the frequency domain (see Box 6.1).

Box 6.1 The Mathematics of Coherence

The coherence between two signals x and y is defined for each frequency λ by the ratio of the cross-spectrum between the two signals and the two autospectra as follows:

$$|R_{xy}(\lambda)|^2 = \frac{|f_{xy}(\lambda)|^2}{f_{xx}(\lambda)f_{yy}(\lambda)}. \quad (1)$$

Here, R_{xy} is the coherency function, f_{xy} is the cross spectrum, and f_{xx} and f_{yy} are the autospectra. Equation (1) is comparable to Pearson's correlation coefficient in the time domain (Box 2.1).

Whereas, coherence provides an estimate of the strength of the coupling between two signals, timing information can be obtained from the phase spectrum, defined by the argument of the cross spectrum:

$$\Phi(\lambda) = \arg\{f_{xy}(\lambda)\}. \quad (2)$$

The argument of a complex number $z = a + ib$ is the angle with the positive real axis in the complex plane in which a is plotted on the real axis and b is plotted on the imaginary axis (Box 4.1).

In the time domain the association between the two signals x and y can be expressed in the cumulant density function which is obtained from the inverse Fourier transform of the cross spectrum f_{xy}. The maximum peak in the cumulant density function provides a measure of time delay between signals.

Coherence can attain values between 0 and 1 (in contrast to correlation which takes values between -1 and 1, see Sect. 2.4.4). When coherence is 1 there is a perfect linear relationship between two signals, whereas when it is 0 there is no relationship at all. It is not as easy as for correlation to visualize what strong and weak coherence mean, but an attempt is made in Fig. 6.5.

As can be observed in Fig. 6.5, for coherence to be high, both amplitude and phase of the two signals must covary with each other. Phase provides a measure of the temporal relationship between two signals and allows to determine whether the first signal is leading or lagging the second (see Sect. 6.4.2.2).

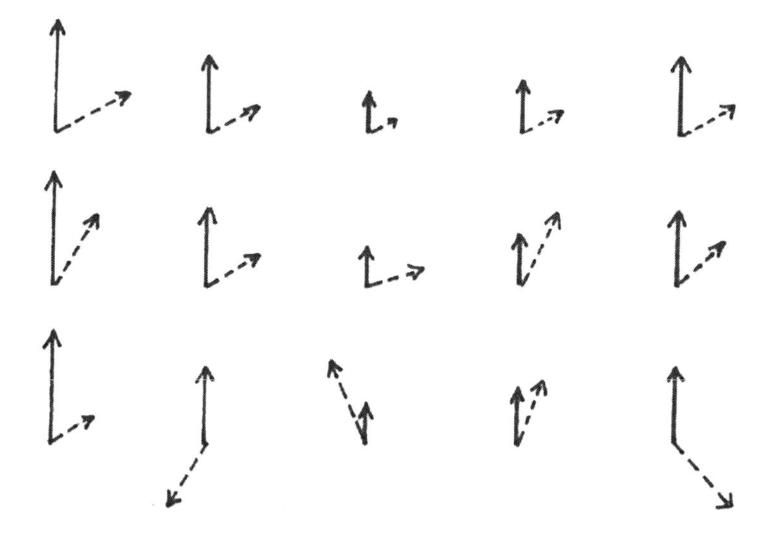

Fig. 6.5 Illustration of *top*: perfect; *middle*: intermediate; and *bottom*: weak coherence between two signals, represented by the *solid* and *dotted arrows* at five different moments in time. The length of the *arrows* represents amplitude, the angle between the *arrows* represents phase. In the first case, the relative length of the arrows and the angle between the arrows remains constant. In the second case, there is some relationship, but not as strongly as in the first and in the last case there is no relationship at all

6.4.2 *Tips and Tricks When Calculating Coherence*

6.4.2.1 Significance

Before interpreting coherence it is important to establish its statistical significance. To obtain an estimate of coherence, the data is typically segmented into N smaller parts of a few (T) seconds, yielding a spectral resolution of $1/T$ (cf. Sect. 3.3.3). Thus, when segments of 2 s are chosen, a spectral resolution of 0.5 Hz results. The number of segments N determines the significance threshold as follows (Halliday et al. 1995):

$$\text{significance threshold} = 1 - \alpha^{1/(N-1)}, \tag{6.1}$$

where α is the significance level. For example, when 100 segments are obtained and a significance level of 0.05 is assumed, coherences larger than $1 - 0.05^{1/99} = 0.03$ are significant and can be interpreted. A significance level of 0.05 implies that 1 in 20 significant coherence peaks is still coincidental. There is always a trade-off in choosing the number of segments between the spectral resolution (better in case of fewer segments) and the significance level (better, i.e., coherence does not need to be as large to be significant, in case of more segments).

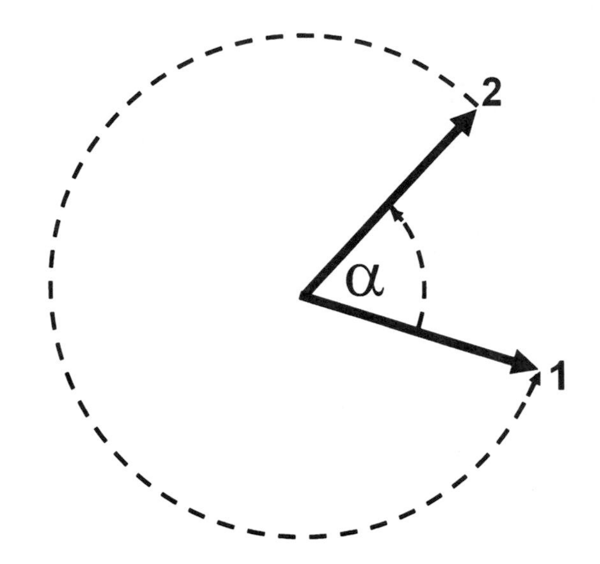

Fig. 6.6 Illustration of the problem in identifying lag or lead between two signals based on a single phase α; either signal 2 is leading signal 1 by phase α or signal 2 is lagging signal 1 by 360°-α. Compare Fig. 6.5 for interpretation of the *arrows*

Even when coherence is significant for a certain frequency, other criteria are often taken into account before interpretation. For example, to avoid interpreting spurious coherence peaks (taking into account that at a significance level of 0.05, 1 in 20 peaks is still coincidental), interpretation is usually limited to those peaks that are significant for at least a few (3–5) consecutive bins.

6.4.2.2 Phase: Calculation and Interpretation

First and foremost, phase should only be interpreted for the frequencies at which coherence is significant, even though the phase spectrum provides a value for phase at every frequency. However, since phase is expressed in units of degree, a single value cannot be interpreted. As an example, a phase of 30° may mean that signal 1 is leading signal 2 by this amount or that signal 2 is actually leading signal 1 by 330° (see Fig. 6.6).

Only when multiple frequencies have the same time-lag, independent of frequency, the time lag can be estimated by regression analysis. To make this more clear, imagine a point traveling over a unit circle. This point will take just as long to progress 30° along the circle at 10 Hz or to progress 60° at 20 Hz (Mima and Hallett 1999). Since the accuracy of the phase estimate (expressed in the frequency-dependent phase variance) depends on the coherence estimate at that frequency, a weighted regression must be performed. The smaller the variance for that frequency (i.e. the more reliable the phase estimate), the larger the weight. The slope of the regression line then provides a measure of time delay. If the slope is positive, the first signal is lagging the second, whereas when it is negative, the first signal is leading the second (see Fig. 6.7).

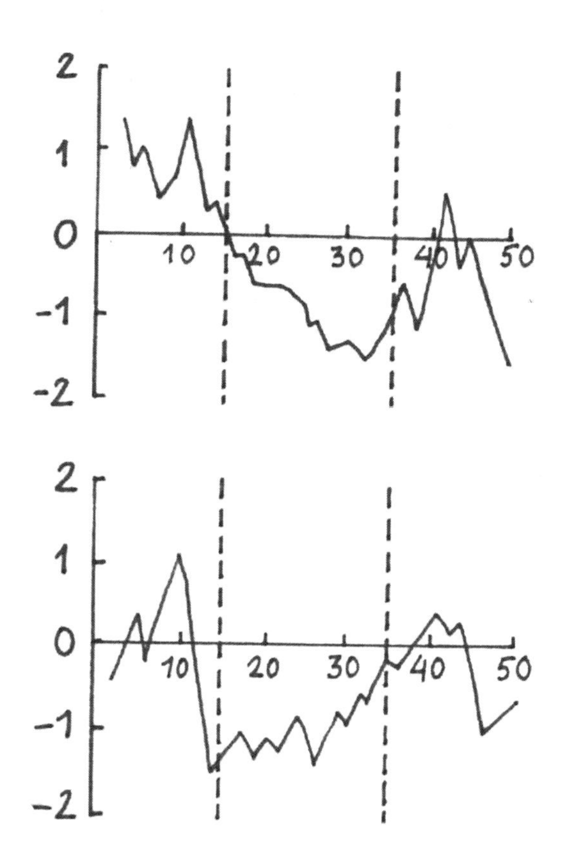

Fig. 6.7 Schematic example of phase spectra. *Vertical scale*: phase in radians, *horizontal scale*: frequency in Hz. For these examples it is assumed that coherence between signals 1 and 2 is significant between 15 and 35 Hz (indicated by the *dashed vertical lines*). In the *top plot*, signal 1 leads signal 2, in the *bottom plot*, signal 1 lags signal 2

6.4.2.3 Nonphysiological Factors Influencing Coherence

EEG is always recorded by comparing the voltage at one electrode against the voltage at another (set of) electrode(s); the reference (see also Sect. 4.2). Since the reference has a large effect on the EEG signal, it also strongly influences corticomuscular or corticocortical coherence. Generally, there are three categories of references; those that include only a few electrodes (e.g. a linked-earlobe reference), neutral references (e.g. a *noncephalic* or common average reference), and current source density estimation references (e.g. the *Hjorth* or *Laplacian* transform of the EEG). A disadvantage of the first category of references is that an artifact in the reference influences all other EEG channels. Theoretically, the second category is optimal, because these references are not influenced by the EEG. However, a noncephalic electrode can still suffer from artifacts (e.g., due to breathing), yielding the same disadvantage as for the references in the first category. The common average reference is only neutral when a sufficiently large number of electrodes, that also cover the lower parts of the head, are included. This may be the case for 128 or 256 EEG channels but not for a standard 10–20 configuration. Finally, the current source density estimation methods filter out widespread waveforms (low spatial frequencies) and emphasize localized waveforms (high spatial frequencies), thereby providing a better spatial resolution of the EEG. However, this may be a disadvantage if the interest is in the lower frequency bands of the signal. Typically, current source density methods are optimal for calculating corticomuscular coherence, whereas neutral references are optimal for calculating corticocortical coherence.

For corticocortical coherence in particular, coherence may be overestimated due to volume conduction and common reference effects (Nunez et al. 1997). Volume conduction implies that the electrical activity of one source can be detected on multiple electrodes, if the source is strong enough. If this is the case, coherence between two distant electrode sites is high just because activity is measured from one and the same source and not because the two brain areas underlying the electrodes are functionally coupled. The common reference effect entails artificially high coherence due to a common signal in the two EEG signals of interest, whenever the common reference is not neutral. A similar increase in coherence results from a common artifact on both EEG signals. Finally, two neighboring electrodes will always show moderate to high coherence due to volume conduction effects, but when the coherence is very close to 1, one should be aware that this may be the result of the electrode paste forming a conducting bridge between the two electrodes.

6.5 EEG-EMG Analysis in Individual Patients

Patient 1

In this patient who suffered from rheumatoid arthritis for more than two decades, EEG was recorded from 20 channels positioned according to the 10–20 system, to investigate a possible cortical generator for her myoclonic jerks. In addition, the electrocardiogram (ECG), electro-oculogram (EOG), and left *extensor digitorum communis* (EDC) EMG were recorded. The EEG was found to be diffusively slow with reasonable differentiation and reactivity (see Sect. 4.2). Epileptic activity, consisting of peaks and peak-wave complexes, was observed over bilateral temporo-occipital areas, right more than left (Fig. 6.8a). Some paroxysmal delta activity was also found, possibly related to episodes of drowsiness (Fig. 6.8b). A simultaneous video recording revealed multifocal myoclonic activity in the face and extremities, which could not be directly coupled to epileptic EEG activity.

To determine whether cortical activity was generating the myoclonic activity, back-averaging was performed. EMG onsets of 131 myoclonic bursts were determined manually. The data was filtered using a 1 Hz (24 dB/oct) high-pass filter to stabilize the baseline and segmented around these onsets (150 ms pre-onset, 50 ms post-onset). Additional artifact rejection was found to be unnecessary. Finally, all segments were averaged. The result is shown in Fig. 6.9.

After averaging, a negative field could be observed over right centroparietal areas, starting around 30 ms before EMG onset and ending around 20 ms after EMG onset, providing evidence for a possible cortical generator of the myoclonic activity.

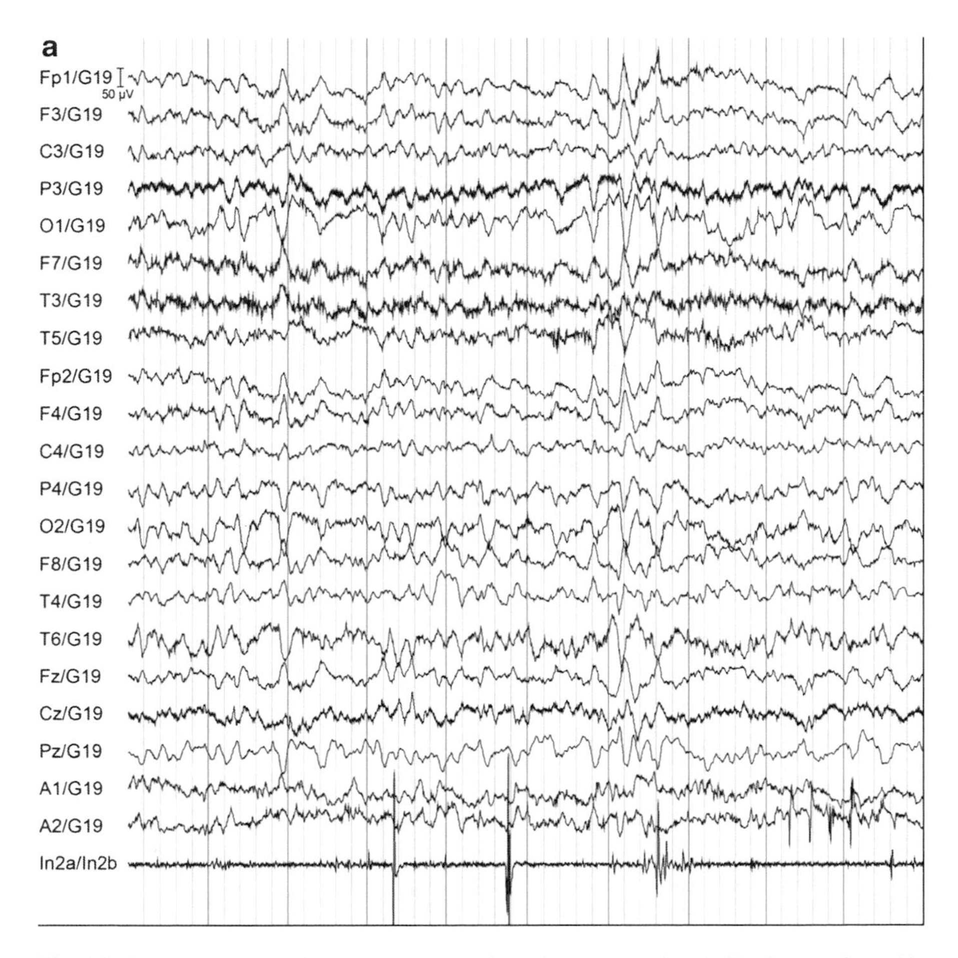

Fig. 6.8 Samples of EEG (10 s each) as recorded in patient 1 (eyes closed). The *bottom channel* is the extensor digitorum communis (EDC) EMG. (**a**) Example of ongoing EEG during which multiple myoclonic bursts occur in the EMG. Not all bursts can be coupled directly to epileptic activity in the EEG. (**b**) Example of a burst of paroxysmal delta-activity (at 2 s)

The absence of a temporal relationship between epileptic EEG activity and EMG onsets, suggests that the cortical myoclonus is not of epileptic nature.

The patient received a single dose of Clonazepam on admission, which decreased the intensity of the myoclonic jerks. Clonazepam is used as an antiepileptic drug, but causes muscle relaxation as a side effect as well, thus not proving an epileptic origin of the myoclonus, either. Since the patient suffered from continuing drowsiness and an increase in her *apneas* as side effects of the Clonazepam, this was unfortunately not a long-term treatment option.

Patient 2

In this patient who suffers from myoclonic jerks, possibly as a side effect of her antiepileptic medication, EEG was recorded using 19 channels and EMG was

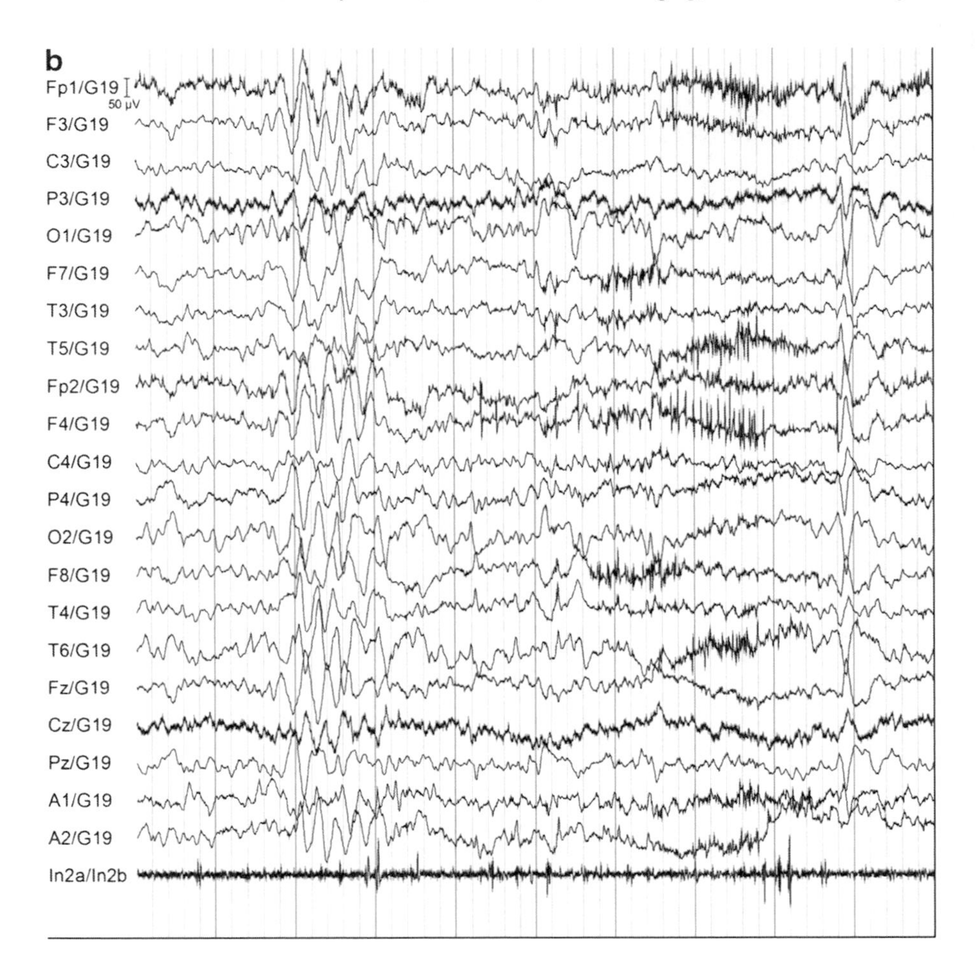

Fig. 6.8 (continued)

recorded on six channels (*flexor carpi radialis*, EDC and *biceps brachii*, bilaterally). The EEG was a bit slow, but well differentiated and reactive. Epileptic activity occurred rather frequently, both multifocally as well as generalized, with a focus of activity in frontotemporal areas. In some cases the epileptic activity coincided with a clear myoclonic burst in the EMG channel, but this was not always the case (see Fig. 6.10a). The EMG showed multifocal, very narrow (50–100 ms) bursts in multiple muscles (see Fig. 6.10b). When the patient was asked to extend the arms and hands, to induce sustained EMG activity, pauses occurred in the EMG infrequently, coinciding with a drop of hand posture, indicating the presence of negative myoclonus as well as positive myoclonus.

To determine a cortical generator of the myoclonic activity, back-averaging was performed in this patient, as well. EMG onsets were determined manually in the EDC channel, resulting in 45 bursts that occurred in relatively artifact-free EEG periods. The data was filtered using a high-pass filter (1 Hz, 48 dB/oct) and segmented

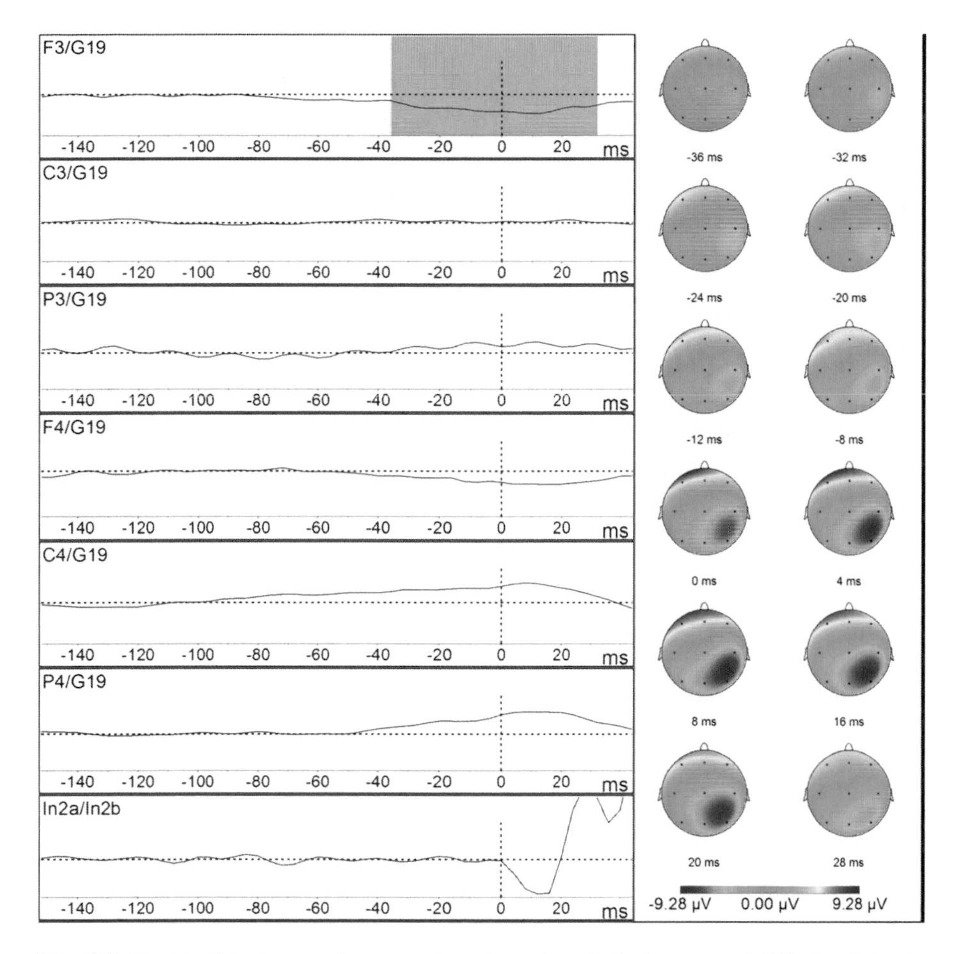

Fig. 6.9 Result of back-averaging procedure in patient 1. *Left*: average EEG signal in six centroparietal channels and average EMG (*bottom channel*), positive down. *Right*: EEG mapping from 36 ms before EMG onset to 28 ms after EMG onset. View from above, nose on *top*

from 100 ms pre-onset to 50 ms post-onset. Before averaging, artifact rejection was applied; all segments in which the amplitude exceeded ±50 µV were removed on a channel-by-channel basis. The final result is shown in Fig. 6.11.

After averaging, a clear cortical potential, with a positive-negative-positive configuration which was most pronounced on left centroparietal electrodes, was found to precede the myoclonic burst, starting approximately 25 ms before the onset of the EMG. In conclusion, there is evidence for a cortical myoclonus which is not only of epileptic origin.

Patient 3

In the 19-channel EEG of this patient, who suffers from tremulous and myoclonic movements and epileptic seizures, spike-wave complexes (see Sect. 4.2) were identified, confirming a form of epilepsy. In addition, electrical stimulation of the median nerve at the wrist at rest evoked an abnormal muscle response (C-reflex), suggesting hyperactivity of the sensorimotor cortex. To further investigate the

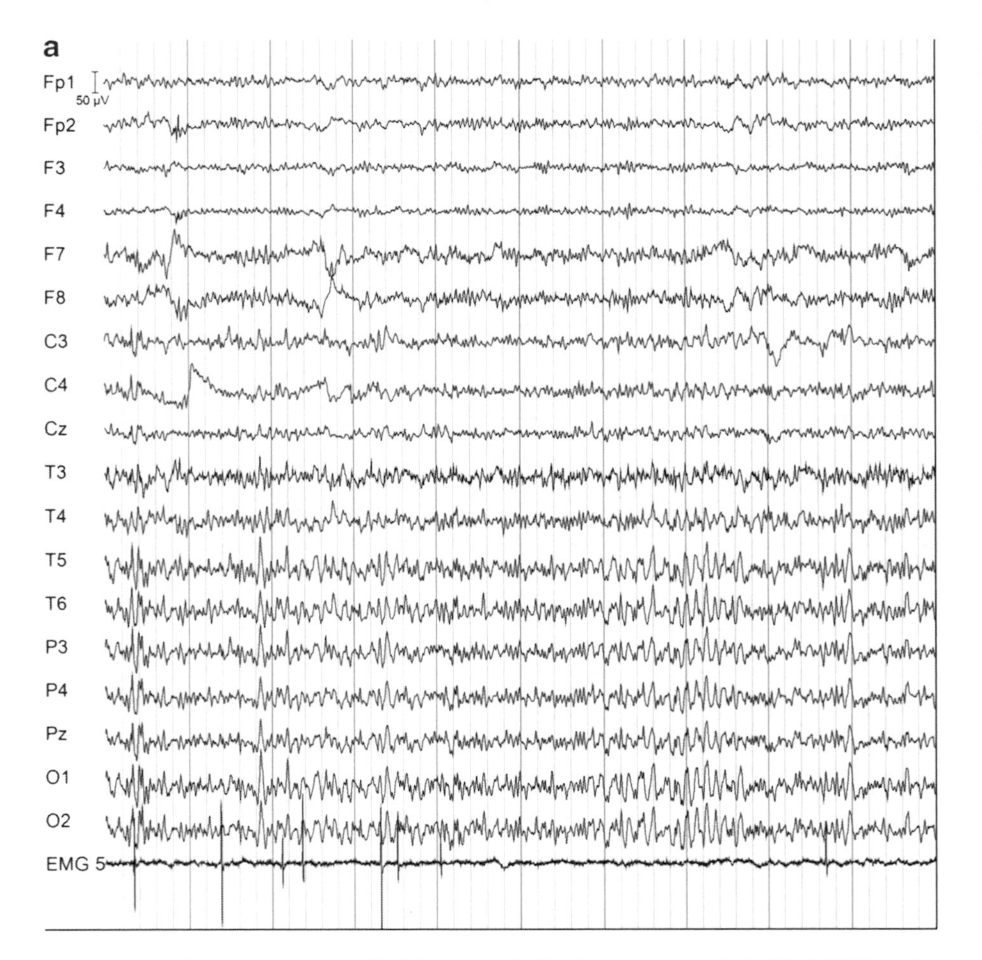

Fig. 6.10 (**a**) Example of ongoing EEG in patient 2. The *bottom channel* is the EDC EMG on the *right side*. (**b**) Example of EMG in patient 2. From *top* to *bottom*: flexor carpi radialis, EDC, and biceps brachii, first left, then right. The bursts occur in all recorded muscles irregularly and are very short-lasting (50–100 ms) implying that the patient suffers from multifocal myoclonus

origin of the myoclonic movements in this patient, EMG from the most involved muscles in the right lower arm was recorded concurrently with the EEG. The bursts of EMG activity were found to be highly irregular (see Fig. 6.12).

In contrast, if this patient would have suffered from ET, the bursts would have been regular (cf. EMG from patient 3 in Sect. 3.4). The irregularity of the myoclonic bursts precludes back-averaging to investigate a cortical origin. Therefore, coherence analysis was used instead (see Figs. 6.13 and 6.14).

Coherence analysis proves that patient 3 indeed suffers from cortical myoclonus; there is significant coherence between the EEG over the sensorimotor cortex and the EMG of the involved muscle(s), the cortical signal leads the muscle signal and by an amount of approximately 16 ms, which is fast but consistent with known conduction times (the time it takes a potential to travel from the cortex to the muscle; cf. Chap. 11). Combined with the finding of epilepsy, this patient was

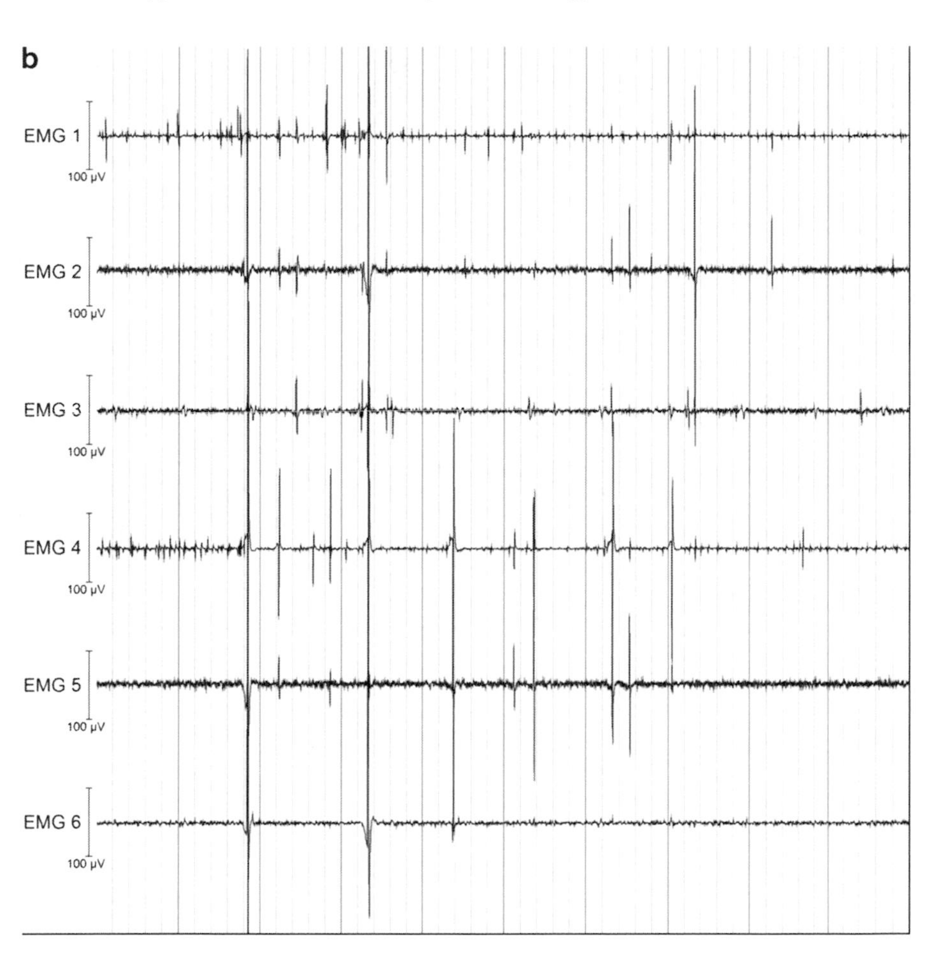

Fig. 6.10 (continued)

diagnosed with familial cortical myoclonic tremor with epilepsy (FCMTE), a rare hereditary disease (van Rootselaar et al. 2005, 2006).

6.6 Other Applications of Coherence Analysis in Neurology

Most applications of coherence analysis in neurology have a more fundamental purpose; coherence is then compared between two groups of subjects. Unfortunately, overlap in coherence values between two groups of patients is often too large to allow clinical diagnostic applications. However, the applications described in this section do provide more insight in neurophysiological processes underlying important neurological diseases using patient-friendly and noninvasive techniques, thereby providing a first step towards clinical use.

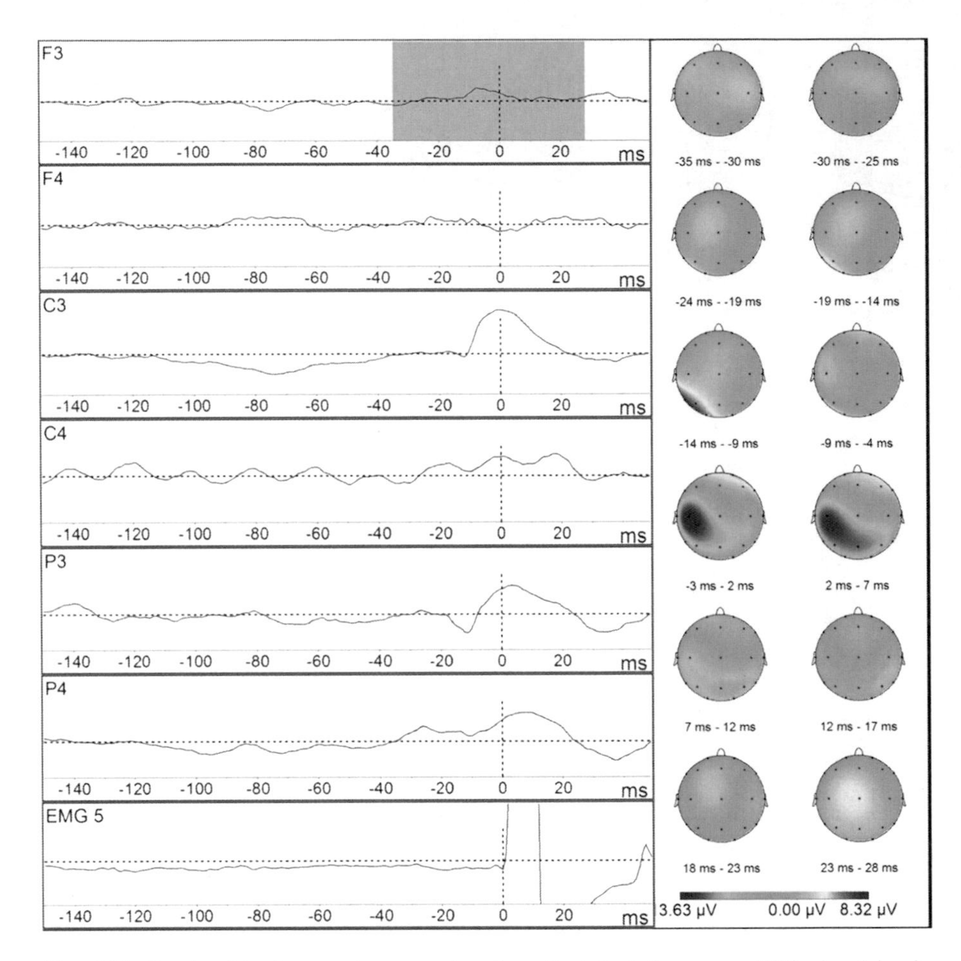

Fig. 6.11 Result of back-averaging procedure in patient 2. *Left*: average EEG signal in six centroparietal channels and average EMG (*bottom channel*), positive down. *Right*: EEG mapping from 35 ms before EMG onset to 28 ms after EMG onset. View from above, nose on *top*

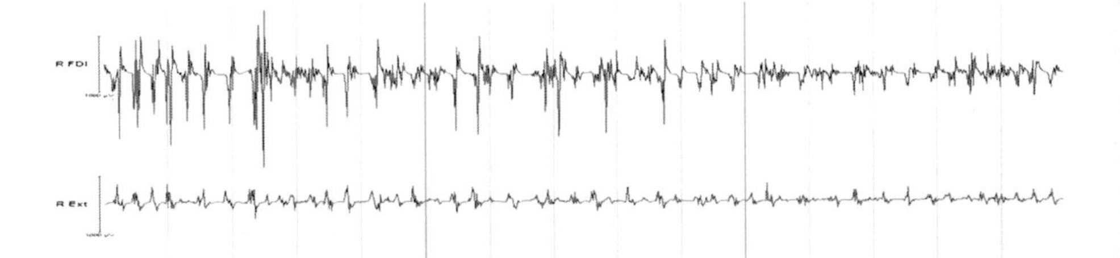

Fig. 6.12 EMG recorded from right first dorsal interosseus muscle (*top*) and right lower arm extensor muscles (*bottom*) in patient 3

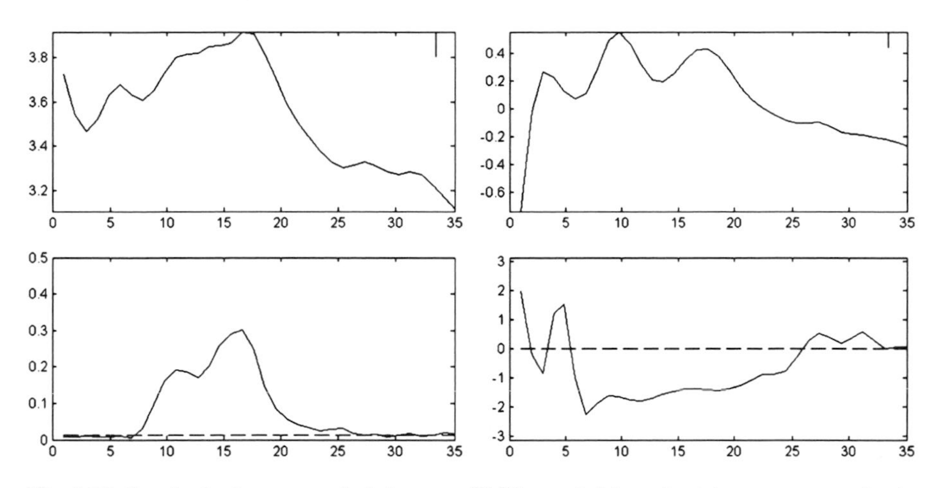

Fig. 6.13 Result of coherence analysis between EMG recorded from the right extensor muscles in the lower arm (rEXT) and EEG channel C3 in patient 3. *Top left*: rEXT spectral power (log plot), *top right*: C3 spectral power (log plot), *bottom left*: coherence between rEXT (*horizontal dashed line*: significance level), and C3 and *bottom right*: phase in radians between rEXT and C3. The frequency in Hz is on the horizontal axis. Phase is positive for the frequency range in which the coherence is significant (7–27 Hz), indicating that rEXT is lagging C3 in this range

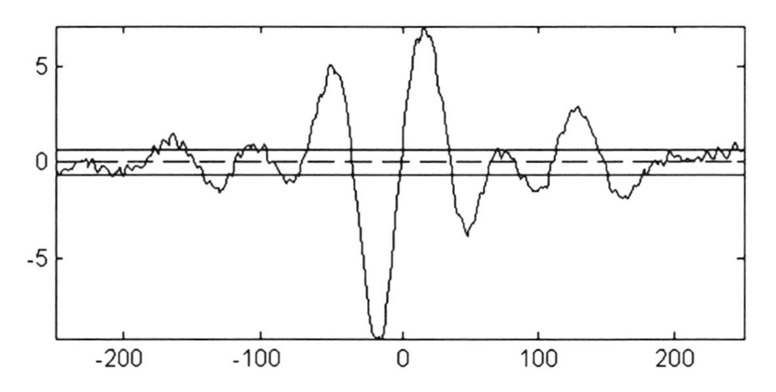

Fig. 6.14 Cumulant density function (see Box 6.1) for patient 3. The maximum peak indicates that there is a delay of approximately 16 ms between the cortical signal at C3 and the muscle signal at rEXT

6.6.1 EEG-EEG Coherence in Alzheimer's Disease

Alzheimer's disease (AD) is the most common cause of dementia, which by itself is the most common neurological disorder among elderly. An important issue in the care for AD patients is to establish the diagnosis as early as possible and in that sense the prediction of the progression of mild cognitive impairment – thought to be a precursor of AD – into AD is of interest. Since AD is a cortical form of dementia, EEG has long been used to try and diagnose AD. It has been found repeatedly that EEG is slower (increase in lower frequency waves) in older individuals and particularly in patients with AD. In addition to EEG slowing, a decrease in the corticocortical coherence between different brain regions has been observed, supporting the view of AD as a cortical disconnection syndrome. Besides anatomical disconnections between cortical areas, reduced cholinergic coupling between cortical neurons may also underlie the decreased coherence in AD. The abnormalities in coherence increase with the severity of the disease (Jeong 2004). Unfortunately, the EEG can basically only distinguish between groups and is less helpful in the diagnosis of AD in individual patients.

6.6.2 EEG-EMG Coherence Analysis for Tremor Diagnosis

As illustrated earlier in this chapter, EEG-EMG coherence analysis can be used to examine the cortical correlates of myoclonic jerks, but it can also help distinguish between tremors that are clinically alike. An example is the movement disorder that patient 3 suffers from: FCMTE which is clinically often mistaken for the more ubiquitous ET (van Rootselaar et al. 2006). Not only can coherence analysis be used to prove an oscillatory coupling between the bursts of muscle activity and neuronal activity in the cortex as shown in Sect. 6.5, but van Rootselaar et al. also showed that the type of corticomuscular and musculomuscular coupling is different between FCMTE and ET during maintained posture of the dominant hand. When comparing healthy subjects to patients with FCMTE and patients with ET in two frequency bands (4–8 Hz around tremor frequency and 8–30 Hz), healthy subjects and ET patients showed normal weak corticomuscular coherence in the higher frequency band (see also Sect. 6.4), whereas both corticomuscular and musculomuscular coherences were very strong in FCMTE patients. In the lower frequency band, both ET and FCMTE patients showed increased musculomuscular coherence, which was absent in the healthy subjects. Hence, coherence in the lower frequency band allows to distinguish between patients and healthy subjects, while coherence in the upper frequency band allows to distinguish between FCMTE and ET patients.

6.6.3 Blood Pressure-Cerebral Blood Flow Coherence in Cerebral Autoregulation

To function adequately and consistently, the brain requires a constant supply of the nutrients that are transported by the blood. Since the arterial blood pressure (BP) varies, the cerebral blood vessels have developed the ability to adapt to these changes using several mechanisms such that cerebral blood flow (CBF) remains constant. The process of changing cerebral resistance in response to a change in perfusion pressure to keep CBF constant is known as cerebral autoregulation (Panerai 2004; van Beek et al. 2008). Coherence analysis can be used to understand the physiological process of cerebral autoregulation and to aid in the diagnosis,

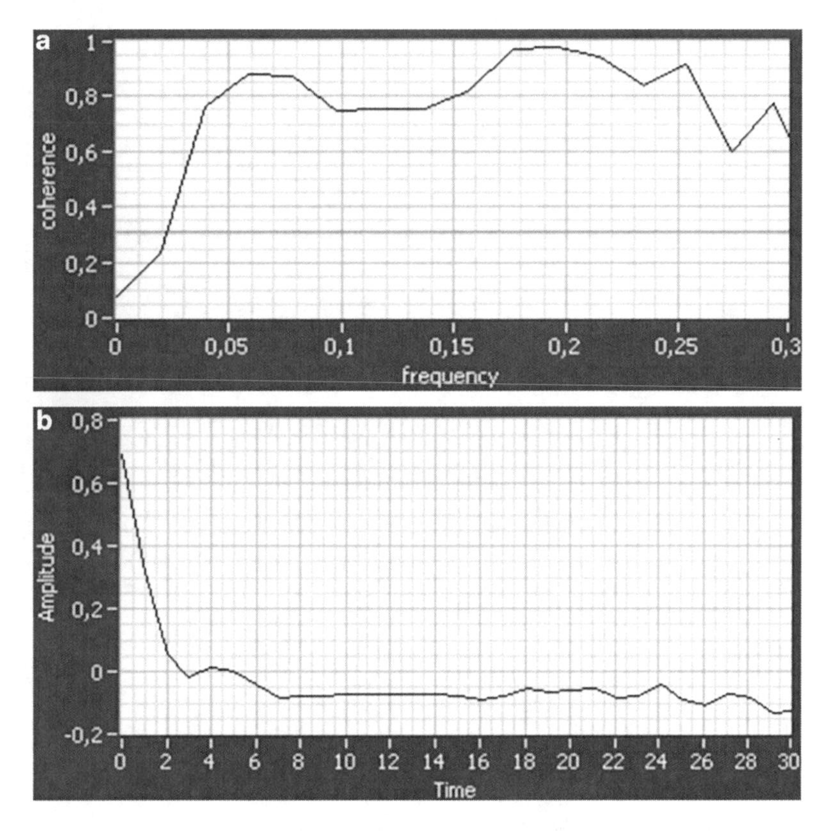

Fig. 6.15 (**a**) Example of coherence between BP and CBF in a healthy subject. Coherence is studied for very low frequencies only (here up to 0.3 Hz), because cerebral autoregulation is a relatively slow process. For the lowest frequencies, coherence is typically negligible. (**b**) Step response, showing a quick return to baseline after a change in BP, indicating well-functioning pressure autoregulation

monitoring, and prognosis of patients suffering from conditions that endanger this process. Although the process of cerebral autoregulation is determined by multiple factors (Panerai 2004), it can be simplified and considered as a dynamical system with BP as input and CBF as output. The behavior of this system is then described by its transfer function (see also Sect. 4.3.1). Typically, estimates of the transfer function are obtained in the frequency domain, employing the inherent variability of the BP (or inducing additional variability in the BP by manipulating the patient's posture). The coherence between BP and CBF provides a measure of the extent to which BP variations at a certain frequency linearly determine variations at that same frequency in CBF. In healthy subjects, coherence should be low for the lowest frequencies because CBF should not follow changes in BP when cerebral autoregulation is functioning effectively. An example of coherence between BP and CBF in a healthy subject is provided in Fig. 6.15.

6.7 Answers to Questions

Answer 6.1
It takes approximately 20 ms for a cortical potential to travel along the spinal cord to a lower arm muscle.

Answer 6.2
The duration of a recording in a patient with myoclonic bursts can only be determined beforehand if the jerks occur regularly and the inter-jerk interval is known. Then the duration should simply be at least 50 times the inter-jerk interval.

Answer 6.3
Suppose the initial discrete representation of the signal is (0, 0, 0, 0, 1, 1, 1, 1). After smoothing with the window presented in the question the discrete representation of the signal is (0, 0, 0, 0.25, 0.75, 1, 1, 1). Hence, the sharp change in the signal has disappeared.

Glossary

Apnea Delay of breathing.

Ataxia Lack of muscle coordination resulting in e.g., low muscle tone (hypotonia), undershoot, or overshoot of intended position (dysmetria) and an inability to perform rapid, alternating movements (dysdiadochokinesia).

Biceps brachii Upper arm muscle, elbow flexor.

Cerebellar ataxia Ataxia due to dysfunctioning of the cerebellum.

Cranial nerves Nerves emerging directly from the brain or brainstem.

Etiology Origin, cause of a disease.

Extensor digitorum communis Lower arm muscle, one of the wrist extensors.

First dorsal interosseus Muscle between thumb and index finger.

Flexor carpi radialis Lower arm muscle, one of the wrist flexors.

Hjorth Similar to Laplacian reference.

Integrated (EMG) Temporal summation of EMG values.

Jerk Sudden, often uncontrolled movement.

Laplacian Weighted average reference in which the contributions of the (surrounding) reference electrodes are weighted according to their distance to the middle input electrode.

Masseter Muscle connecting the upper and lower jaw; its function is to aid in chewing.

Multifocal myoclonus Brief, involuntary twitching of a muscle or a group of muscles occurring independently in different body parts.

Noncephalic Not related to the head, here, a reference electrode that is not attached to the scalp but, e.g., to the sternum.

Obstructive sleep apnea syndrome Sleep disorder characterized by regular cessation of breathing caused by a blockage of the airway (e.g. by the tongue or other tissues).
Orbicularis oculi Muscle around the eye; its function is to close the eyelid.
Perception deafness Hearing loss due to a dysfunction of the inner ear or hearing pathways.
Polyneuropathy Neuropathy (see Glossary Chap. 2) affecting multiple nerves.
Rectified (EMG) Absolute value of the EMG, i.e. positive values remain positive and negative values become positive (by removing the minus sign).
Rheumatoid arthritis Systemic inflammatory autoimmune disease affecting many tissues and organs but mostly the joints that contain a lubricating (synovial) fluid.
Smoothed (EMG) Running average of EMG amplitude, sometimes the weights are not the same for every EMG value. Smoothing is a form of low-pass filtering.
Sternocleidomastoid Muscle in the frontal part of the neck; its function is to rotate and flex the head.

References

Online Sources of Information

http://www.neurospec.org. Library of Matlab routines to calculate multivariate Fourier analysis of time series (including coherence), maintained by David Halliday

Papers

Barrett G (1992) Jerk-locked averaging: technique and application. J Clin Neurophysiol 9:495–508

van Beek AHEA, Claassen JAHR, Olde Rikkert MGM, Jansen RWMM (2008) Cerebral autoregulation: an overview of current concepts and methodology with special focus on the elderly. J Cereb Blood Flow Metab 28:1071–1085

Cassim F, Houdayer E (2006) Neurophysiology of myoclonus. Neurophysiol Clin 36:281–291

Engel AK, König P, Kreiter AK, Schillen TB, Singer W (1992) Temporal coding in the visual cortex: new vistas on integration in the nervous system. Trends Neurosci 15:218–226

Halliday DM, Rosenberg JR, Amjad AM, Breeze P, Conway BA, Farmer SF (1995) A framework for the analysis of mixed time series/ point process data – theory and application to the study of physiological tremor, single motor unit discharges and electromyograms. Prog Biophys Mol Biol 64:237–278

Jeong J (2004) EEG dynamics in patients with Alzheimer's disease. Clin Neurophysiol 115:1490–1505

Maurits NM, Scheeringa R, Van der Hoeven JH, De Jong R (2006) EEG coherence obtained from an auditory oddball task increases with age. J Clin Neurophysiol 23:395–403

Mima T, Hallett M (1999) Corticomuscular coherence: a review. J Clin Neurophysiol 16:501–511

Nunez PL, Srinivasan R, Westdorp AF, Wijesinghe RS, Tucker DM, Silberstein RB, Cadusche PJ (1997) EEG coherency: I: statistics, reference electrode, volume conduction, Laplacians, cortical imaging, and interpretation at multiple scales. Electroencephalogr Clin Neurophysiol 103:499–515

Panerai RB (2004) System identification of human cerebral blood flow regulatory mechanisms. Cardiovasc Eng 4:59–71

van Rootselaar AF, van Schaik IN, van den Maagdenberg AM, Koelman JH, Callenbach PM, Tijssen MA (2005) Familial cortical myoclonic tremor with epilepsy: a single syndromic classification for a group of pedigrees bearing common features. Mov Disord 20:665–673

van Rootselaar AF, Maurits NM, Koelman JH, van der Hoeven JH, Bour LJ, Leenders KL, Brown P, Tijssen MA (2006) Coherence analysis differentiates between cortical myoclonic tremor and essential tremor. Mov Disord 21:215–222

Varela F, Lachaux JP, Rodriguez E, Martinerie J (2001) The brainweb: phase synchronization and large-scale integration. Nat Rev Neurosci 2:229–239

Chapter 7
Psychogenic Movement Disorders, Bereitschaftspotential, and Event-Related Potentials

After reading this chapter you should know:

- Why the Bereitschaftspotential can help differentiate between organic and psychogenic movement disorders
- How the Bereitschaftspotential can be calculated from an EMG-EEG recording
- How evoked and event-related potentials differ
- How analysis parameters influence event-related potentials and the Bereitschaftspotential in particular
- How event-related potentials can be used in the evaluation of neurological disorders

7.1 Patient Case

Patient 1
A 22-year-old male patient is suffering from jerky movements in his right upper leg since 2 years. These movements only occur during rest and never during action, but are otherwise position-independent. The frequency and duration of these movements have gradually increased over time and now they occur approximately once every 10 s. The frequency increases further as a result of stress or sleep deprivation. The patient reports that he can not willingly suppress the movements. Thinking that there may be an epileptic origin of the movements, antiepileptic medication has been tried in the past but to no avail. Initially, his treating neurologist thinks that he may be suffering from *propriospinal myoclonus*. During the neurological examination, which yields no other abnormalities, it seems that the myoclonic movements decrease in frequency and are even absent sometimes.

N. Maurits, *From Neurology to Methodology and Back: An Introduction to Clinical Neuroengineering*, DOI 10.1007/978-1-4614-1132-1_7,

A tremor recording (polymyogram; see Chap. 3) indeed shows myoclonic-like contractions of the upper leg muscles, although it seems that the patient can be distracted from the myoclonic movements, which would be an indication for a *psychogenic* origin. However, the polymyographic recording also shows that the muscle activity during myoclonus sometimes only lasts for as short as 100 ms, which is very short for an intentional movement, thus indicating that the movement may be unintentional nevertheless. In this patient, the combination of clinical examination and polymyography is not yet sufficient to decide with certainty whether or not the movements have an *organic* origin.

7.2 The Bereitschaftspotential: Recording the Intention to Move by Simultaneous EEG-EMG

Quite often in clinical practice patients – such as the patient introduced in the previous section – with peculiar movements that do not straightforwardly meet the criteria for known (organic) movement disorders, are encountered. In these cases, the patient's history, the clinical neurological examination, the response to (placebo) medication, and the absence of other underlying disorders may point towards a psychogenic origin of the movement disorder. Often, a laboratory test for a psychogenic movement disorder is lacking. Fortunately, in some of these cases it is possible to make use of the specific EEG activity preceding voluntary movement, which was discovered in 1964 by Kornhuber and Deecke. They recorded EEG and EMG simultaneously and – computer software for *online* averaging not being available at that time – stored all data on magnetic tape. By cleverly playing the tape backward they were able to derive an average of the EEG activity preceding movement, in a manner similar to the back-averaging procedure described in Chap. 6. In this way, they discovered an EEG component, before EMG onset, which they named the Bereitschaftspotential (BP). This potential is an event-related potential (ERP) and one of the motor-related cortical potentials (MRCPs) also known as readiness potential. The BP was later divided into more components. A complicating factor in the latter matter is that different authors use different terminology and from as little as two, to as many as five different components have been distinguished within the BP. Fortunately, for clinical practice this distinction is not so important. What matters is just the presence or absence of the BP.

The BP starts approximately 2–1.5 s before the onset of movement, is maximal over centroparietal areas (Fig. 7.1), symmetrically, and widely distributed over the scalp, although *topographical* focality can be found in the late phase. The late (part of the) BP, starting around 400 ms before EMG onset, is usually maximal over movement-associated contralateral motor areas. The exact onset of the BP and its amplitude differ highly from subject to subject and depend on the particular conditions of movement. If someone is asked to perform a self-paced movement, say approximately once every 5 s, preparation time is typically longer than in more natural

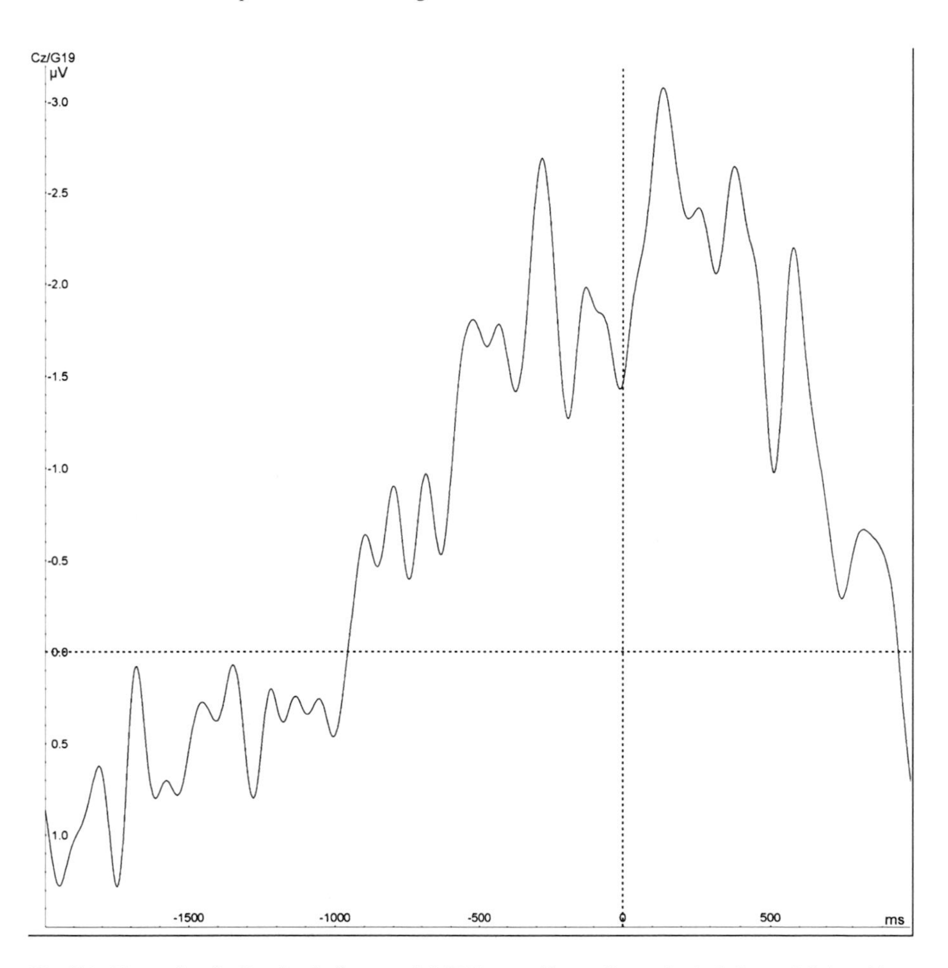

Fig. 7.1 Example of a Bereitschaftspotential (BP) recording at Cz: early (−1.5 to −0.5 s) and late (−0.5 to 0 s) BP. The *dashed line* indicates (foot) movement onset

circumstances and the BP onset will be earlier. If an external stimulus, such as the position of a clock hand, is used to trigger the movements (cued condition), there is less internal preparation and the BP will be diminished or may even be absent.

Interestingly, when a movement disorder has a psychogenic origin, it is often possible to derive a BP. It should be taken into account however, that the BP indicates certain brain mechanisms for generating movement, as seen with ordinarily voluntary movements, but by itself does not indicate that a movement is voluntary (Hallett 2010). Yet, before purely involuntary movement, the slow EEG shift that constitutes the BP should not occur. On the other hand, an absent BP in a patient does not prove an organic origin of the aberrant movement, since absence could also be due to many other physiological (e.g., in the case of more stereotype and automated movements), psychological, and technical reasons. Yet, a BP may be helpful in the diagnosis of psychogenic movement disorders.

7.3 Calculating Event-Related Potentials

7.3.1 Are Event-Related Potentials Different from Evoked Potentials?

In general, ERPs are small changes in electrical brain activity that are recorded from the scalp and that are brought about by some stimulus. In this sense, they do not seem very different from the evoked potentials (EPs) that were introduced in Chap. 5 and in fact, the terms are sometimes used interchangeably. It is possible to distinguish ERPs from EPs, though. ERPs are typically *endogenous*; they are generated by cognitive processes, have a widespread distribution on the scalp, and long latencies (up to several hundred ms). *Exogenous* EPs are generated directly by external stimuli and have short latencies (typically up to approximately 100 ms). Additionally, ERP amplitudes are typically higher than EP amplitudes, allowing for fewer repetitions of the stimulus to arrive at a sufficient signal-to-noise ratio (SNR, see Box 5.1). Furthermore, spectral properties of ERPs and EPs differ; ERPs generally have more spectral power at lower frequencies, because they represent slower changes in cortical potentials. Another important difference is that ERPs are not that much influenced by physical properties of the stimuli that are used to evoke them: the stimulus frequency or stimulus modality (e.g., visual or auditory stimulation) may differ, but the ERP will largely be the same. On the other hand, ERPs are much more susceptible to changes in psychological state or mood and an effect known as *habituation* can easily occur. For example, in case of the BP, the level of intention, the complexity of the movement and force with which the movement is executed will influence its amplitude, whereas preparatory state and speed of movement execution will influence its onset latency. These properties make ERPs less reliable and less reproducible than EPs, which is probably why they are less often used in diagnosis of neurological disorders. Yet, besides the above mentioned BP, there are some other interesting applications of ERPs for clinical practice (see Sect. 7.5).

Question 7.1 In Chap. 5, it was shown that to get a sufficient signal-to-noise ratio (SNR) for an EP, stimulations must typically be repeated 100–500 times. Suppose that a typical ERP component has an amplitude of 20 μV. When 500 repetitions are needed to obtain an EP component with an amplitude of 2 μV, how many repetitions would you need to get a reliable estimate of the amplitude of this ERP component?

7.3.2 Steps in Offline Calculation of Event-Related Potentials

In several aspects, the calculation of ERPs is the same as that of EPs. First and foremost, the "signal + noise" model is also adopted (see Sect. 5.3.1). The subsequent steps that are taken to get from an ongoing EEG recording with event markers to an average ERP are similar to those needed to calculate an EP, but more caution must be taken when choosing parameters for each calculation step and some additional steps are made. There is also more variation in the time interval that is considered in ERPs; this can be before or after the stimulus, but also before or after a (motor) response, whereas in EPs usually only the time interval after the stimulus is considered. Furthermore, often more than one stimulus is involved in an ERP paradigm, leading to several ERPs that can be compared in the analysis. A simple example is the P300 ERP, more extensively discussed in Sect. 7.5.3, in which both target and standard stimuli are used. Finally, for scientific purposes, ERPs are nowadays often recorded using multiple scalp electrodes (up to as many as 256). This allows to use spatial topographies in addition to temporal properties to better distinguish the different ERP components. The calculation of the BP is very similar to the calculation of a jerk-locked back-average in (cortical) myoclonus, as discussed in the previous chapter.

The steps that are usually taken to calculate an ERP are:

1. Filtering
   By first applying filters, strange effects on the edges of segments are circumvented and some artifacts (with frequencies outside the pass-band, see Sect. 4.3.2) are already removed. In contrast to EPs, for which often only a high-pass filter is used, ERPs also contain low frequencies and a typical pass-band may be 0.1–30 Hz. More on the peculiarities of this step can be found in Sect. 7.3.4.1.
2. Ocular (eye movement) artifact correction
   Especially in ERP paradigms, stimulus-locked eye movements (ocular artifacts) can be quite common. Some people may blink whenever a stimulus is presented or, even when instructed to focus on a central fixation point, some people will exhibit horizontal eye movements whenever a stimulus is presented in the peripheral visual field. In that case, eye movements may be preserved in the average ERP. Under these circumstances it is advisable to first remove ocular artifacts. Several methods are available to do so, such as the Gratton et al. (1983) method or independent component analysis (ICA). These methods are typically available in EEG analysis software packages and will not be explained in detail here.
3. Global artifact correction
   During this step, EEG channels that only contain artifacts (such as noise because of a loose electrode connection or slow *DC* shifts due to a wobbly electrode-scalp contact) or that are flat (because they are e.g., switched off by the amplifier) can be removed from the analysis. When recordings are done with respect to an average

reference the average needs to be redefined after electrodes have been removed from the set-up.

4. Segmentation
   For ERPs more attention needs to be paid to this step than for EPs, simply because there are more options to choose from (stimulus- vs. response-locked and pre- and post-event). Furthermore, since some of the ERPs are rather slow (in the order of several hundred milliseconds), the interval for analysis must be chosen sufficiently long, without overlap with previous or later intervals, also because part of the pre-stimulus interval is used subsequently for baseline calculation in step 6.
5. Artifact correction per trial or segment
   When, after filtering and ocular artifact removal, signal changes such as large amplitude fluctuations remain, these are most likely artifacts as well. Most EEG analysis packages provide a way to (semi)automatically remove these artifacts based on an evaluation of their (change in) amplitude (e.g., so-called detrending).
6. Baseline correction
   As for EPs, the baseline must be long enough to give a good estimate of the noise level in the recording. A rule of thumb is to use 20% of the total segment length as a baseline, which can be quite long in case of ERPs. Note that particularly for ERPs, you need to be sure that there is no remaining activity from the previous trial in the baseline. This can be quite hard to ascertain because ERPs often are the result of cognitive processes that can last long.
7. Averaging
   The last remaining step is the simplest: all remaining filtered and corrected segments need to be averaged. In principle, since both operations are *linear*, the steps of averaging and baseline correction may also be executed in the reverse order.

7.3.3 *Quantifying and Interpreting ERPs and the BP in Particular*

As for EPs, ERP components are most commonly quantified in terms of their (onset or peak) latency and (peak-to-baseline or peak-to-peak) amplitude. In addition, again similar to EP peaks, ERP components are identified by their *polarity* (positive = P, negative = N) and approximate latency in milliseconds (N100, P300, N400) or by their order of appearance in time (N1 = first negative component, P3 = third positive component). When peaks can not be so clearly identified, a mean amplitude can be determined over a time interval centered around the expected peak component latency. This time interval should be small enough not to include (parts of) other components.

In contrast to EP peaks, that usually signify physiological activity in a well-defined and limited area of the brain (such as the N20 SEP peak originating in Brodmann area 3b, in the primary somatosensory cortex), ERP components typically concern brain activity resulting from cognitive or strategic processes due to activity in multiple and widespread brain areas. This implies that when ERP components are compared between conditions or (groups of) subjects, it may well be that not only the amplitude or latency of the component changes, but that a component is maximal at a different electrode or that the entire shape of the component is altered. In the latter case, or when the effect of an experimental manipulation on the ERP component can not be hypothesized beforehand, other analysis methods are required. For these purposes, changes in an ERP component can be assessed in a more exploratory fashion by taking into account the entire shape of a component or the entire ERP waveform. In that case, usually not just a few isolated electrodes, peaks, and a limited time window are considered, but the data set is considered in its entirety. As an example, a simple approach that is less sensitive to small changes in component properties between conditions or subjects is to determine the area under the curve (AUC) over a particular time interval instead of the peak amplitude only. By dividing the AUC by the length of the time interval the mean amplitude over this time interval results again. By comparing mean amplitude over successive time windows (of e.g., 100 ms) between conditions or groups and using appropriate statistical techniques, it is possible to quantify any observable differences. More complex approaches that take into account ERP activity over multiple electrodes and/or multiple time points at once, are mapping (Sect. 7.3.3.1), principal or independent component analysis (ICA; not explained in detail here), and source localization (Chap. 8).

7.3.3.1 Mapping

Originally, most ERPs were recorded using only a few scalp channels. In case of the BP, recording from C3, Cz, and C4 should suffice to get a rough measure of brain activity in the primary motor areas. Similarly, to record the P300 (see Sect. 7.5.3) only electrodes Fz, Cz, and Pz are typically used. Now that multichannel recordings are feasible, it has become possible to consider the variation over time of the voltage on a certain electrode, as well as variations over time in voltage distributions over the entire scalp. This adds a spatial aspect to the analysis of ERPs. A voltage distribution on the scalp is also referred to as a (topographic) map.

Especially when there is a lot of EEG data available (as in 32-, 64- or 128-channel recordings) it can be more insightful to make use of mappings. In a topographic mapping, the voltage at a certain moment in time at an electrode is expressed as a color value in a 2D view of the scalp. The value of intermediate, nonsampled, points on the scalp is determined by *interpolation* (see Box 7.1). The advantage of mapping is of course that the voltage at multiple spatially divided points at a certain moment in time can be observed and evaluated simultaneously. Besides colors, contour lines that connect points with the same voltage value (or, in case of EEG, equipotential lines),

Box 7.1 One-Dimensional Interpolation Methods

Interpolation allows to calculate the value of a new data point between two existing data points. Several interpolation methods exist. The simplest uses only two adjacent data points to calculate the interpolated value, whereas more complex methods require more than two data points to do this. Furthermore, more complex methods typically yield smoother data, having its associated mathematical advantages (such as differentiability). Many methods determine a function that fits the existing data points, allowing to calculate the value of all intermediate values straightforwardly. One-dimensional interpolation can be generalized to multidimensional interpolation (as necessary for topographic mapping), but this will not be further discussed here.

Linear interpolation
In this method, the fitting function is simply a straight line between two adjacent existing data points. For a point (x,y) between (x_0,y_0) and (x_1,y_1) its y-value can be calculated from

$$y = y_0 + (x - x_0)\frac{y_1 - y_0}{x_1 - x_0}. \quad (1)$$

Polynomial interpolation
This method is a generalization of linear interpolation, in the sense that a linear function is a first-order polynomial. It can be shown that a polynomial of degree at most n can be fit through $n + 1$ datapoints (x_i, y_i), $i = 0,\ldots,n$ uniquely. Since an nth order polynomial has $n + 1$ unknowns, finding the polynomial fitting function amounts to solving an $n + 1$-dimensional system of linear equations.

Spline interpolation
Spline interpolation uses low-degree polynomials in each interval between data points, and chooses the polynomial pieces such that they fit smoothly together. Splines can be linear, cubic, or of even higher order. Cubic spline interpolation can be thought of as taking a piece of flexible wire and hammering a nail through this wire at all data points in the plane. The resulting shape of the wire is the cubic spline interpolated fitting function. The advantage of using splines is that no superfluous oscillations result between interpolated data points, as may happen when using higher order polynomial fitting.

can be used as well. This type of mapping is very similar to the more well known geographic topographic maps in which contour lines connect points with the same altitude and similar colors indicate e.g., areas with the same type of soil or land use.

A disadvantage of an EEG topographical mapping is that it suggests that the EEG has actually been measured over the entire head, whereas in reality only a

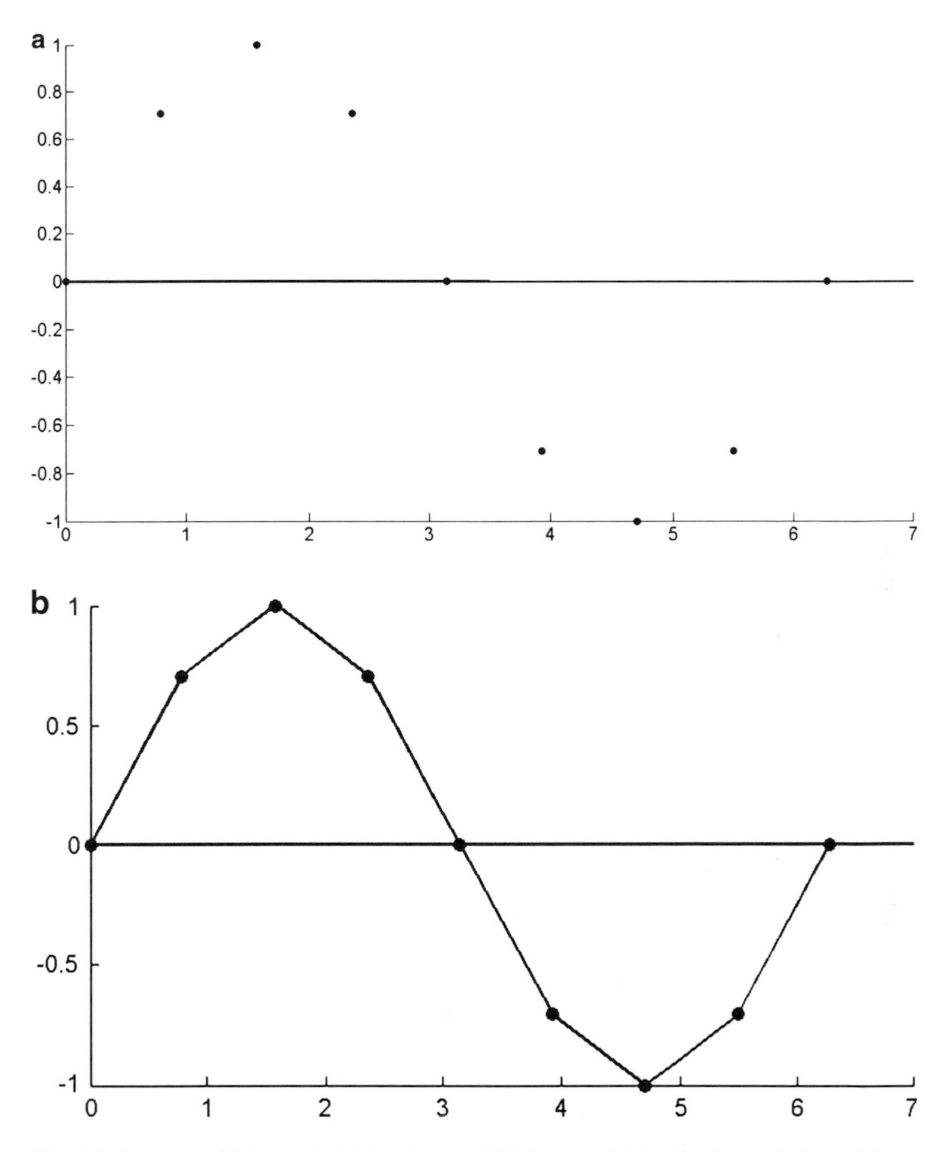

Fig. 7.2 Example of (**a**) sampled data points and (**b**) linear and (**c**) spline interpolations of these data points

limited number of spatially divided points (the electrode positions) has been sampled. This suggestion becomes even stronger when smooth (polynomial or spline) spatial interpolation methods (see Box 7.1) are used. In this sense, an advantage of using the most simple linear interpolation method, in which e.g., a point half way two points with the values 0.5 and 1 gets the value 0.75 and a point at 3/5 the value 0.8, is that the interpolation remains visible in the nonsmooth contourlines (see Fig. 7.2).

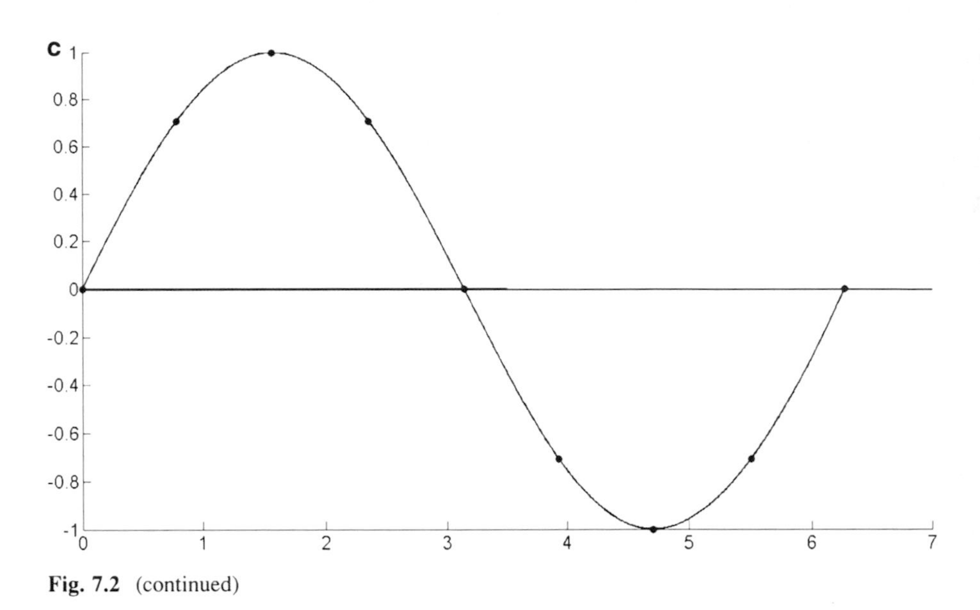

Fig. 7.2 (continued)

Similar to the sampling of an EEG or EMG signal in time (see Sect. 3.3.2) there is a Nyquist rule to determine the minimum spatial sampling frequency for EEG. As an example, suppose that the shortest wave length of interest in the EEG signal is 5 cm, then the highest spatial frequency would be 0.2 cm^{-1} and the spatial Nyquist frequency would be twice the highest signal frequency, i.e., 0.4 cm^{-1}. This means that for this particular example the signal would be adequately sampled spatially when electrodes are placed at an interelectrode distance of 2.5 cm (0.4 electrodes per cm). It has been shown that all spatial frequencies that can be present in a scalp EEG recording can be adequately sampled when the distance between electrodes is approximately 2 cm. This is compatible with 64–128 electrodes in a single recording. This does not imply however that all EEG recordings need to use at least 128 electrodes: this fully depends on the purpose of the recording and on the particular EEG signal under investigation.

Finally, a mapping can be made in many different ways. The mapping is influenced by the type of interpolation that is used, by the colors that are used and by the presence or absence of contour lines. In Fig. 7.3, three different representations of the same scalp voltage distribution are shown as an example of how interpolation influences the appearance of an EEG mapping.

Mappings can be used to display data in the time as well as in the frequency domain. In this way a topographic mapping can also be used to visualize that alpha power is mostly present in occipital regions. Mappings can also be displayed at regular time intervals as “snapshots” in a movie of brain activity or as an actual movie.

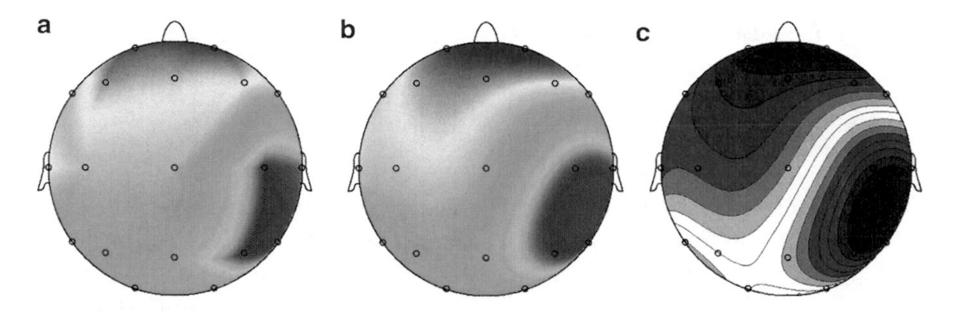

Fig. 7.3 Different representations of the same scalp voltage distribution. (**a**) Linear and (**b**) spline interpolation using *rainbow colors* and (**c**) spline interpolation using discrete colors with contour lines. Electrodes are indicated by *small circles*. In all figures *warm colors* indicate positive voltages whereas *cool colors* indicate negative voltages

Question 7.2 What would a mapping of alpha activity look like?

7.3.3.2 Quantifying the BP

The BP is a peculiar ERP to quantify since it does not have a maximum or minimum at all. Instead, it consists of a slowly increasing negative potential over centro (parietal) electrodes. Depending on the involved limb, maximum activity is found over more lateral or more medial electrodes (C3 for right hand, Cz for feet, C4 for left hand, etc.). For clinical purposes it is not so important to determine the actual size of the BP but, as mentioned earlier, its presence or absence. A BP can be said to be present when its distribution is consistent with a generator in the motor area associated with the observed movement (i.e., a negative maximum over central areas), when its amplitude increases gradually from approximately 2 to 1.5 s before EMG onset until EMG onset, and when this increase in amplitude is higher than variations in the base line (earlier than 2 s before EMG onset). If one would like to quantify the BP, even though this is not often done in literature, the most appropriate way would probably be to use mean amplitude over consecutive time intervals before EMG onset. Determining the onset of the BP may be especially important for its late part, which is related to activity in the contralateral motor areas associated with the moving body part.

7.3.4 Tips and Tricks When Calculating ERPs and the BP in Particular

7.3.4.1 Applying Filters

The BP is a slowly increasing potential, making it difficult to remove slow drifts in the raw EEG signal that are due to bad electrode contacts, sweating, or movement. After all, simply applying a high-pass filter will also remove part of the BP signal itself. An alternative option is to use a DC-correction method to remove the drift, but this has the same disadvantage. Instead, it may be better to remove drifts in the signal on a trial-per-trial base manually. When enough trials remain, drift-containing trials can just be removed.

7.3.4.2 Removing Artifacts

As mentioned before, ERPs are calculated by averaging the brain response to many stimuli. In doing this, the assumption is that the response of the brain to the stimulus is always the same, whereas other "background" activity of the brain differs from trial to trial and will average out when a sufficient number of trials is taken into account (the "signal + noise model"). What is not considered in this assumption is the presence of artifacts. Some artifacts, such as stimulus-related eye blinks (Sect. 4.3), always occur at approximately the same time after stimulus onset (i.e., they are stimulus-locked) and always have the same polarity. These artifacts will not disappear by averaging and will severely distort the shape of the ERP. Eye blinks are an example of artifacts of physiological origin, as are muscle and electrocardiographic artifacts. Instrumentation artifacts originate from sources outside the body. Both types of artifacts have been described in more detail in Sect. 4.3. Of particular importance for EMG are 50 Hz mains noise (see Fig. 7.4), but also slow drifts in the signal due to movement of electrodes or wires or due to badly placed electrodes. What makes noise is its high amplitude compared to the ERP amplitude, its frequency outside the ERP frequency band or its disturbing effect on the ERP waveform.

The most effective way to remove noise and preserve the ERP signal without distortions depends on the type of noise and the type of ERP. In general, for effective removal of artifact by filtering, it is important that the spectra of the noise and the signal do not overlap (cf. Fig. 4.11). However, even better than removing artifacts by filtering (Sect. 4.3) is to avoid artifacts.

Preventing and Removing Mains Noise

Mains noise (also known as electromagnetic interference) can sometimes be prevented by better placement of electrodes and improvement of electrode impedance. Furthermore, repositioning the patient may help. Also, when many

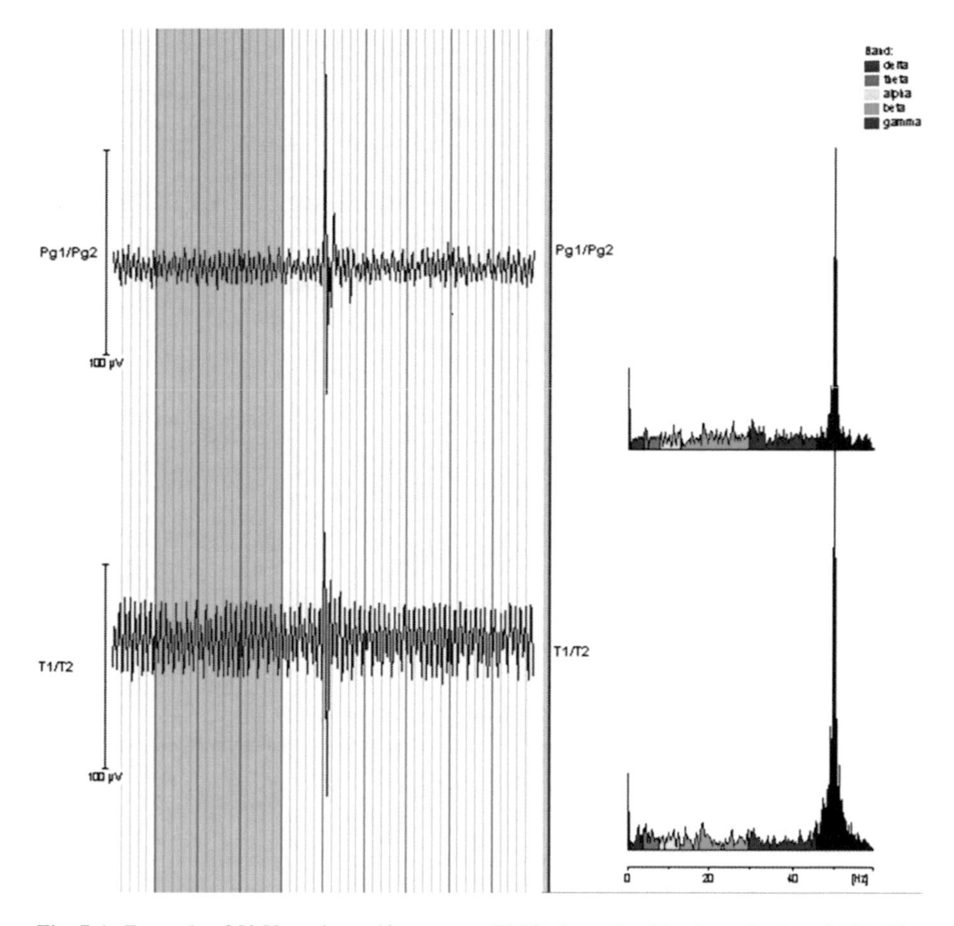

Fig. 7.4 Example of 50 Hz mains artifact on two EMG channels with a burst (in the *middle*). The 50 Hz peak is clearly visible in the spectra on the *right*

electrical devices are present in the recording area, signal quality may improve when electrode wires or cables are repositioned. Switching off any nonrelevant electrical devices helps and crossing of electrical cables with electrode wires should be avoided. However, if mains noise cannot be avoided, an effective way to remove it is to use a low-pass (smoothing) filter. In case of EEG recordings this can be done without many consequences for the signal itself, since most of the EEG power is below 30 Hz and mains noise has a frequency of 50 (or 60) Hz. However, if the mains noise frequency is right in the middle of the frequency band of interest (as may be the case for EMG signal that has power up to 250 Hz), a simple low-pass filter throws away too much of the signal. In that case using a band-stop filter around the mains frequency may be an option.

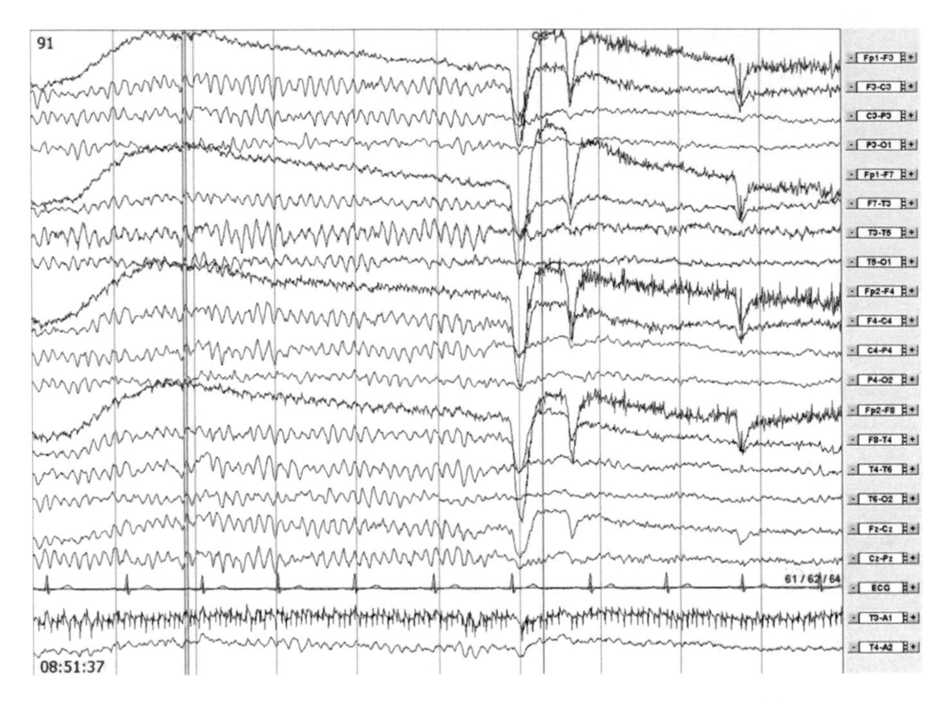

Fig. 7.5 Illustration of how closing the eyes increases alpha power and reduces blinks. Here, EEG was recorded during eyes closed (first half of the page), when the EEG is dominated by alpha-waves. After the eyes are opened (middle of the page), alpha disappears and the patient blinks more often

Preventing and Removing Muscle Artifact

Muscle artifact on EEG channels is often due to tension in the neck muscles (occipital electrodes), frowning (frontal electrodes), or clenching of the jaw (temporal electrodes). By asking the subject to relax, drop their jaw, or open the mouth slightly and by making the subject as comfortable as possible, either by lying down or by using a chair with a head-rest, muscle artifact can often be prevented or diminished. Removing muscle artifact by filtering is more difficult because muscle activity has a broad spectrum (cf. Fig. 4.7).

Preventing and Removing Eye Blinks

The number of eye blinks varies considerably between subjects. To prevent eye blinks subjects should first be instructed to blink as little as possible. It should be made clear however, that "not blinking" may not become a second task, thereby possibly influencing performance on the main task. Depending on the task and the signal of interest, it may be possible to simply let subjects close their eyes. This eliminates eye blinks, but will increase the alpha power in the EEG signal, which is often undesirable (Fig. 7.5).

In many subjects the number of eye blinks is significantly diminished when they are instructed to keep their eyes open, but look downward.

Removing eye blinks can be done in many ways, but filtering is not typically one of them because of overlapping signal and artifact spectra. If the number of eye blinks is limited, simple removal of trials with eye blinks is the best option. This can be done automatically, by rejecting all trials in which the amplitude on the frontopolar channels exceeds a certain level, or manually, which has the advantage that trials with less clear or smaller blinks can also be removed.

Alternatively, eye blinks can be removed by subtracting part of a separately recorded EOG (electro-oculogram) signal from each EEG signal. In contrast to the removal of trials, this type of correction is performed on unsegmented EEG data. Several methods can be used to calculate the fraction (propagation factor) of the EOG that spreads to each EEG signal. An often used approach is the linear regression method designed by Gratton et al. (1983), but many more methods have been proposed and used since (Hoffmann and Falkenstein 2008). A possible problem with regression methods is that the EEG signal is distorted, especially in frontal regions, where the effect of the correction will be largest. Other alternatives for ocular artifact correction are dipole source modeling methods and ICA. The latter method often works quite effectively when sufficient data with representative eye blinks is available.

Question 7.3 What filter would you use to remove mains noise in an EEG recording?

7.3.4.3 Detecting EMG Onsets

In calculating the BP, the reliable detection of EMG onsets, either automatically or manually (see Sect. 6.3.3 for general (dis)advantages of each approach) forms a crucial step. When there is not much EMG activity outside movements and the onset of the movement is brisk, EMG onset detection is relatively easy and can often be done automatically. When movement onset is slower or when there is more continuous EMG activity, EMG onset detection can become troublesome. Before using any automated method for EMG onset detection it is advisable to apply a (10 Hz) high-pass filter to remove slow drifts in the signal and to *rectify* the signal. It is important not to use low-pass (smoothing) filters on the EMG signal when onset of activity needs to be detected. Depending on the width of the smoothing window and the steepness of the EMG onset, the actual EMG onset and the onset of smoothed activity can then shift considerably in time.

EMG onset detection can be automated, which is fast and can be done using different methods (see Sect. 6.3.3). The simplest method is to use a level trigger. In that case a marker is set whenever the EMG value becomes higher than a preset value. To avoid getting multiple markers within one EMG burst, usually a temporal offset (or delay) between triggers is imposed. In case of the BP this would be several seconds. Alternative methods make use of more complex criteria. An example is a method in which a marker is set whenever the EMG amplitude becomes larger than a multiple of the variability in the baseline. The advantage of this type of method is of course its objectivity. A disadvantage may be that this type of method breaks down when movements are more complex or when there are slow drifts in the baseline that cannot be removed easily. For best performance it is important to always check the results of an automated detection procedure. For a BP, manual EMG onset detection may be best when different movements are present in the dataset, which cannot be differentiated easily in an automated fashion. In that case, it may also help to have additional video footage to ensure that the observed EMG activity is indeed related to the pathological movement. The effect of the different methods and settings on the resulting BP is not necessarily very large as long as the movement is consistent, as can be seen in Fig. 7.6.

7.4 Bereitschaftspotential in an Individual Patient

To be able to calculate a BP from task-unrelated jerky or shock-like movements, it is important that the onset of each movement can be well identified on EMG and that the movements are far enough apart to be able to observe the development of a BP, i.e., at least 2 s apart. Furthermore, the EEG must be relatively artifact-free. This implies that if the movement is accompanied by large movement artifacts in the EEG, it may still not be possible to derive a BP. Further, at least 50–100 repetitions of the same movement are needed to obtain a sufficient SNR in the ERP. Finally, in clinical practice this must all be achieved within a reasonable recording time (30–60 min). But if all these conditions are met, the BP can be of diagnostic help as the example of Patient 1 shows.

Patient 1

The EMG was recorded from two of the quadriceps muscles bilaterally: m. rectus femoris and m. vastus medialis. For the EEG, 21 electrodes were used at standard 10–20 positions. Just before or during the myoclonic movements, no epileptic phenomena were observed in the EEG. Frequently occurring (>100 in a recording of 35 min) myoclonic, shock-like movements were seen in the EMG, that lasted between 100 and 300 ms. Movement artifacts sometimes made exact identification of the burst onset difficult. In this particular example, considerable mains artifact (50 Hz noise) was found on the EMG channels, which was first filtered out using a notch (band-stop) filter (see Sect. 4.3.2). Thereafter, EMG onset markers were set using an amplitude criterion (−30 μV) on the right m. vastus medialis EMG channel using a minimum interval between markers of 2 s. Setting markers manually was tried in addition, but these approaches lead to more variable onsets and a lower BP amplitude. After segmentation of the data from 2,000 ms before marker onset to 1,000 ms after marker onset, segments were removed in which the amplitude on Fp1 or Fp2 was larger than −100 or 100 μV, thereby effectively

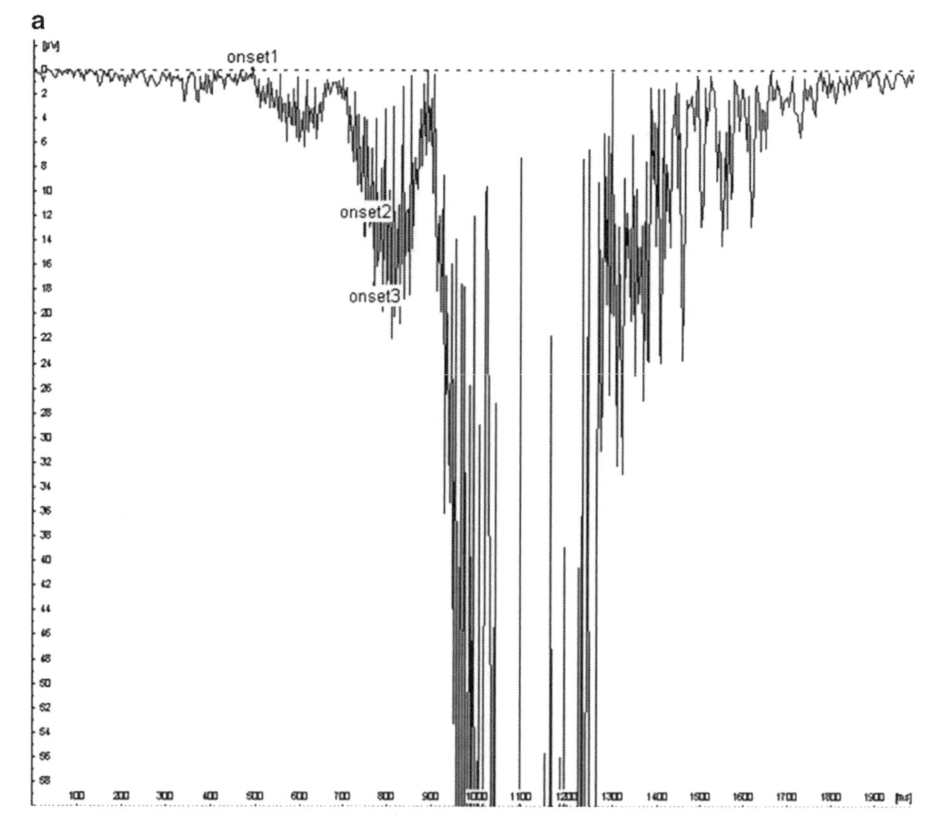

Fig. 7.6 (**a**) Example of an EMG burst (positive down) due to a voluntary brisk right wrist extension movement. Two different methods have been used to detect EMG onset: manual detection (onset1) and a level trigger (at 10 μV: onset2 and at 15 μV: onset3). (**b**) The BP resulting from each of the three methods applied in (**a**). Each of the BPs results from an average of 75 trials. The manual detection method is indicated in *black* and the level trigger method is indicated in *red* (10 μV) and *green* (15 μV). In this example, the automated detection methods seem to perform better (as indicated by a *smoother curve*), probably because there was considerable variability in the first part of EMG activity, which made detection of the first onset of EMG difficult. Setting a higher level trigger implies that not the first onset of EMG is detected, but a moment slightly later into the EMG burst is marked. This has a small effect on the latency of the maximum in the BP

removing the influence of eye blinks on the data in a simple way. Finally, the remaining 67 segments were used to determine the average BP.

A clearly distinguishable BP was found in this patient (see Fig. 7.7b). Despite the relatively short duration of the EMG bursts (which were as short as 100 ms) this result points toward a psychogenic origin of the myoclonus.

In this particular patient the EEG-EMG recording and the subsequent analysis to determine a BP helped to provide a more solid basis for a diagnosis of psychogenic movement disorder. On the other hand, it is important to realize that the absence of a BP is not helpful and can be consistent with both a psychogenic and an organic origin of a movement.

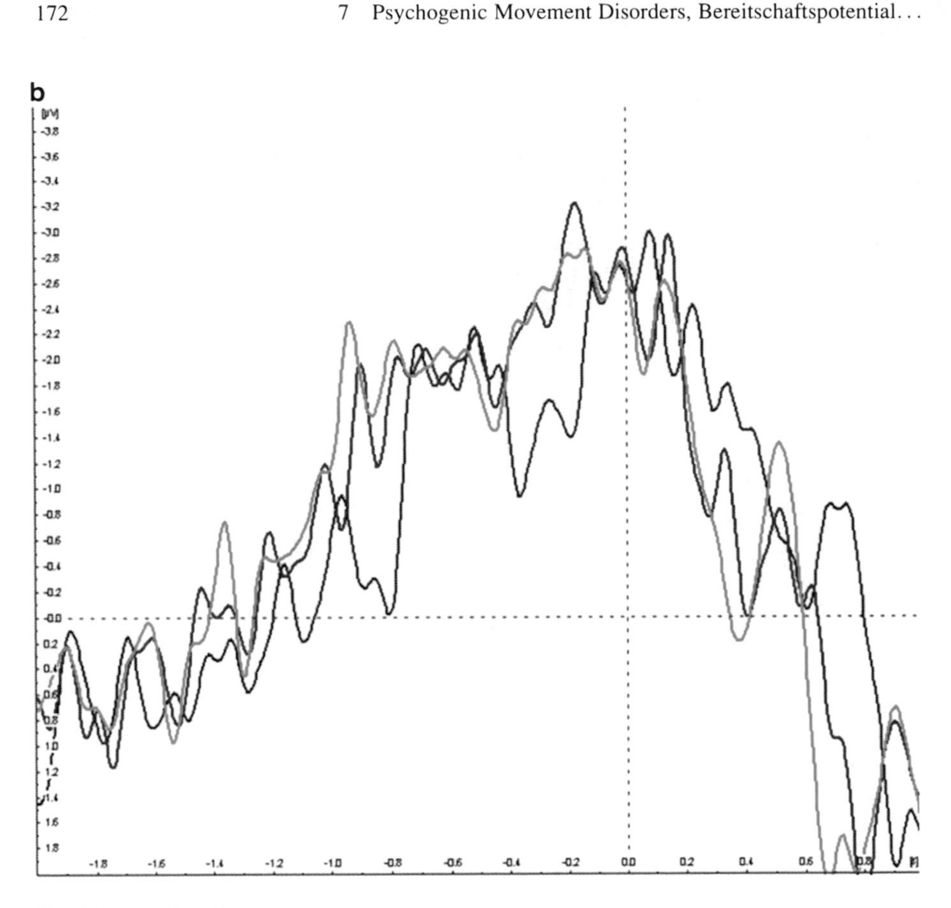

Fig. 7.6 (continued)

7.5 Other Applications of Event-Related Potentials in Neurology

As mentioned in Sect. 7.3.1 ERPs are susceptible to changes in psychological state or mood which makes them generally less reliable and less reproducible than EPs. This is probably why ERPs are less often used during the diagnostic process of neurological disorders. Yet, there are some examples of ERPs, besides the BP, that are promising in this respect. Whereas EPs are used to evaluate the trajectory from sense organ to associated primary sensory cortex, ERPs are suited to investigate the neurophysiology of cognitive functions. Since several neurological diseases are associated with cognitive dysfunction (such as Alzheimer's disease and other dementias, but also Parkinson's disease and other *parkinsonisms*) and these diseases become more ubiquitous with the aging of our population there is a demand for sensitive and objective neurophysiological tools to measure changes in cognitive function. In the next sections, some examples of ERPs with clinical potential are discussed. An overview of applications of MRCPs in movement disorders was recently given by Colebatch (2007).

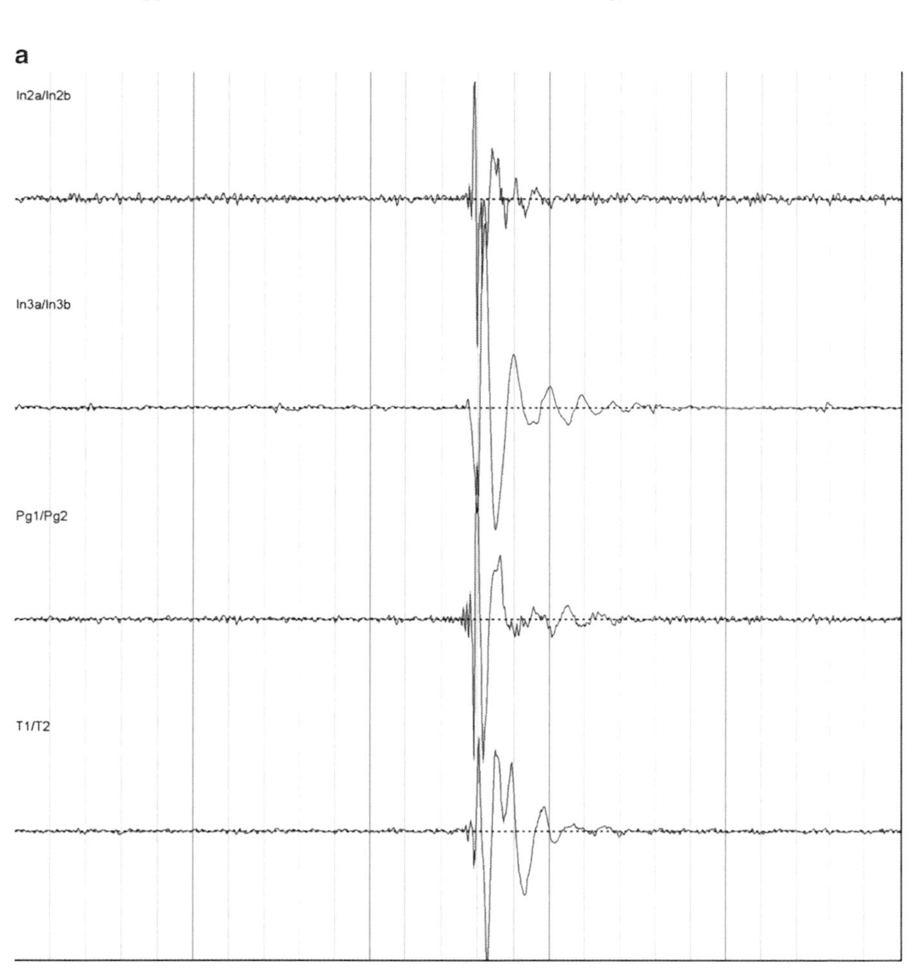

Fig. 7.7 Results of EEG-EMG recording in patient 1. (**a**) Example of an EMG burst on (first right then left) m. rectus femoris and m. vastus medialis. The time interval between two light *green lines* equals 200 ms. (**b**) BP at Cz. The time interval that is displayed ranges from −2,000 to 1,000 ms with respect to the EMG onset marker. At other electrodes the amplitude of the BP was smaller, consistent with an origin in the leg area of the primary motor cortex. (**c**) Topographic mapping of BP distribution on the scalp at (from *left* to *right*) $t = -528$ ms, $t = -308$ ms and $t = -104$ ms

7.5.1 Lateralized Readiness Potential (LRP)

The asymmetric distribution of the BP associated with unilateral body part movements has been studied separately as the lateralized readiness potential (LRP) since the 1980s. The LRP is usually calculated as the difference between the potentials recorded at C3 and C4 before EMG onset and is thought to reflect response selection and motor programming or preparation processes for unilateral movement in the brain. Often, the LRP is derived from the EEG recorded during a

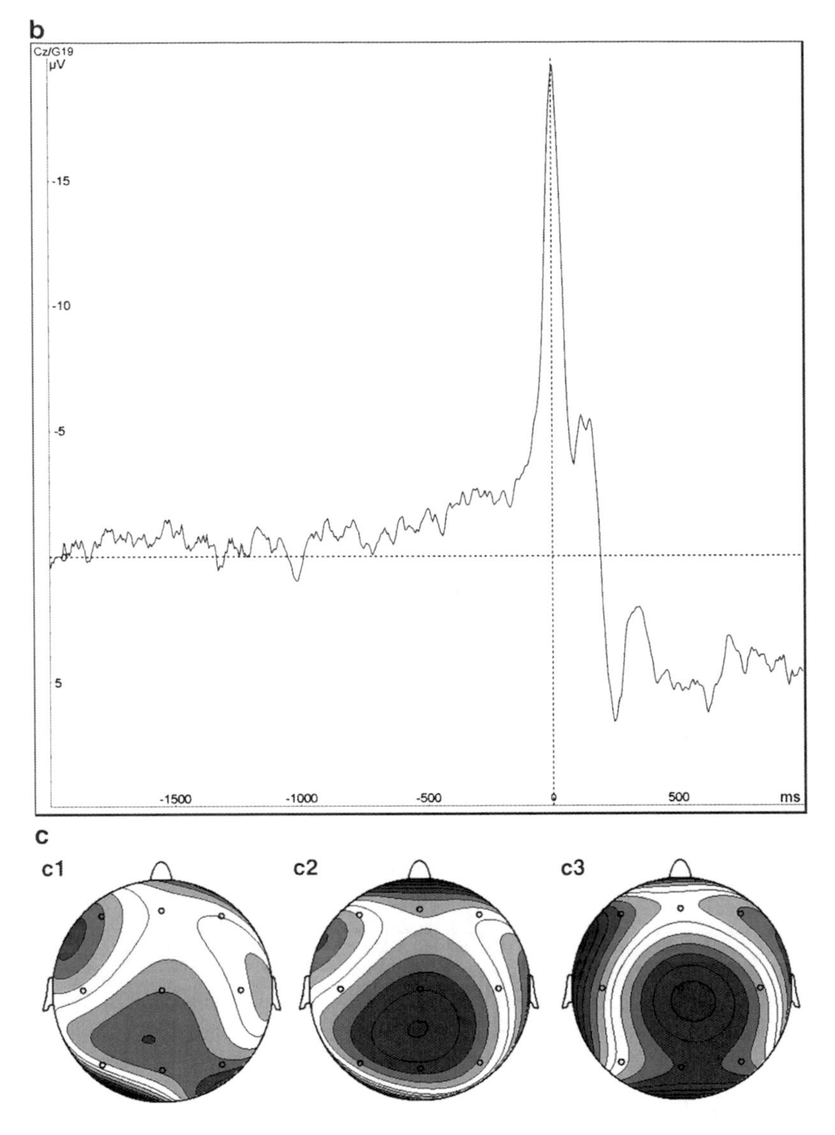

Fig. 7.7 (continued)

choice-reaction task, in which subjects have to give a response either with the left or right hand, depending on stimulus presentation and on the task at hand. This has to be done as fast and accurately as possible.

To compare the LRP between conditions or between groups of subjects it may be of interest to determine its onset, which would be a measure of the beginning of the preparation to make a motor response. In comparison to determining the latency of a waveform peak it is typically more difficult to determine the latency of a waveform onset. This is illustrated by the fact that if the baseline is not very stable, eyeballing the onset is already quite difficult. Yet, several methods have been proposed to identify the LRP onset. In (2000) Mordkoff and Gianaros showed that available methods fall into three categories: criterion-based, baseline-deviation, and regression-based methods. Without going into much detail here, the latter (and most recent) methods are shown to be least prone to errors in estimating LRP onset. In regression-based methods, a straight line is fit through the curve during the baseline period and another through the curve rising up to the peak. The latency of the intersection point of the two lines then signifies LRP onset.

There are several scientific studies that have shown deviant brain activation in patients with Parkinson's disease during choice-reaction tasks, likely due to basal ganglia dysfunction, resulting in failure of efficient inhibition. However, so far, results on the LRP in patients still lack the specificity and sensitivity necessary to be useful in clinical practice. Of interest may be the fact that some recent ERP studies have shown that directing attention and movement preparation produce similar frontocentral and occipitotemporal lateralized potentials fields. It could be that combined attention-motor tasks may later prove to be of more value in the diagnosis and understanding of Parkinson's disease than motor tasks alone.

7.5.2 Contingent Negative Variation (CNV)

The contingent negative variation (CNV), first described in (1964) by Walter et al., is a slow negative wave that develops in the interval between a warning stimulus and a "go" or cue stimulus after which a (unilateral) movement has to be made. The CNV is thought to reflect several processes, such as anticipation for the next signal and preparation for the upcoming motor response. The CNV has a frontocentral maximum a few hundred milliseconds after the warning cue which is thought to originate from activity in premotor areas. When the period between the warning and the cue stimulus is lengthened it is possible to observe that the CNV actually consists of two negative waves, one directly following the warning stimulus (reflecting its processing) and another, similar to the BP, preceding the cue stimulus. In Parkinson's patients, a decreased CNV is quite consistently found.

7.5.3 Oddball Paradigm (P300)

An often used protocol resulting in a relevant ERP when studying cognition, is the oddball paradigm. In such a protocol, one and the same stimulus is presented repeatedly, but at infrequent intervals a different stimulus (the oddball) is presented. If this oddball is presented relatively infrequently (in 10–20% of the cases), a specific ERP component develops approximately 300 ms after stimulus presentation. This P300 or P3 component was first described by Sutton et al. (1965). The P300 component can be evoked by visual, auditory as well as sensory stimulation and under very different circumstances (in attentive as well as inattentive conditions, with or without actual oddballs or other distracting stimuli). Most investigators now agree that the P300 component is actually built up by two separate waveforms, the P3A and P3B. The P3A waveform is earlier and has a more frontocentral distribution whereas the P3B waveform is later and has a more centroparietal distribution (Fig. 7.8). Nowadays, when referring to the P3 or P300 component, usually the P3B is meant. It is not known exactly what brain areas are involved in generating the widespread P3, although many have been implicated in the past decades. Nor is it clear what specific cognitive processes underlie the P3 waveform. However, the influence of several manipulations on P3 amplitude and latency is now well understood. The P3 wave is not only very sensitive to target stimulus probability, but also to task difficulty, invested effort, and to many biological variables (such as the amount of sleep or coffee consumption). This implies that there is high variability in P3 amplitude and latency between but even within healthy subjects, making clinical application difficult (Fig. 7.8).

Soon after its discovery, evaluation of the P300 potential as a possible diagnostic test for cognitive dysfunction was initiated. And indeed, differences between normal subjects and groups of patients with dementia or head injury were reported. These results fitted nicely with cognitive theories on P300 latency as an index for certain mental processes, such as stimulus processing speed. In head injury patients and patients with movement disorders, increased, decreased as well as similar P3 activity as in healthy subjects have been found. Altogether, sensitivity and specificity in a clinical setting have thus been highly variable.

Research in our own group (Elting et al. 2005) showed that by using source localization approaches (see Sect. 8.3), the varying contribution of P3A and P3B waves to the P3 waveform can be identified efficiently, thereby reducing at least one source of variability. This approach helped to show that the increased P3 latency that is often found in patients with head injury was actually due to a decreased contribution of the P3A wave to the P3 waveform. Yet, these results by themselves have not unequivocally helped to improve sensitivity and specificity of P3 parameters in head injury.

7.6 Answers to Questions

Answer 7.1

In Box 5.1 it was explained that to increase the SNR by a factor of N, the number of trials must be increased by a factor N^2. Thus, when the amplitude of the signal is a factor 10 higher, it suffices to use a factor $N^2 = 100$ fewer trials. In this case, 500/100 = 5 repetitions would be enough to estimate the ERP with the same reliability as the EP.

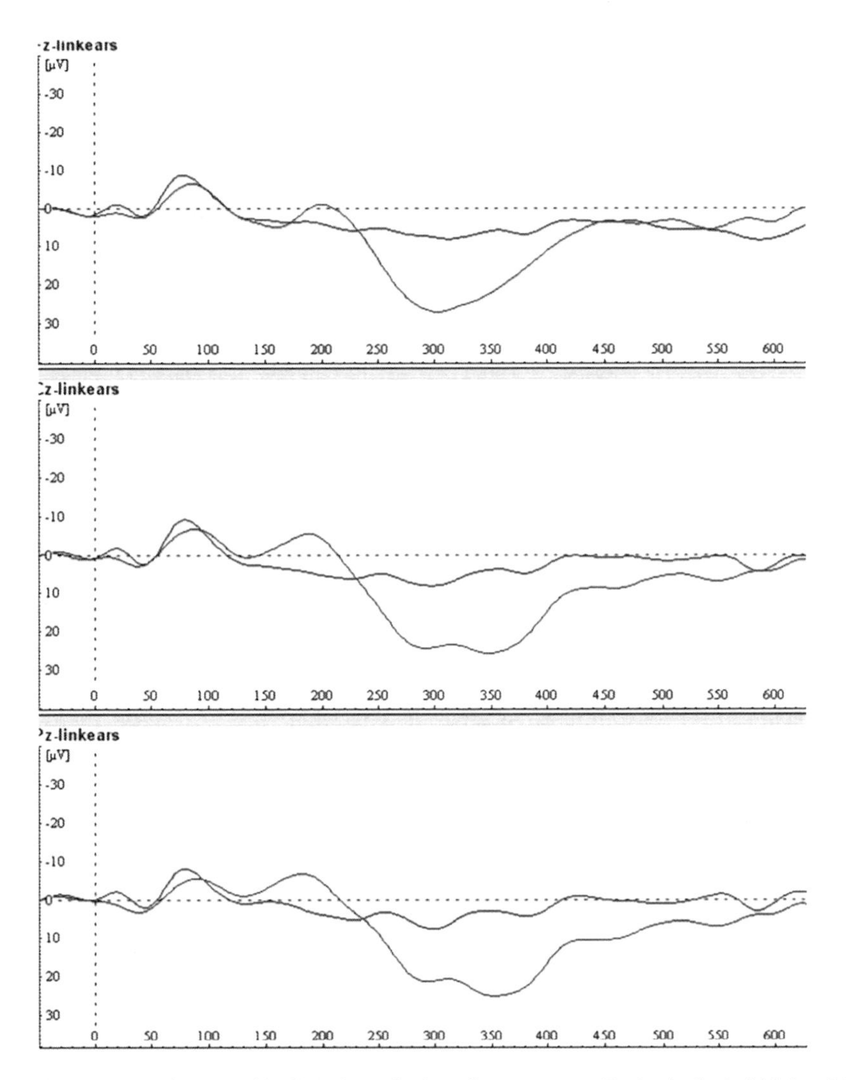

Fig. 7.8 Example of target (*black*) and standard (*red*) averages on Fz (*top*), Cz (*middle*) and Pz (*bottom*) electrodes resulting from an auditory oddball paradigm. The P300 wave in the target average is clearly visible. It deviates from the standard average between 250 and 400 ms after stimulus presentation (*dotted vertical line* at 0 ms). In this example, the P3A (around 300 ms) and P3B (around 350 ms) components can also be discerned visually and result in a bifurcated P300 waveform, particularly on Cz and Pz. Notice the large P3 amplitude of 20–30 µV (compared to the much lower EP amplitudes in Sect. 5.4). Furthermore, this example also shows that the target average deviates from the standard average as early as 150 ms after stimulus presentation: this is the N2 component. The difference in N2 between the target and the standard averages is also referred to as mismatch negativity and is thought to be related to task-irrelevant detection of deviant stimuli

Answer 7.2
If rainbow colors were used, a mapping of alpha activity would show red colors around occipital electrodes, possibly extending to more parietal and central areas, depending on the subject. Blue colors would be shown around frontal electrodes.

Answer 7.3
Since most of the EEG activity is limited to frequencies lower than 35 Hz and mains noise frequency (in Europe) is 50 Hz, a low-pass filter at 40 or 45 Hz should work as long as the transition from pass-band to stop-band is steep enough. A higher order Butterworth filter could do the trick (see Chap. 4).

Glossary

DC Direct current (in contrast to alternating current [AC]): unidirectional flow of electric current. Also known as the constant, zero-frequency, or slowly varying local mean value of a current.

Endogenous Arising from within (the brain).

Exogenous Opposite from endogenous: coming from outside (the brain).

Habituation Decrease in response to a stimulus after repeated regular exposure to that stimulus over a duration of time.

Interpolation Constructing new datapoints within the range of existing data points according to a mathematical method (e.g., linear, polynomial or spline interpolation).

Linear In mathematics a linear operation f is an operation that obeys the following two criteria: $f(x + y) = f(x) + f(y)$ and $f(ax) = af(x)$ where a is a scalar.

Myoclonus Sudden, quick, shock-like, single, or repetitive involuntary movements without loss of consciousness.

Online Executing a calculation or an operation on a dataset while this dataset is being recorded (in contrast to offline analysis).

Organic Affecting or relating to an organ, of physiological origin.

Parkinsonism Neurological syndrome characterized by tremor, hypokinesia (slow or decreased movement), rigidity (increased muscle tension when moved passively), and postural instability. This syndrome is found in Parkinson's disease, but also e.g., in corticobasal degeneration and progressive supranuclear palsy.

Polarity Sign: positive or negative.

Propriospinal Relating to the spinal cord and specifically, denoting those nerve cells and their fibers that connect the different segments of the spinal cord with each other. Propriospinal myoclonus: jerks of the trunc and/or limbs, usually due to a generator in the spinal cord. Typical is a delay of the jerk in the different involved parts of the body consistent with the distance traveled by the electric signal along the slow conducting propriospinal nerve fibers.

Psychogenic Disease for which no structural or anatomical abnormality can be identified (in contrast to an organic disease).

Rectify A signal is rectified by retaining all positive values and by replacing all negative values by positive values of the same size.

Topography As in: topographic (brain) mapping. Spatial variations in scalp voltage, often plotted in two dimensions in a top view of the head.

References

Online Sources of information

http:\crbooks.google.co.uk. Evoked potentials in Clinical Medicine, edited by K.H. Chiappa and published by Lippincott-Raven, 1997. Of particular interest is Chapter 15 Endogenous event-related potentials, by B.S. Oken

http://en.wikipedia.org/wiki/Bereitschaft_potential. Overview of the origin and derivation of the BP

http://www.cis.hut.fi/projects/ica/book/intro.pdf. Introduction to the more mathematical concepts of ICA from the book by Hyvärinen A, Karhunen J and Oja E. (2001) Independent Component Analysis, Wiley

Books

Jahanshahi M, Hallett M (eds) (2003) The Bereitschaftspotential. Movement-related cortical potentials. Kluwer Academic, New York

Handy TC (ed) (2005) Event-related potentials. A methods handbook. MIT Press, Cambridge

Luck SJ (2005) An introduction to the event-related potential technique. MIT Press, Cambridge

Papers

Colebatch JG (2007) Bereitschaftspotential and movement-related potentials: origin, significance and application in disorders of human movement. Mov Disorders 22:601–610

Elting JW, van der Naalt J, van Weerden TW, De Keyser J, Maurits NM (2005) P300 after head injury: pseudodelay caused by reduced P3A amplitude. Clin Neurophysiol 116:2606–2612

Gratton G, Coles MG, Donchin E (1983) A new method for off-line removal of ocular artifact. Electroencephalogr Clin Neurophysiol 55:468–484

Hallett M (2010) Physiology of psychogenic movement disorders. J Clin Neurosci 17:959–965

Hoffmann S, Falkenstein M (2008) The correction of eye blink artifacts in the EEG: a comparison of two prominent methods. PLoS One 3(8):e3004

Key AP, Dove GO, Maguire MJ (2005) Linking brainwaves to the brain: an ERP primer. Dev Neuropsychol 27:183–215

Mordkoff JT, Gianaros PJ (2000) Detecting the onset of the lateralized readiness potential: a comparison of available methods and procedures. Psychophysiology 37:347–360

Shibasaki H, Hallett M (2006) What is the Bereitschaftspotential? Clin Neurophysiol 117:2341–2356

Sutton S, Braren M, Zubin J, John ER (1965) Evoked-potential correlates of stimulus uncertainty. Science 150:1187–1188

Walter WG, Cooper R, Aldridge VJ, McCallum WC, Winter AL (1964) Contingent negative variation: an electric sign of sensorimotor association and expectancy in the human brain. Nature 203:380–384

Chapter 8
Brain Tumor, Preoperative Function Localization, and Source Localization

After reading this chapter you should:

- Know why source localization is useful preoperatively
- Know what the principles of source localization and its methods are
- Know the basic mathematical concepts underlying source localization (additional material)
- Understand the basics of the most common solution methods
- Know how the quality of a source model can be evaluated
- Know how source localization can be used in other neurological applications

8.1 Patient Case

Patient 1

This 61-year-old female patient suffers from repeated episodes of word-finding difficulties. These problems persist for several hours and then disappear again. The first diagnosis is that these episodes concern *TIAs*, originating in the left hemisphere, and the patient is sent home with blood-thinning medication. Two months later, the patient is admitted to the hospital after an epileptic seizure, during which the word-finding difficulties occur again, but this time in conjunction with a right-sided *paresis*, incontinence for urine, and rhythmic limb movements. An EEG shows a left parietal epileptogenic focus. CT scans only show an old infarction in the right *internal capsule*, but no other abnormalities, particularly not in the left hemisphere. Another 4 months later, a similar seizure occurs and this time an MRI is made which shows a left parieto-occipital lesion which could be either an old infarct or a *glioma*. In the next few months, problems with word-finding and clumsiness with the right hand are continuously present, although with variable severity. There are no problems with sensibility, swallowing, or walking.

N. Maurits, *From Neurology to Methodology and Back: An Introduction to Clinical Neuroengineering*, DOI 10.1007/978-1-4614-1132-1_8,

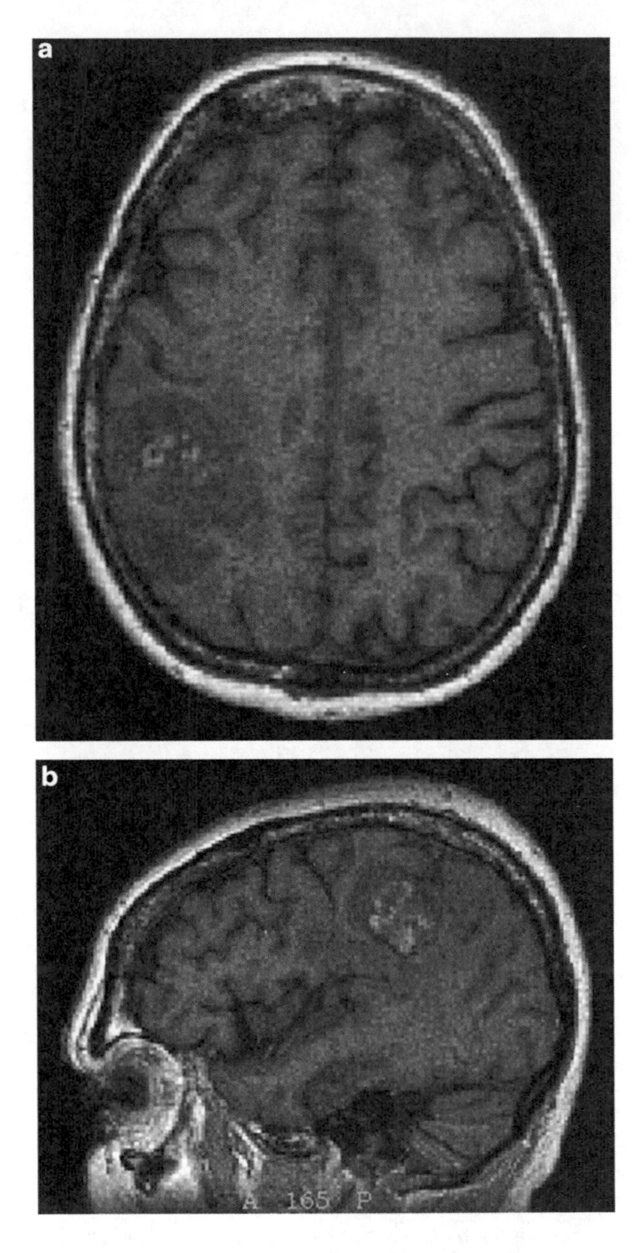

Fig. 8.1 Some MR images of patient 1. (**a**) *Axial* image (left side is left hemisphere), (**b**) *sagittal* image (left is front). The tumor is clearly visible in the left hemisphere as a hyperintensive (whiter) mass, surrounded by hypointensive (darker) fluid

The patient does not report any headache or nausea either. When she suffers from a major seizure again a few months later, she is admitted to hospital and a thorough clinical investigation follows.

Another EEG shows normal rhythms over the right hemisphere, but slow activity left-temporally, spreading to left frontal areas, consistent with a lesion with locally irritative phenomena. Another CT scan, this time employing a radiocontrast agent sensitive to changes in tissue, indicates a mass left parietally, which could be a tumor or a metastasis of the endometrial cancer that the patient suffered from 1 year earlier. An MRI scan (see Fig. 8.1) supports the CT finding and allows to better localize the mass to the left pre- and postcentral *gyri*.

Since it is important for treatment to know the exact diagnosis, a biopsy is taken from the mass in the brain and pathological analysis shows that it is most likely an astrocytoma, a tumor originating from astrocytes, a type of glial cells. Yet, to arrive at a definite diagnosis, the neurosurgeon decides that a *craniotomy* is necessary, during which more malignant tissue will be removed for pathological diagnosis.

8.2 Preoperative Function Localization: Identifying Eloquent Brain Areas

In the case of patient 1, the neurosurgeon decides to remove the tumor, but there is an important risk involved in that. If brain areas are removed that are essential for daily life functions (such as motor, sensory or visual processing or speech), the patient may recover from surgery with better life expectancy, but with important handicaps, such as paresis or *aphasia*. In case of this patient, the tumor is indeed located in an area which is crucial for motor and sensory functioning (the primary sensorimotor area around the central *sulcus*), as well as language (Wernicke's area). There is thus a high risk that these functions will deteriorate when, during excision of the tumor, the healthy brain areas are accidentally damaged. Therefore, before surgery, the neurosurgeon tries to determine the position of these so-called eloquent areas with respect to the tumor, so that they can be avoided during surgery if possible and the patient will hopefully recover from surgery with as few side effects as possible. In a healthy brain, the primary somatosensory cortex is easily identified as the postcentral gyrus. However, in this patient, the tumor has pushed the surrounding brain tissue aside, hindering the identification of the somatosensory cortex directly from anatomical brain images.

8.3 EEG Source Localization

Some of the eloquent brain areas can be localized using EEG source localization methods. These methods can be used to find the number, localization, and activity of electrical sources in the brain generating an EEG as recorded on the scalp. Besides localizing eloquent brain areas, another application is the localization of an epileptogenic focus on the basis of e.g., interictal sharp waves as observed in the EEG (see Sect. 4.2).

8.3.1 Principles

The mathematics of EEG source localization consist of solving an inverse problem: given the electrical potential distribution on the scalp, what are the generating

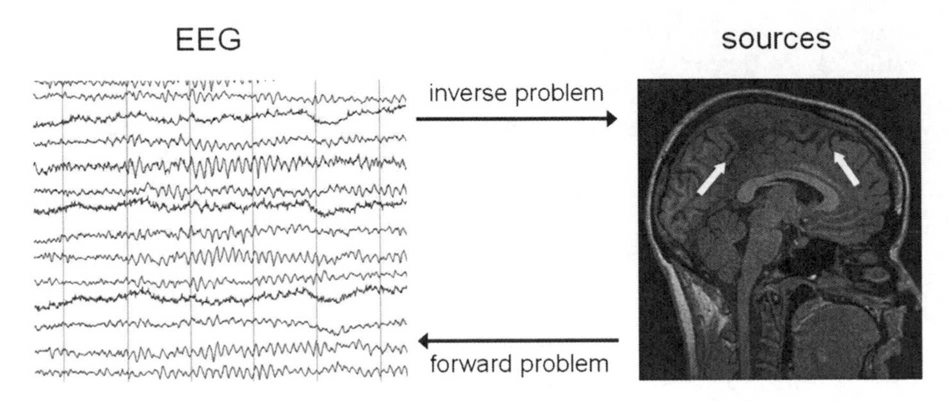

Fig. 8.2 Schematic of the relation between the inverse and the forward problem in EEG source localization. Potential sources are indicated by *white arrows*

sources in the brain? Already in 1853 Hermann von Helmholtz, a German physician and physicist, showed that this inverse problem has no unique solution: there are multiple configurations of sources that generate the same potential distribution. Fortunately, the forward problem does have a unique solution that can also be calculated in most cases. This means that given some sources, there is only one potential distribution on the scalp that is generated by these sources (see Fig. 8.2). Maxwell's and Ohm's laws allow to derive a mathematical relation between the sources and the potential distribution (see Box 8.1).

Equation (7) in Box 8.1 shows that solving the forward problem is equivalent to solving a Poisson problem. In principle, the potential V (the EEG as measured on the scalp) can be calculated from the Poisson equation, if the primary current $\mathbf{J}^{\mathrm{p}}$ and the conductivity σ of the head are known. In some cases an *analytical solution* exists, but mostly computers and numerical mathematical approaches are needed to find an approximate solution to the Poisson equation. An analytical solution to the Poisson equation exists for the case in which σ is position independent and the medium is infinite and *isotropic*:

$$V(\mathbf{r}_0) = -\frac{1}{4\pi\sigma} \int\int\int_{volume} \frac{\nabla \cdot \mathbf{J}^{\mathrm{p}}}{|\mathbf{r} - \mathbf{r}_0|} d\mathbf{r}. \tag{8.1}$$

Here, the volume integral represents a summation over all current sources in the volume. Unfortunately, in the real world of EEG source localization this solution cannot be used because in the head conductivity depends on position (it is much lower for bone than for brain tissue and cerebrospinal fluid), the head is finite and it is not isotropic. Because of this and the ill-posedness of the problem (there is no unique inverse solution), it is necessary to make additional assumptions.

The first assumption that is made concerns the current sources. Typically, it is assumed that the current sources can be represented by one or more dipoles. A dipole consists of two charges of opposite sign that are of the same size and positioned

Box 8.1 Mathematics of the Electromagnetic Inverse Problem

Maxwell's equations describe how the electric and magnetic fields that are the result of a current in the brain are related. They can be used to derive the mathematical equation that needs to be solved in the context of EEG source localization.

Since the speed of electromagnetic waves resulting from potential changes in the brain is very high ($O(10^5)$ m/s), these potential changes can be detected almost instantaneously everywhere on the skull. Therefore, it suffices to consider only the stationary (time-independent) Maxwell equations for the electric field **E** and the magnetic field **B**:

$$\nabla \cdot \mathbf{E} = \frac{\rho}{\varepsilon_0}$$

$$\nabla \times \mathbf{E} = 0,$$

$$\nabla \cdot \mathbf{B} = 0,$$

$$\nabla \times \mathbf{B} = \mu_0 \mathbf{J}. \tag{1}$$

Here ρ is the charge density, ε_0 is the dielectric constant (*permittivity*) of vacuum, μ_0 is the electric *permeability* of vacuum, and **J** the current density. Note that **E**, **B**, and **J** are vector valued functions defined for every point **r** in space.

Because of the second of the equations in (1), the electric field can be written as the gradient of a potential field V:

$$\mathbf{E} = -\nabla \mathbf{V}. \tag{2}$$

Ohm's law states that the current density **J** results from the primary current $\mathbf{J}^{\mathrm{p}}$ and a passive volume current proportional to the electric field **E**:

$$\mathbf{J} = \mathbf{J}^{\mathrm{p}} + \sigma \mathbf{E} \tag{3}$$

and by substituting **E** according to (2) we arrive at:

$$\mathbf{J} = \mathbf{J}^{\mathrm{p}} - \sigma \nabla \mathbf{V}, \tag{4}$$

where σ is the *conductivity* of the medium (in the case of EEG: brain, cerebrospinal fluid, skull and scalp) which depends on the position in space. For EEG, $\mathbf{J}^{\mathrm{p}}$ is due to the activity of neurons in or very close to the cell itself, whereas the volume current flows everywhere in the medium. The source of the EEG is found by localizing the primary current. Thus, $\mathbf{J}^{\mathrm{p}}$ can be considered the driving "battery" of the recorded EEG.

(continued)

Box 8.1 (continued)

Finally, by taking the divergence of (4):

$$\nabla \cdot \mathbf{J} = \nabla \cdot \mathbf{J}^{\mathrm{p}} - \nabla \cdot (\sigma \nabla \mathbf{V}), \tag{5}$$

taking the divergence of the last of the equations in (1), using that the divergence of a rotation is zero:

$$\nabla \cdot \nabla \times \mathbf{B} = \nabla \cdot \mu_0 \mathbf{J} = \mu_0 \nabla \cdot \mathbf{J} = 0 \tag{6}$$

and by combining (5) and (6) we arrive at:

$$\nabla \cdot \mathbf{J}^{\mathrm{p}} = \nabla \cdot (\sigma \nabla \mathbf{V}). \tag{7}$$

This is a Poisson equation for the primary current (the source) which can in principle be solved from the measured potential $\mathbf{V}$ (the EEG) when the conductivity σ of the medium is known.

infinitely close to each other. The current distribution of a dipole at position $\mathbf{r}_Q$ and with strength $\mathbf{Q}$ is mathematically expressed as:

$$\mathbf{J}_{dip}(\mathbf{r}) = \mathbf{Q}\delta(\mathbf{r} - \mathbf{r}_Q). \tag{8.2}$$

Here, $\delta(.)$ is the Dirac delta function, which can be thought of as being zero everywhere except at position zero where it is infinite, in such a way that its integral is 1. These dipoles are physically interesting because it can be shown that the potential due to multiple current sources in a volume can always be written as an infinite series of monopoles, dipoles, tripoles, etc. In most cases, only the dipoles are sufficient to approximate the actual current sources. The dipoles are also physiologically interesting as a model for populations of neurons. Remember that the scalp EEG is generated by groups of parallel oriented pyramidal neurons in the outer layer of the cortex (Chap. 4). When a neuron is excited by other neurons through postsynaptic potentials, current starts to flow along the cell membrane. Since current only flows in closed loops, the intracellular current along the membrane induces an extracellular volume current in the opposite direction. This current loop can be modeled by a combination of a current sink and a current source, i.e. as a so-called "equivalent" current dipole (ECD; see Fig. 8.3). An ECD represents the activity in a few square centimeters of cortex.

The second assumption that needs to be made concerns the volume conductor, which encompasses the shape of the head and its compartments (bone, cerebrospinal fluid, brain) and the conductivities of the different tissues. The simplest assumption is to view the head as a sphere. The advantage of this assumption is that the *analytical solution* to the Poisson problem for a dipole inside a homogeneous sphere is known. The potential distribution of a single dipole in a spherical head model is displayed in Fig. 8.4.

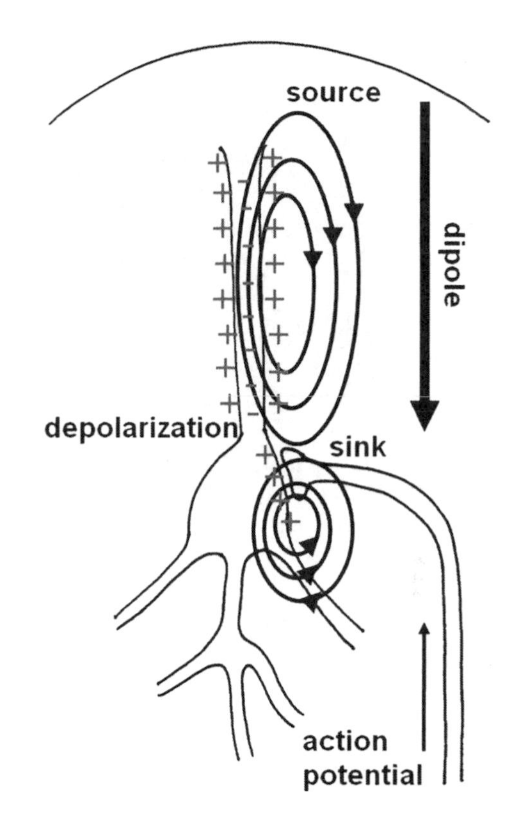

Fig. 8.3 Schematic of pyramidal cell in the cortex with a depolarizing current flowing along the membrane. The loop resulting from the intracellular and extracellular currents can be viewed as resulting from a current source and a current sink which can be modeled by a current dipole

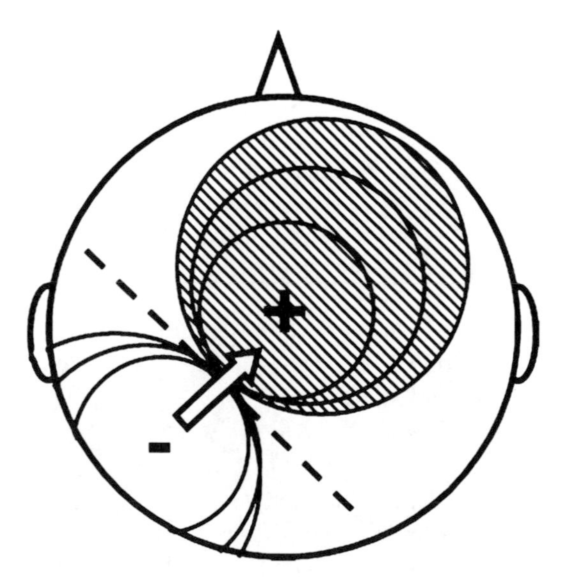

Fig. 8.4 Potential distribution of a single dipole, oriented parallel to the scalp and positioned just below the zero equipotential line (*dashed*), in a spherical head model

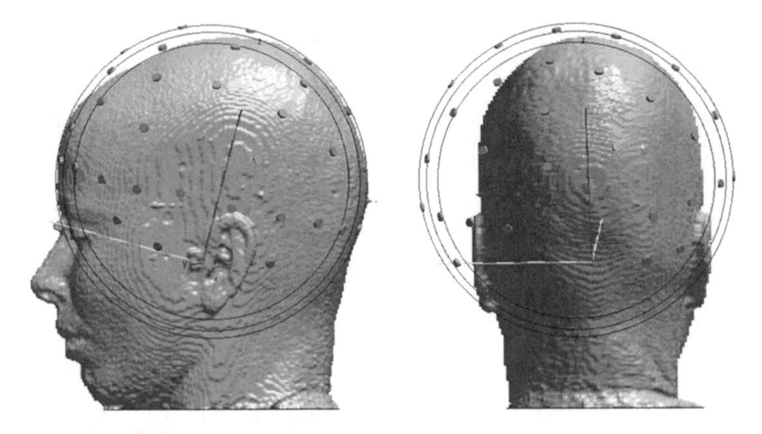

Fig. 8.5 Multilayer spherical model fitted to an anatomically correct head. Notice the good fit on top and bad fit at the sides and bottom

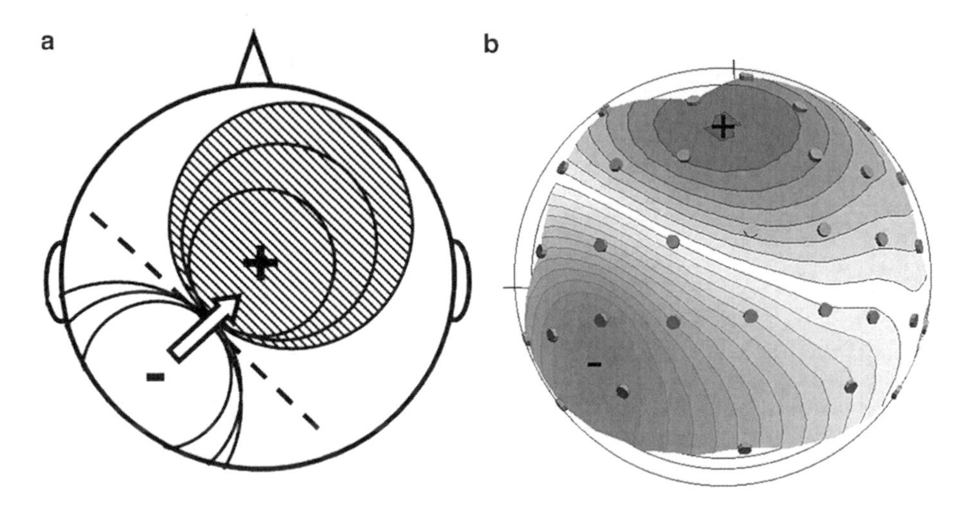

Fig. 8.6 (**a**) Potential field resulting from a tangential dipole in a spherical head model. (**b**) Measured potential resulting from a sensory evoked potential

> *Question 8.1* What is another ECD model that can explain the potential distribution in Fig. 8.4?

The solution for multiple dipoles in a sphere can easily be found by adding the solutions for the single dipoles since the Poisson problem is linear. This assumption may seem an oversimplification, but for some cases it actually works quite well. A sphere can be a close fit to the human head shape on top, although it is worse for the bottom or sides of the head (see Fig. 8.5). Yet, when localizing activity in more *dorsal* areas of the brain, such as the sensorimotor areas, a spherical head model may suffice.

Box 8.2 Solving the Poisson Problem

To represent the boundaries between the different tissues in the head in a computer, they are covered with triangles (triangulation; see Fig. 8.7). The triangulations are very useful because the potential on the scalp, which is expressed as a volume integral (8.1), can also be expressed as an integral equation on each of the boundaries. When this integral (the solution to the Poisson equation) is approximated numerically, the first step is to limit the number of points on the scalp where the solution needs to be approximated. The triangulations provide an intuitive way to do that: the potential is calculated only at each of the vertices. Since the number of points for the solution is now limited, the continuous surface integral can be translated into a large matrix equation that needs to be solved and there are a multitude of methods available to do that. This approach of solving a surface integral is called the boundary element method.

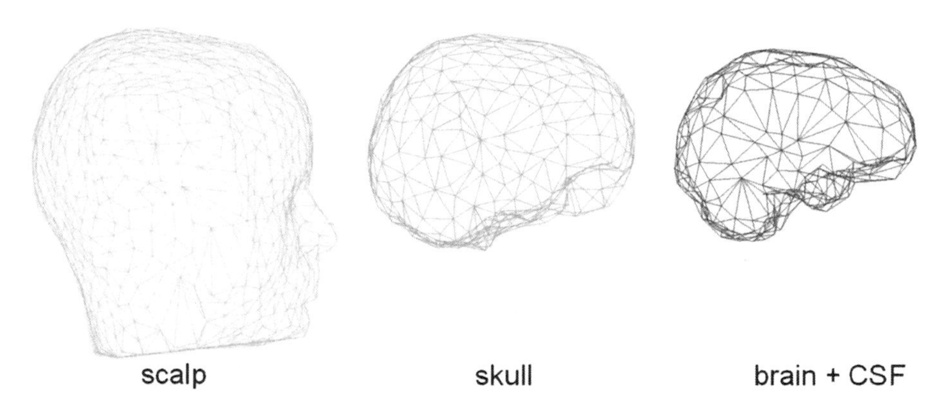

Fig. 8.7 Individual volume model for boundary element method as derived from the MRI. Each boundary is covered with *triangles*, which allows a rather accurate representation of the actual anatomy. From *left* to *right*: scalp, skull, brain, and cerebrospinal fluid

In Fig. 8.6, the potential resulting from a tangential dipole (the orientation when the dipole is located in the wall of a sulcus) in a spherical head model is indicated, next to a measured potential field resulting from a sensory evoked potential, showing that this simple model can be a good approximation of reality under some circumstances.

Question 8.2 What would the potential field of a radial dipole (a dipole located on a *gyrus*) in a spherical head model look like?

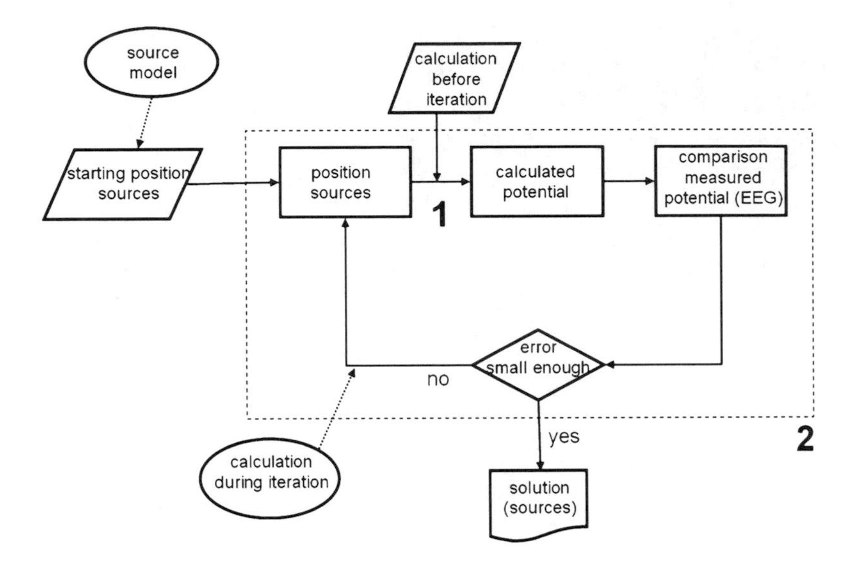

Fig. 8.8 Diagram of the iterative approach to solve the inverse problem. *1* Forward problem; *2* inverse problem

A better approximation of the head as a volume conductor is a sphere with multiple layers for the scalp, the skull, and the brain with the cerebrospinal fluid. An analytical solution for the potential on the sphere resulting from a dipole inside the brain layer still exists, although it is more complicated. For realistic dipole models of EEG activity in other than dorsal areas, it is necessary to use anatomically more correct models. Although this necessity was realized early on, solution of the Poisson problem in these more irregular volume conduction models requires numerical mathematical approaches and powerful computers which were not available for a long time. Nowadays, a model is typically used which is based on the average anatomy as obtained from MR images in several subjects. This approach allows comparison of dipole model results between different people which is important for fundamental investigations of brain organization. If however, a dipole model is required for an individual patient, the individual anatomy must be taken into account. Usually, an individual volume model or realistic model is obtained by segmenting an MRI of the patient, which identifies the boundaries between brain/CSF and skull, between skull and scalp, and between scalp and the outside world. Numerical mathematical methods are then used to find a solution to solve the Poisson problem (see Box 8.2).

So far, the focus has been on how the forward problem can be solved, but remember that the intention was to solve the inverse problem. The inverse problem involves estimating the current sources in the brain, given the potential distribution on the scalp. The solution can be approximated using an iterative approach (see Fig. 8.8): the forward problem is solved for a set of dipoles and a certain volume conductor and the resulting calculated potential distribution is compared to

Box 8.3 Deriving an Estimate for the Number of Sources Using SVD, PCA, or ICA

SVD on a matrix with the recorded potential values (electrodes in rows and time points in columns or vice versa) results in a number of spatial components or topographies, such that the first component explains the largest part of the variance in the data. The second component is (mathematically) perpendicular to the first and explains the largest part of the remaining variance, etc. In general there is no one-to-one correspondence between the SVD components and the activity of the actual dipole sources because of the physiologically artificial need for the components to be mathematically perpendicular, but the number of important SVD components is usually a good indication of the number of sources. PCA is comparable to SVD except that the components can be rotated with respect to each other. ICA performs best, because it does not require the components to be mathematically perpendicular, but only to be independent.

the measured potential distribution. Depending on the error, the set of dipoles is adapted (its positions, orientations and/or strengths) and the procedure is repeated until the error is small enough.

The error is usually calculated as the difference between the calculated and the recorded potential distribution in some measure, such as the sum over all electrodes and time points of the squared differences between the measured and the calculated potential distribution.

Question 8.3 What is another useful measure for the error that needs to be minimized when solving the inverse problem?

An important step in solving the inverse problem is the initial estimate for the dipole model (the number of dipoles, their positions, orientations and strengths): the better this initial estimate approximates the final solution, the easier it is to find it. Therefore, it is best to make an educated guess. If something is already known about the likely sources of the activity this can be incorporated in the dipole model. As an example, in case of a somatosensory evoked potential (SEP), one would expect activity related to the earliest scalp component in the contralateral postcentral gyrus (the primary somatosensory area). Furthermore, information about the position of sources could be obtained from other modalities such as functional MRI. Finally, several methods can be employed to derive the number

Box 8.4 Repeated Calculation of the Forward Problem: The Lead Field Matrix

Instead of repeatedly solving a new matrix equation for the forward problem in a realistic head model, some calculations can be done once, before the iterative process, which greatly reduces calculation time.

As indicated before, the Poisson equation is linear. This allows writing the potential V_i at an electrode i as a linear combination of the primary current density $\mathbf{J}^{\mathrm{p}}$:

$$V_i = \int \mathbf{L}_i(\mathbf{r}) \cdot \mathbf{J}^{\mathrm{p}}(\mathbf{r})\, dv. \tag{1}$$

Here, $\mathbf{L}_i$ is the lead field operator which depends on the geometry and conductivities of the head model. When the current source is a dipole with strength $\mathbf{Q}$ at position $\mathbf{r}_Q$ (1) can be rewritten using (8.2) to:

$$V_i(\mathbf{Q}, \mathbf{r}_Q) = \int \mathbf{L}_i(\mathbf{r}) \cdot \mathbf{Q}\delta(\mathbf{r} - \mathbf{r}_Q)\, dv = \mathbf{Q} \cdot \mathbf{L}_i(\mathbf{r}_Q). \tag{2}$$

The lead field operator can thus be obtained by calculating the potential at every position on the scalp, resulting from a dipole with strength $\mathbf{Q}$ at every possible position $\mathbf{r}_Q$ in the head model. When $\mathbf{L}_i$ is known, it can be reused every time the potential must be calculated in the same volume conductor due to a different dipole configuration. For a spherical model this is easily done, for a realistic head model however, this is a cumbersome calculation.

The discretized version of the lead field operator, the lead field matrix, has as many rows as there are electrodes and as many columns as there are possible source locations. Thus, an entry of the lead field matrix corresponds to the potential at one recording site due to one specific source. The number of possible source locations can be enormous for a spatially detailed head model, which would effectively inhibit the calculation of the lead field matrix. The reciprocity principle of von Helmholtz provides a way out (Weinstein et al. 2000). It states that given a dipole $\mathbf{Q}$, and a need to know the resulting potential difference between two points A and B, it is sufficient to know the electric field $\mathbf{E}$ at the dipole location resulting from a current I, placed between points A and B:

$$V_A - V_B = \frac{\mathbf{E} \cdot \mathbf{Q}}{-I}. \tag{3}$$

This can be used to easily calculate $\mathbf{L}$. So, rather than iteratively placing a source in every element and computing a forward solution at the electrodes, the process can be inverted by placing a source and sink at pairs of electrodes, and for each pair compute the resulting electric field in all of the possible

(continued)

Box 8.4 (continued)

source locations. The reciprocity principle can then be used to reconstruct the potential differences at the electrodes for a source placed in any possible location. In practice, this is done by making one of the electrodes the ground electrode, putting a unit current sink at that position and putting a unit current source at the position of another electrode. The forward solution can then be computed which results in a potential field, and by taking the gradient the electric field **E**, defined at every position in the volume conductor, is obtained. By computing $\mathbf{E}/-I$ at every possible source position, a row in the lead field matrix results.

Box 8.5 Nonlinear Optimization with the Levenberg–Marquardt Method

To understand the Levenberg–Marquardt algorithm, it is convenient to visualize the error function that needs to be minimized as a hilly landscape, in which the valleys represent lower errors and the peaks higher errors. The purpose of optimization is to find the deepest valley without getting stuck in a local minimum, i.e., a valley which is not the deepest. In the steepest descent method a step downhill (in the direction opposite to the local gradient) is made which is so big that a bigger step would result in going uphill again. Finding this optimal step size is rather inefficient and the minimum is approached in a zigzag manner. The Levenberg–Marquardt algorithm improves upon the steepest descent method by also taking the local curvature of the error landscape into account. This helps decide the optimal step size in the direction opposite to the local gradient.

of sources from the data itself such as principal component analysis (PCA), independent component analysis (ICA), and singular value decomposition (SVD) (see Box 8.3).

Since the forward problem must be solved repeatedly, it is also important to consider how this can be done as efficiently as possible, without having to repeat the same calculation for each iteration. A solution is to calculate the so-called lead field matrix efficiently beforehand and reuse the matrix for every calculation (see Box 8.4).

The last step in the iteration process that needs to be considered is the iterative method itself. This entails finding new source configurations from one iteration to the next, as long as the difference between the measured and the calculated potential distribution is still considered too large. This type of problem is called a nonlinear optimization problem, which is encountered in many other fields as well. In source localization, the Levenberg–Marquardt algorithm is mostly used instead of the simpler but less efficient steepest descent method (see Box 8.5).

8.3.2 Alternative Methods

In the previous section, a limited number of equivalent current dipoles was used as a model for the current sources underlying the EEG as measured on the scalp. One of the risks of the described approach is that a suboptimal solution may be found when the nonlinear optimization method does not result in a global, but in a local minimum of the error function.

An approach to overcome the problem of local minima is the use of a scanning method. As a bonus, these methods do not need to know the number of sources beforehand. The scanning methods search for optimal dipole positions throughout the source volume by systematically visiting all nodes of a discrete grid within the head model. Likely source locations are then determined as those for which a metric computed at that location exceeds a given threshold. The most common scanning methods are the linearly constrained minimum variance (LCMV) beam forming and multiple signal classification (MUSIC) and their variants. MUSIC suffers from a few problems. One of these is that it usually detects multiple peaks in a 3D-volume of the head, each of which may correspond to either a different ECD or to a local minimum in the error function. The extended version of MUSIC, recursively applied and projected (RAP)-MUSIC helps find the true sources by recursive estimation of multiple sources.

Another class of source localization methods that do not require an a priori assumption on the number of sources concerns the distributed source localization methods. Distributed source localization estimates the amplitudes of a dense set of dipoles distributed at fixed locations within the head volume, usually limited to the gray matter. These methods are based on reconstruction of the brain electric activity in each of the fixed locations. The number of these points is typically much larger than the number of measurement points (electrodes) on the scalp surface. Again, the inverse problem that needs to be solved here has no unique solution and thus assumptions need to be made to find a single optimal solution. The different methods described in literature (Michel et al. 2004) differ in their choice and implementation of these assumptions. The constraints can be of a mathematical nature, but biophysical or physiological knowledge or findings from other modalities may also be incorporated. Some examples are the Minimum Norm method and its improvement the Weighted Minimum Norm method, Low Resolution brain Electromagnetic Tomography (LORETA) and its improvement standardized LORETA (sLORETA) and local autoregressive average (LAURA).

8.3.3 Tips and Tricks When Performing Source Localization

Most source localization methods will yield a solution when the difference between the model solution and the measured EEG potential distribution (the error) is smaller than a predetermined threshold in a certain *metric*, or when the error does not decrease significantly from one iteration step to the next anymore. Yet, because of the problem of local minima (when limiting our discussion to the ECD source localization method), even a good solution from the point of view of the method, can be a bad solution from the point of view of (patho)physiology. The results of source localization may therefore be hard to interpret, since a priori information on the number, positions, orientation, and activity of the sources is typically not available. Furthermore, the EEG signal itself is measured at a limited number of positions (electrodes) only, the signal is smeared on the scalp with respect to the signal on the cortex because of the presence of the bony skull and there is a lot of noise (background EEG, not originating from the source of interest). The question thus remains how the reliability of a certain source solution can be judged. Here, some practical approaches are discussed.

8.3.3.1 Optimizing the Signal-to-Noise Ratio

First and foremost, the quality of the source solution improves when the signal-to noise ratio (SNR) of the EEG is higher. This can be intuitively understood as it cannot be expected that a source that yields activity which drowns in the background EEG can be resolved. Depending on the application, there are some approaches that can be taken to optimize the SNR. When the signal of interest can be measured several times (as in the case of interictal epileptic peaks or evoked potentials; see Chaps. 4 and 5), averaging helps improve the SNR. In these cases, the noise (or background EEG) is averaged out, helping the signal of interest, which is assumed to be the same every time, to stand out. Care must be taken to only average EEG activity that originates from the same source. This may be difficult for epileptic spikes that may originate from multiple locations. In those cases, similar spikes may be selected on the basis of their spatial topography.

In principle it also helps to measure the activity from more locations, although care must be taken that the quality of the EEG signals is optimal on every included electrode. It is better to take fewer electrodes into account than use additional electrodes with a bad quality signal. At some point, it does not help to add additional electrodes: when the spatial resolution of the EEG signal has been resolved (sampling at more than twice the spatial Nyquist frequency, see also Sect. 5.3.4.1) more electrodes only oversample the signal. For EEG, the optimal interelectrode distance is in the order of 2–3 cm, which can be reached with 64- or 128-channel electrode caps.

8.3.3.2 Stability of the Solution

Another method to assess the reliability of the model solution is to rerun the process for a different model (other types or number of dipoles) or for the same model but with a different initial solution. If the same model solution is found even when the initial solution is perturbed, evidence for the solution becomes stronger. Similarly, when the source model consists of multiple dipoles, the quality of the solution should be checked for models in which each one of the dipoles is successively removed and the forward solution is calculated again. The solution should also be stable for small changes in the duration of the time interval that is considered and for changes in the exact dipole model (*stationary*, rotating or moving in time) that is used. Another option is to split the EEG dataset in two (if it is large enough) and to redo the source localization for the separate datasets. If the solutions remain stable (the same) under all these types of manipulations, one can be fairly sure that the solution is (patho)physiologically reliable.

8.3.3.3 Overlapping Evidence

A reliable source solution should be more or less independent of the method that is used, although it is known from studies that compare the solutions for different methods that some of them have specific biases. As an example, the Minimum Norm method tends to place the sources more superficially in the brain and LORETA solutions are more blurred. Yet, the location of the sources should largely coincide and not, e.g., shift hemispheres when calculated with a different method.

Second, depending on the type of EEG data that is used to find sources, it may be possible to have an a priori hypothesis on their anatomical localization. For example, if the EEG activity related to the first cortical component of a SEP needs to be localized, it may be expected in the primary somatosensory cortex, contralateral to the stimulated limb.

Question 8.4 Where would you expect to find the sources of the first cortical component in an auditory evoked potential?

If the anatomy is not disturbed due to a pathophysiological process, it is usually possible to identify such an area of interest on an individual's MR image. If the anatomical localization of the studied brain process is not known, e.g., because the process involves multiple and variable brain areas as is the case for many cognitive processes, brain areas identified using other neuroimaging approaches (functional MRI, positron emission tomography [PET]) may be used for reference. It should be noticed however, that some EEG sources can be "silent" for fMRI and vice versa, because of the different physiological processes underlying signal generation.

Finally, the location of sources should be in line with available knowledge obtained from other functional imaging modalities. For evoked or event-related potentials, a requirement is that the experiment has been performed in the same manner, using the same parameters, in both modalities.

8.3.3.4 Goodness-of-Fit Measures

Most software packages provide a numerical measure to assess the reliability of the source model which is the goodness-of-fit. An often used measure of goodness-of-fit is explained variance. Explained variance provides a measure of the discrepancy between the measured EEG and the potential distribution predicted by the model. The higher the explained variance, the better the model. Explained variance can be expressed as a percentage. In that case, a value of 100% would indicate a perfect fit of the model to the data. Good solutions typically have at least 80–90% explained variance, but models with lower values may be acceptable when there is other independent evidence for their reliability (see Sect. 8.3.3.3). Probably not surprisingly, the higher the SNR of the EEG data, the higher the explained variance of the model. Explained variance is identical to the square of the correlation between the observed values and the values predicted by the model. Note that, although explained variance and goodness-of-fit are also used in the context of (statistical) correlations, a strong correlation (more than 50% explained variance) would not correspond to a good model in the context of source localization. Sometimes the goodness-of-fit is expressed as residual variance, which is the complement of explained variance (residual variance = 100 – explained variance).

8.4 Preoperative Function Localization in an Individual Patient

To perform source localization in an individual patient several requirements need to be fulfilled. First, an individual head model is needed, based on an anatomical MRI which is then segmented into the number of head model compartments that is needed, e.g., brain with CSF, skull, and scalp. Second, a suitable EEG dataset is required and third, the exact 3D-position of the electrodes with respect to the head must be known. The latter provides information on the spatial origin of all EEG data. If these three requirements (see Fig. 8.9) are met, source localization can be performed.

Patient 1

Since patient 1 suffers from somatosensory disturbances, one of the eloquent areas that should be localized before her tumor is excised is the somatosensory area. In an anatomically normal brain, the primary somatosensory cortex is easily identified as the postcentral gyrus. However, in this patient, the tumor has caused the

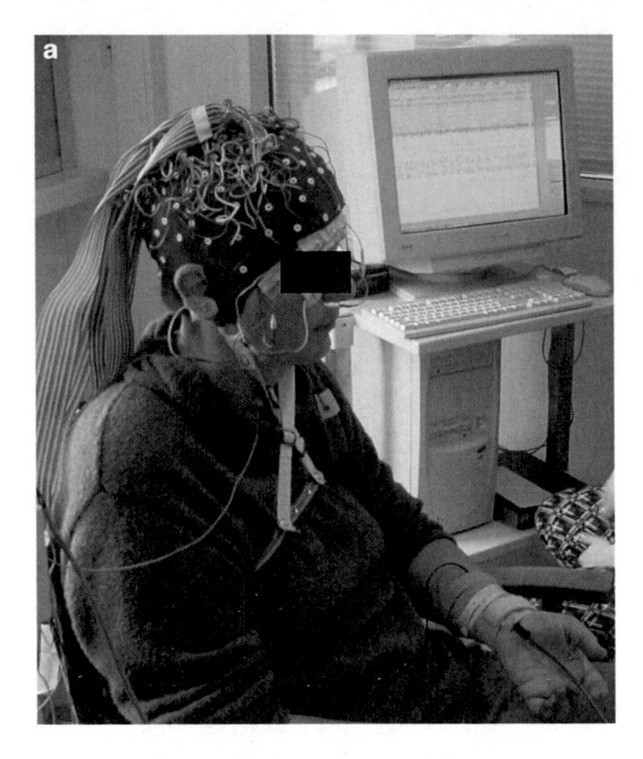

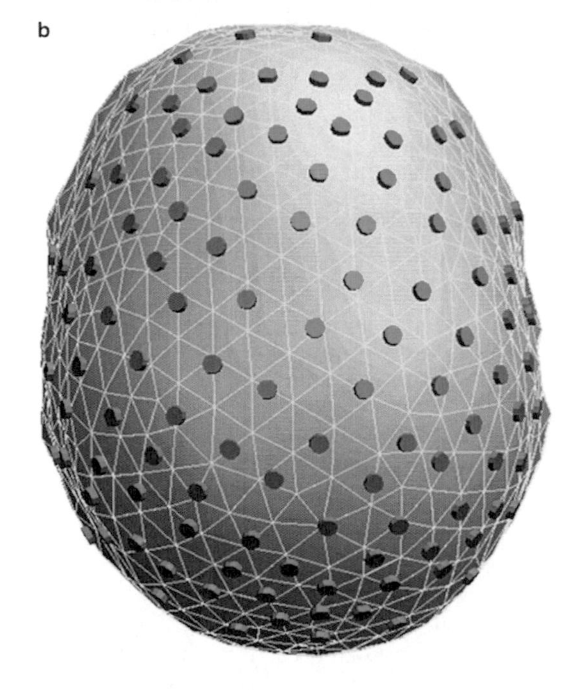

Fig. 8.9 The three requirements for individual source localization. (**a**) Evoked potential recording, (**b**) electrode positions in 3D-space, and (**c**) individual head model

Fig. 8.9 (continued)

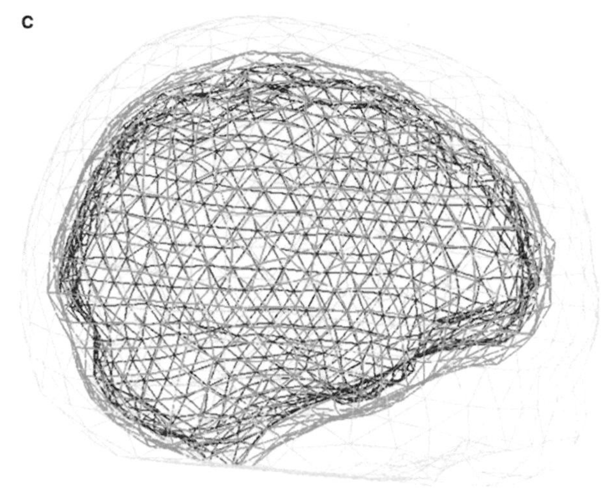

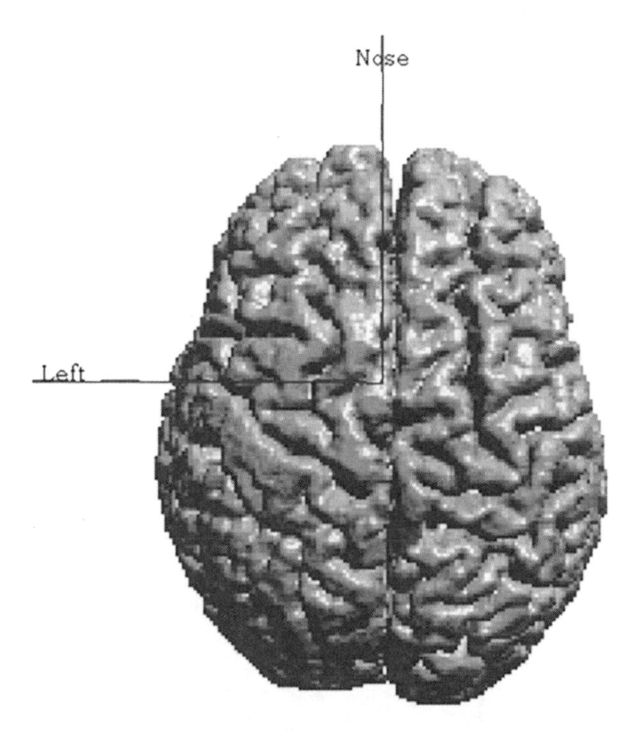

Fig. 8.10 The brain of patient 1 as segmented from the anatomical MR image. Notice the shallow sulci and the shifted central sulcus in the left hemisphere

surrounding brain tissue to shift (see Fig. 8.10), hampering the identification of the somatosensory cortex directly from the anatomy.

An alternative way to localize the postcentral gyrus is to perform source localization on a preoperatively obtained SEP recording. Since patient 1 suffers particularly from somatosensory problems in her hand, a *median nerve* SEP recording was performed, employing 128 EEG channels to optimize the signal-to-noise ratio.

Question 4.2 Why would increasing the number of EEG channels help improve the signal-to-noise ratio?

The median nerve was stimulated 500 times at 2.8 Hz with small electrical shocks at the wrist, just strong enough to evoke a motor twitch in the thumb, ensuring that all sensory nerve fibers were evoked. The patient's individual MRI was preprocessed to identify the boundaries of scalp, skull, and brain employing segmentation methods, resulting in an individual head model with three layers which was then triangulated. The 3D-localization of the electrodes was obtained using a tracking system and the electrode positions were *co-registered* with the head model using three anatomical landmarks (the nasion and left and right preauricular points). These positions were both tracked and visible in the MRI.

The first cortical component of the SEP (the N20, arriving in the primary somatosensory cortex approximately 20 ms after stimulation at the wrist) was employed for source localization. This component usually has a good signal-to-noise ratio, making it suitable for source localization. Because it is also known that the N20 is generated in *Brodmann area* 3b, a well-defined region in the primary somatosensory cortex, we modeled the source of the N20 by a single dipole. This is also a suitable assumption because the potential distribution of the N20 is similar to the potential distribution of a single ECD in a sphere, as can be seen in Fig. 8.11.

This procedure resulted in an explained variance of 86% for right median nerve stimulation and a position that was robust to changes in the initial position of the dipole. As can be seen in Fig. 8.12, the dipole was positioned just anterior and dorsal to the tumor, consistent with the sensory difficulties the patient had.

In this patient, functional MRI was additionally exploited to localize eloquent areas, providing independent evidence for localization from a different modality than EEG. Her right hand was stroked with a brush for several seconds. In addition, the patient voluntarily contracted and relaxed her fist to obtain an indication of the location of the primary motor cortex. Figure 8.13 illustrates that the ECD identified from the SEP data and the somatosensory area as identified from the fMRI data, are very close together. The positions do not overlap completely, likely because finger-stroking may also involve some reactive finger motor activity and is less specific than median nerve stimulation.

During craniotomy, the identified eloquent areas were avoided as much as possible employing *neuronavigation* techniques (see Fig. 8.14).

The preoperative results were first verified during surgery. Unfortunately, because the preoperatively identified areas were localized rather deep, superficial electrical stimulation of the cortex did not result in a motor response, nor could SEPs be recorded directly from the cortex using a small electrode grid. The surgery

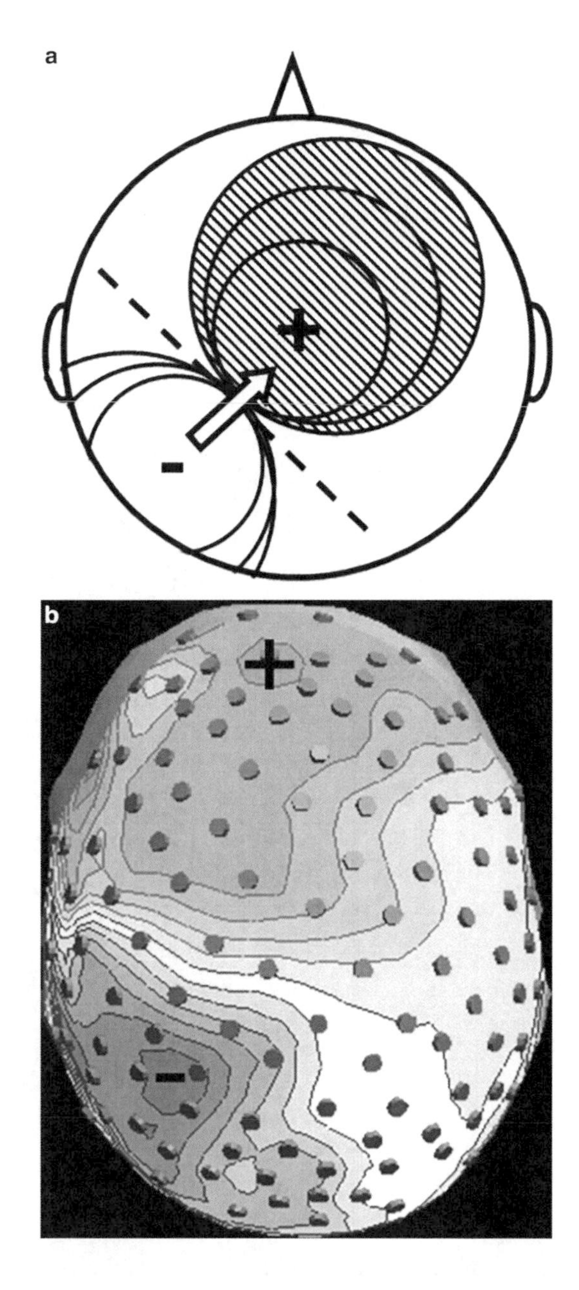

Fig. 8.11 (**a**) Potential distribution of an ECD in a spherical head model. (**b**) Potential distribution of N20 median nerve SEP component for patient 1

was continued anyway, taking into account the preoperative localization results as much as possible. When the tumor was resected, the surgeon succeeded in evoking a motor response by directly stimulating a brain area rather deep in the central sulcus (and not on the gyrus as would normally be expected). Directly

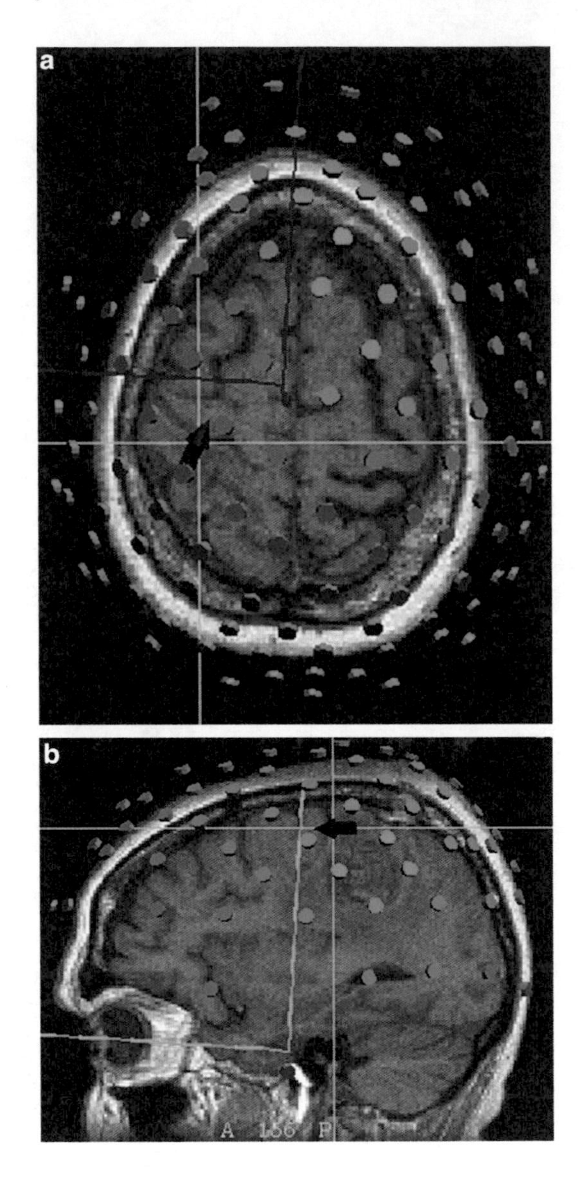

Fig. 8.12 Position and orientation of the dipole with the best fit to the SEP data at peak N20 activity, with respect to the anatomical MRI of patient 1. All 128 electrodes are also indicated. (**a**) Axial orientation, (**b**) sagittal orientation

postoperatively, the patient had good motor control of arms and legs although she complained about a tingling feeling in the right hand. One day after surgery her paresis increased again, as did her language problems. One week after surgery she could be discharged from the hospital; the severity of her complaints had abated again to the level that she had preoperatively, indicating that the surgery had caused minimal additional harm. Without preoperative function localization, results could have been much worse.

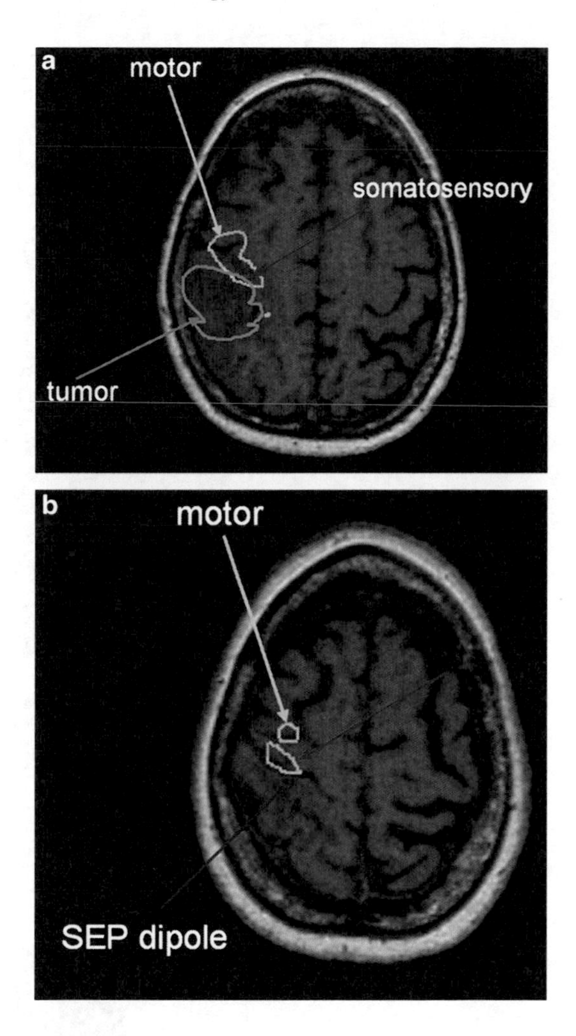

Fig. 8.13 Position of the dipole with the best fit to the SEP data at peak N20 activity, with respect to the fMRI activation. The tumor and areas involved in motor and somatosensory activity are also indicated. (**a**) Axial slice on the level of the tumor, (**b**) axial slice just above the tumor

8.5 Other Applications of Source Localization in Neurology

8.5.1 Epileptogenic Focus Detection

In a subgroup of patients with epilepsy, pharmacological therapy is not sufficient to control their seizures. In these refractory epilepsies, surgery may be considered to remove the epileptic focus, for therapeutic and hopefully curative purposes. Preoperatively, the epileptic focus should be determined accurately, to make surgery efficient, preserve eloquent brain areas as much as possible, and optimize the effect of surgery. The gold standard for focus localization is electrocorticography; EEG recorded directly from the cortex. This implies that another surgical procedure – in

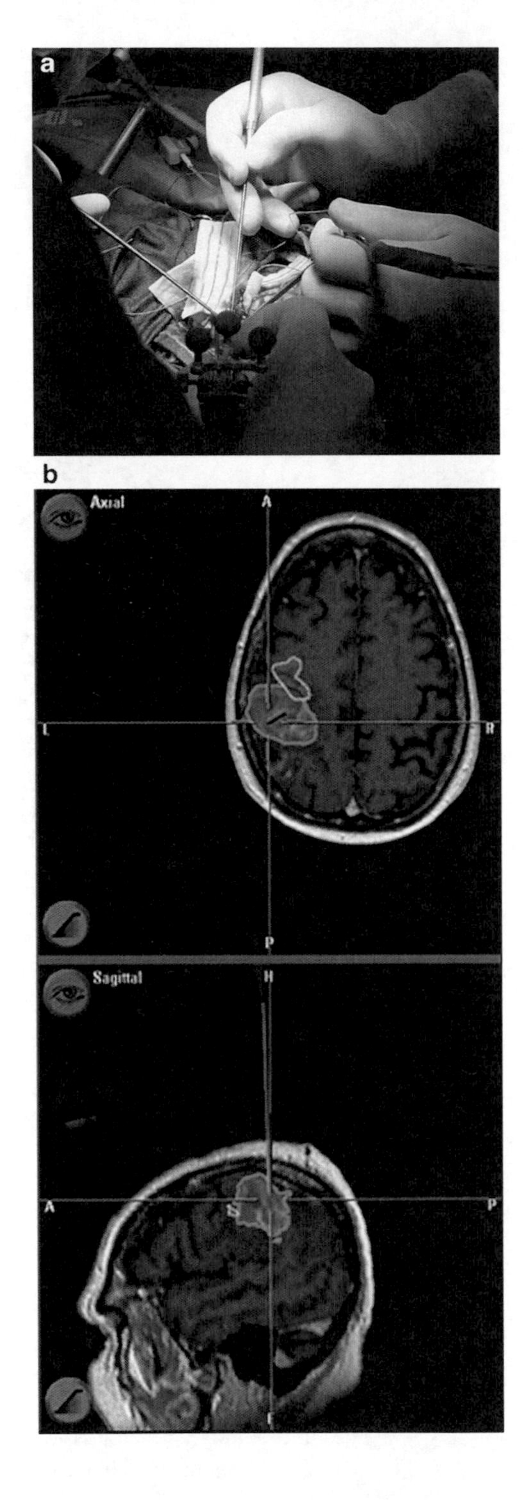

Fig. 8.14 Tumor resection using neuronavigation in patient 1. (**a**) Reflective spheres allow a camera to recognize the position of the neurosurgical tools. (**b**) The position of the tool is integrated in a visual display of the patient's preoperatively obtained MRI with the results of fMRI and SEP drawn in

which a large electrode grid is placed directly on the cortex – precedes the one in which the focus will be removed. But since it is impossible to cover the entire cortex with electrodes, even for this procedure it is important to localize the epileptic brain area as optimally as possible.

Traditionally, clinical interpretation of the EEG and localization of epileptic activity proceeds visually by inspection of waveforms and assuming that activity at a certain electrode mostly represents activity in the cortex directly beneath it. This approach is a bit too simplistic. As an example, consider the tangential dipole in a spherical head model illustrated in Fig. 8.6. In that case, the maximum and minimum of the EEG activity are not found directly above the dipole. EEG source localization allows to localize EEG activity more accurately. During preoperative evaluation of an epilepsy patient who is a candidate for surgery, averaged interictal spikes or sharp wave activity may be used for EEG source localization.

Typically, epileptic spikes are modeled by ECDs. Single spikes usually have a signal-to-noise ratio that is too small to localize them, which is why spikes are averaged before localization. Since not all spikes will originate from the same source, it is important to only average spikes with the same potential distribution on the scalp. The advantage of the interictal activity is that it has the same source as the ictal activity, but without the clinical symptoms (seizures) that would prohibit reliable recordings.

All methods mentioned in Sects. 8.3.1 and 8.3.2 have been applied to localize epileptic activity: from equivalent current dipole methods which are particularly good at localizing the "center-of-mass" of the epileptic activity to distributed methods that take into account that spikes may originate from an area with a certain spatial extent instead of from one focal area (Ebersole and Hawes-Ebersole 2007).

8.5.2 Evoked and Event-Related Potentials in General

In this chapter, localizing the generator of the first cortical component of the SEP has been discussed in some detail because of its application in patient 1. However, in principle any other EEG signal with a sufficiently large SNR can be localized. Because averaging procedures applied to derive an evoked or event-related potential (see Chaps. 5 and 7) improve the SNR, any EP or ERP component is theoretically suited to apply localization procedures. Generally, (early cortical) EP components are easiest to localize because they represent primary processing and are mostly generated in well-defined brain areas, allowing a single dipole model. The resulting model can also be more easily validated for EPs because of our knowledge of the anatomy of e.g., visual and auditory processing. In contrast, ERP components typically represent higher order (cognitive) processing, involving multiple brain areas, making localization notoriously difficult. Yet, in the context of scientific research, source localization has been applied for EPs and well as for ERPs.

The applicability of EP and ERP source localization for neurology has so far been limited however. Some examples of clinical applications of ERP source localization in neurology (diagnosis of Alzheimer's disease, pathophysiological studies of brain injury) exist, although they are more common in psychiatry (brain mechanisms involved in schizophrenia or depression).

8.6 Answers to Questions

Answer 8.1
Another model that could give the same potential distribution consists of two dipoles oriented in parallel on each side of the one dipole at a certain distance and not necessarily at the same depth, with a different strength.

Answer 8.2
The potential field of a radial dipole would be maximal (negative or positive, depending on dipole orientation) immediately above the dipole and drop off quickly with increasing distance from the dipole.

Answer 8.3
Another useful measure of the error could be the sum of the absolute differences between the calculated and recorded potentials over all time points and positions.

Answer 8.4
The first cortical component of an AEP is known to originate from the primary auditory cortex, which would normally be located at the dorsal surface of the temporal lobe.

Answer 8.5
Increasing the number of EEG channels increases the chance of capturing the most salient changes in the signal by better spatial sampling. For example, if we would have only ten channels, chances are that the maximum of the activity is actually in between several electrodes, thus suggesting a different potential distribution. For most purposes, 64 channels are actually enough, because scalp EEG has a limited spatial resolution due to the low-pass filtering properties of the skull.

Glossary

Analytical solution Closed form expression for a variable that is solved from an equation.

Aphasia Acquired language disorder involving difficulties understanding and/or producing language.

Astrocytoma Brain tumor originating from brain cells called astrocytes.

Axial Imaging direction resulting in images that view the brain from the top or bottom.

Central sulcus Fold in the cerebral cortex separating the frontal and the parietal lobe. Primary motor function is located on the gyrus in front of it, primary sensory function on the gyrus behind it.

Conductivity Material property that determines how hard it is for electricity to travel through it.

Co-registration To manipulate two image datasets from the same person but obtained in different modalities, such that the coordinate systems align.

Coronal Imaging direction resulting in images that view the brain from the front or back.

Craniotomy Surgery in which a bone flap is removed from the skull to be able to access the brain.

Dorsal Top, contrast to ventral (bottom).

Endometrial Originating from the inner lining of the uterus.

Glioma A brain tumor originating from the glial cells that support the nerve cells.

Gyrus Ridge on the cerebral cortex.

Internal capsule White matter close to the basal ganglia.

Isotropic The same in all directions.

Median nerve Nerve in the upper limb innervating several muscles in the forearm and hand.

Metric A mathematical distance function, e.g. in 2D-space a metric may be Cartesian distance calculated as $\sqrt{x^2 + y^2}$.

Neuronavigation Computer-assisted technology allowing navigated movement in the skull during brain surgery.

Paresis Partial loss of movement, weakness.

Permeability Material property that determines how much magnetization is obtained when a magnetic field is applied.

Permittivity Material property that determines how much electricity is transmitted.

Sagittal Imaging direction resulting in images that view the brain from the left or right.

Sensorimotor Involving sensory and motor function.

Signal-to-noise ratio Measure used to quantify how much a signal is corrupted by noise; often calculated as the power of the signal divided by the power of the noise (in decibels).

Stationary No changes over time.

Sulcus Fold in the cerebral cortex.

TIA Transient ischemic attack: a temporary period of insufficient blood supply to a part of the brain.

References

Online Sources of Information

http://psyphz.psych.wisc.edu/~shackman/SourceLocalizationMethodology.htm. Links to the most common methods/software packages for EEG/MEG source localization, reviews and sources of method cross-validation

http://en.wikipedia.org/wiki/Maxwell%27s_equations. Description of Maxwell's equations

Software

http://fieldtrip.fcdonders.nl/start. FieldTrip is a free Matlab software toolbox for MEG and EEG analysis that includes several methods for ECD and distributed source localization

http://sccn.ucsd.edu/eeglab/. Homepage of EEGLAB freeware

http://www.uzh.ch/keyinst/loreta. Website to download (s)LORETA freeware and find all information on the method

Papers

Ebersole JS, Hawes-Ebersole S (2007) Clinical application of dipole models in the localization of epileptiform activity. J Clin Neurophysiol 24:120–129

Michel CM, Murray MM, Lantz G, Gonzalez S, Spinelli L, Grave de Peralta R (2004) EEG source imaging. Clin Neurophysiol 115:2195–2222

Weinstein D, Zhukov L, Johnson C (2000) Lead-field bases for electroencephalography source imaging. Ann Biomed Eng 28(9):1059–1065

Chapter 9
Neuromuscular Diseases, Ultrasound, and Image Analysis

After reading this chapter you should:

- Understand how an ultrasound image is built up from transmitted and received sound waves
- Understand the relation between sound frequency, speed, and wavelength and its consequences for the spatial resolution of ultrasound images
- Understand why muscle ultrasound imaging is useful in the differential diagnosis of neuromuscular diseases
- Know how typical myopathies and neuropathies can be distinguished using ultrasound
- Know how to analyze a muscle ultrasound image quantitatively

9.1 Patient Cases

Patient 1

Since 1 year this male 45-year-old patient is not able to walk on his toes. In addition the heels of his shoes show an asymmetric wear pattern. His symptoms are so minor that he did not seek medical advice until the abnormalities were noticed when he had to be tested for a new driver's license. After tests showed that he suffered from an isolated paresis of both tibial anterior muscles, his general practitioner made an MRI of the lumbar vertebra, thinking that his complaints could be related to a nerve entrapment in this region. The MRI was normal, but the blood level of creatine kinase (CK) was extremely high, indicating muscle cell damage. When the patient was sent to the neurologist, absent Achilles (ankle jerk) reflexes were found while his other reflexes were normal, indicating a peripheral problem. He had no sensory

N. Maurits, *From Neurology to Methodology and Back: An Introduction to Clinical Neuroengineering*, DOI 10.1007/978-1-4614-1132-1_9,

disturbances. Although the neurologist thinks that this patient may suffer from a distal *myopathy*, further investigations are needed to confirm this diagnosis.

Patient 2

This female 13-year-old patient suffers from progressive muscle weakness in both her arms and legs since a few months. She cannot lift her arms well, and combing her hair is difficult. She cannot ride a bicycle for longer distances anymore and has trouble getting up from the floor. She has no cramps or pain, and she has no problems talking or swallowing. Her vision has not changed. Her complaints progressed quickly over a period of only 4 weeks, but seem to be stable now. The neurologist finds muscle weakness in the arms (deltoid, triceps, and biceps muscles) and legs (quadriceps, hamstrings, and calve muscles). She has a *Trendelenburg gait* and is not able to hop on one leg. Getting up from a low chair is only possible with support. Her reflexes are absent and sensibility seems intact. Her complaints indicate that she suffers from a problem in multiple nerves.

Patient 3

This female 2-year-old patient suffers from progressive muscle weakness, particularly of the proximal muscles which has impeded her motor development. She started crawling when she was 18 months old and has trouble keeping her head up when she does. She can only stand and walk a few steps when supported. She has no problems swallowing, talks well, and her fine motor skills are normal for her age. The neurological investigation confirms muscle weakness both *axially* and in her extremities. Her lower leg and foot muscles are atrophic. Her Achilles reflexes are absent and her patellar (knee jerk) reflexes can hardly be elicited. Her plantar (foot sole) reflex is normal. She does not seem to have sensory abnormalities. Because her early motor development is not entirely clear, the neurologist thinks she may suffer from either a *neurogenic* problem or a *congenital* myopathy.

9.2 Ultrasound Imaging to Assess Muscle Structure

Although it is likely that patients 1–3 will all be further examined by surface and needle electromyography to diagnose nerve conduction or muscle problems (see Sects. 2.2 and 2.3), other methods such as ultrasound imaging can also be used to assess muscle structure and composition.

An ultrasound image results from sound waves that are sent into the body, where they are reflected by anatomical structures and received again as echoes. Ultrasound imaging, compared to, e.g., computed tomography (CT) or magnetic resonance imaging (MRI), is inexpensive, noninvasive, patient-friendly, portable, and can thus be performed at the bedside. Furthermore, it is easy to use and easily accepted by young children. Ultrasound imaging is probably best known from its application in prenatal investigations of the fetus, but it has already been known for decades that diseased muscles have a different appearance on ultrasound images than normal muscles (see Sect. 9.3.3). Furthermore, muscle ultrasound can be performed during rest and during contraction, allowing to investigate symptoms that only occur when the muscle is actively used; it allows both static and dynamic imaging enabling the visualization of *fibrillations* and the muscle can be easily visualized in multiple planes, both *longitudinally* and *transversally*. Finally, muscle ultrasound can be used to guide muscle biopsies, ensuring that the biopsy is taken from a diseased part of the muscle.

9.3 From Sound to Anatomical Images

To visualize the anatomy, a *transducer* (see Fig. 9.1) emits high-frequency sound waves that are partially reflected at the boundaries between different tissues and received by the same transducer.

The time between emitting and receiving the sound wave is recorded and used to determine the distance between the transducer and the reflecting boundary between tissues, taking the velocity of sound in tissue into account. By emitting sound waves along the entire length of the transducer, a two-dimensional picture can be derived of the anatomical structure of the tissues directly underneath the transducer. To understand the details of this technique, it is necessary to provide some background in the physics of sound.

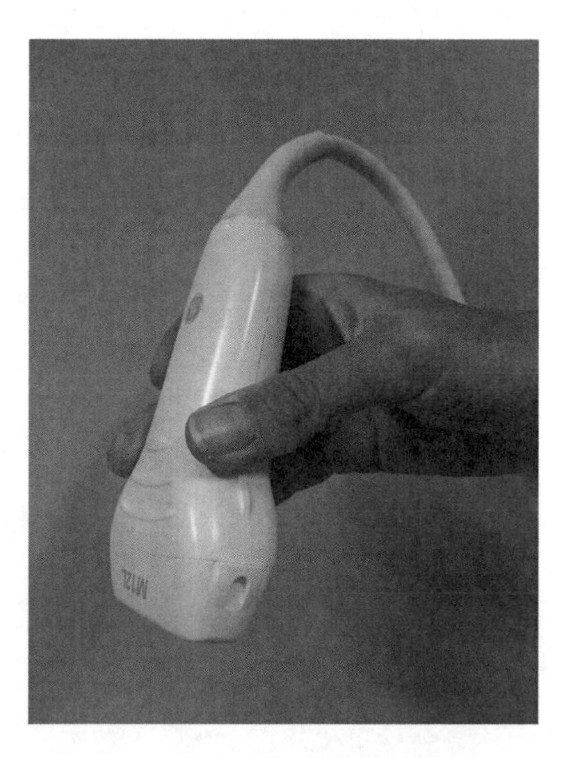

Fig. 9.1 Example of a hand-held transducer used for muscle ultrasound investigations

9.3.1 Physics of Sound

The physics of sound is based on the physics of waves. A wave is a systematic disturbance of matter that changes its position in space as a function of time. Concrete examples of this rather formal definition are waves that develop because a drop of water falls in a puddle or because the end of a rope is moved up and down. These are examples of transversal waves: the wave travels perpendicular to the direction in which the matter (water, rope segments) moves. In longitudinal waves, matter and waves move in the same direction. The latter phenomenon can be observed in a spiral spring of which a few loops are pushed against each other and then released. Diagnostic ultrasound waves, which are high frequency sound waves in liquids and soft tissues, are longitudinal.

Any sound wave can be characterized by a few parameters: its speed (c, in m/s), frequency (f, in Hz), amplitude (u_0, in m), and wavelength (λ, in m). These parameters are not independent but related as follows:

$$c = \lambda \cdot f. \tag{9.1}$$

Note that instead of the letter v, which is usually reserved for speed, the speed of sound is indicated by the letter c. The speed of a sound wave depends on the mechanical properties of the material it travels through, such as density and elasticity which in turn depend on temperature and pressure. Sound velocities differ highly between different tissues: from approximately 300 m/s in air to 4,000 m/s in bone. In muscle, sound velocity is somewhat in between these values. The frequency is independent of material properties, whereas the wave length does depend on these through its relation with speed. The time it takes material particles to complete one wave cycle is called the period (1/frequency). The maximum difference with the neutral position of a material particle during a period is the amplitude.

Sound waves are classified by frequency; infrasonic waves have a frequency below the human hearing threshold (<20 Hz), audiosonic waves have frequencies that can be heard, and ultrasonic waves (ultrasound) have frequencies above the human hearing threshold (>20 kHz). For diagnostic ultrasound MHz waves are typically used.

9.3.2 Sound Waves in Biological Tissues

During diagnostic ultrasound investigations, sound waves are sent into the body that have a certain energy. The echo that is received has considerably lower energy due to *attenuation*. Attenuation is caused by absorption, lateral reflection, and scattering. Absorption is a loss of energy due to particle vibration; energy is then transformed into heat. Lateral reflection causes part of the emitted sound waves to be reflected in a direction other than the transducer. Scattering is the phenomenon in which the sound is dispersed in all directions by a particle, effectively causing the particle to act as sound source itself. The energy is then absorbed somewhere else in the tissue.

The ultrasound image is derived from received sound wave reflections. These reflections arise at the boundaries between tissues in which sound has different acoustical properties, i.e., acoustical impedance. Acoustical impedance Z_a is determined by the sound pressure p, the particle velocity v_p, and the surface area S through which the sound wave travels, as follows:

$$Z_a = \frac{p}{v_p S}. \tag{9.2}$$

Low acoustical impedance thus implies that it is easy for a sound wave to travel through the tissue. Since the human body is rather inhomogeneous, ultrasound reflections will occur often: at boundaries between soft tissues (e.g., skin and fat), at boundaries between soft tissue and bone, and at boundaries with air cavities (lungs, intestines). When the inhomogeneity ("the other tissue") is large compared to the wavelength of the sound, the wave is reflected. When the inhomogeneity is smaller, sound is scattered. This implies that by choosing a different ultrasound frequency (and thereby wave length), the spatial resolution of the image is determined. Small structures can only be seen when the wave length is small enough, or the frequency high enough. On the other hand, high frequency sound is more absorbed (and thus attenuated) by biological tissue than low frequency sound, implying that high frequency ultrasound can penetrate less deep into the tissue.

The combination of all these effects causes the echo of more superficial structures to be stronger than that of deeper structures. Ultrasound hardware partially compensates for this effect by additional amplification of echoes that are received later: so-called "time-gain compensation." But a trade-off between optimal spatial resolution and optimal depth remains. In practice, 3.5–14 MHz transducers are typically used for muscle ultrasound, to achieve an optimal balance between spatial resolution and penetrance.

Question 9.1 What is the size of the smallest spatial structures that can be visualized by 10 MHz ultrasound, assuming a speed of sound in the tissue of 2 km/s?

Different types of structural ultrasound imaging exist that are each suited for different applications. One of the older types which is not typically used anymore is A-mode ultrasound. In this mode a single transducer scans along only one line through the body. The echoes are then plotted as a function of depth. The most common type of ultrasound imaging is B-mode ultrasound, in which a linear array of transducers (built into one larger transducer) scans a plane through the body that results in a two-dimensional image (see Fig. 9.2).

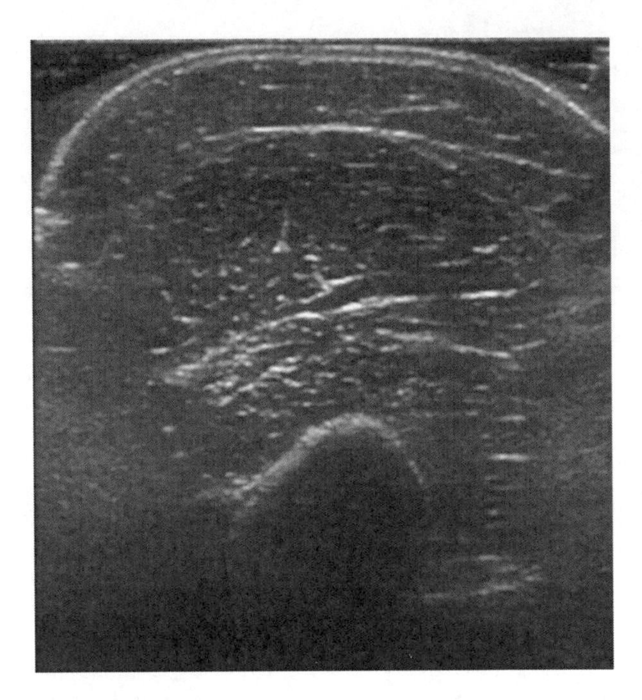

Fig. 9.2 Example of a transversal B-mode ultrasound image of a healthy biceps muscle

The last type of structural ultrasound scanning is M-mode scanning. In M-mode a rapid sequence of A- or B-mode scans is made to derive the position over time of the moving boundaries of (part of) an organ or a blood vessel.

9.3.3 Diagnostic Muscle Ultrasound and Image Analysis

Healthy muscles generate few echoes because they have a highly organized internal structure with few boundaries that could reflect sound. Thus, when scanned transversally, healthy muscles are very dark on an ultrasound image, with the exception of evenly spread small echoes that result from the normal *fibrous* and fatty (fibroadipose) structures of the connective tissue (perimysium) surrounding bundles of muscle fibers (fascicles) (see Fig. 9.2), giving the muscle a homogeneously speckled appearance. When scanned longitudinally, the same structures yield a striped pattern (see Fig. 9.3). The connective tissue around the muscle (epimysium) yields a stronger echo than the perimysium.

The presence of fibroadipose structures varies from muscle to muscle so that, e.g., triceps muscles yield smaller echoes than biceps muscles. In Fig. 9.2 the muscle is imaged at a 90° angle with respect to the bone. The appearance of the image will change when the insonation angle is changed, because part of the reflected sound will no longer be received by the transducer, resulting in lower echo intensity. Because bone yields the most intense echoes, the bone echo suffers most from a non-perpendicular insonation angle. Hence, the correct insonation angle can be

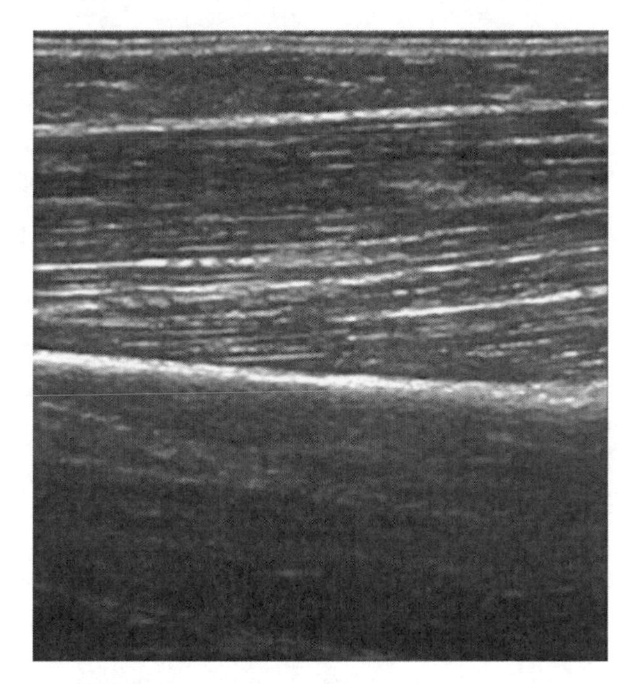

Fig. 9.3 Example of a longitudinal B-mode ultrasound image of the same healthy biceps muscle as in Fig. 9.2

achieved by changing the angle of the transducer until a maximally intense bone echo is obtained. In healthy subjects it is thus generally easy to distinguish blood vessels, (subcutaneous) fat, muscle, and bone, because of their different *echogenicity* (generally: blood vessel < fat < muscle < bone). Muscle thickness can also be obtained easily from the transversal ultrasound image by using the *calipers* available from the ultrasound equipment. Dynamic ultrasound imaging allows to visualize the thickening of the muscle belly in a transversal image and the shortening of fascicles in the transversal image during isometric contraction.

The simplest use of (B-mode) ultrasound in evaluating muscle disease is size measurement. Both muscle atrophy and hypertrophy can easily be determined as long as reference values are available. More interesting are the changes in echogenicity due to neuromuscular disorders. Depending on the disease, the muscle ultrasound image can be more echogenic, more heterogeneous, and the bone echo may disappear. These changes result from underlying changes in muscle composition. Due to increased fat and fibrous content and inflammations, the number of reflective boundaries in the muscle and thereby its echogenicity increases. At very high echogenicity sound cannot even penetrate through the entire muscle anymore, leading to a loss of bone echo. The changes in muscle composition further blur the distinction between connective tissue and muscle, making the image more homogeneously echogenic.

It has been known for a long time that more progressed muscle disease is accompanied by more homogeneously echogenic (whiter) ultrasound images and the increase in echogenicity had been expressed in semi-quantitative scales (such as that developed by Heckmatt et al. 1982: see Box 9.1).

Box 9.1 Heckmatt's Visual Grading of Ultrasound Image Echogenicity

Grade I	Normal
Grade II	Increased muscle echogenicity with distinct bone echo
Grade III	Marked increased muscle echogenicity with a reduced bone echo
Grade IV	Strongly increased muscle echogenicity with absent bone echo

However, subtle changes in echogenicity due to disease progression or treatment cannot be represented on such a scale. Furthermore, even though it is known that myopathies generally cause homogeneously increased echogenicity whereas *neuropathies* generally cause more heterogeneously increased echogenicity (see Fig. 9.4), quantitative assessment is required to reliably make this distinction.

Since an ultrasound image is a digital black-and-white image consisting of pixels with a value between 0 (pure black) and 255 (pure white) when 8-bit grey values are used, it is rather straightforward to determine muscle echogenicity (or density). By selecting the muscle and calculating the average grey value of the selection, a numeric value for echogenicity is obtained. A disadvantage is that this value strongly depends on equipment settings (such as the type of transducer, gain, time-gain compensation, focus, and contrast) and the actual muscle selection that is analyzed. Therefore, it is advised that each laboratory determines its own reference values using fixed settings (or a preset). Furthermore, since color values can change depending on settings and the format in which the image is stored, images should be calibrated before analysis. In our hospital the image is calibrated such that all grey values are present, knowing that the image background is pure black and the letters are pure white. Note that reference values for echogenicity should be obtained for different age groups as echogenicity decreases in the first years of life and then increases into old age (Maurits et al. 2004, Pillen 2009).

Question 9.2 What will change in an ultrasound image when time-gain compensation is switched off?

It is less straightforward to determine the heterogeneity (or inhomogeneity) of an ultrasound image and different methods can be used (see Box 9.2).

Question 9.3 When considering the images in Fig. 9.4, can you order them according to echogenicity and heterogeneity?

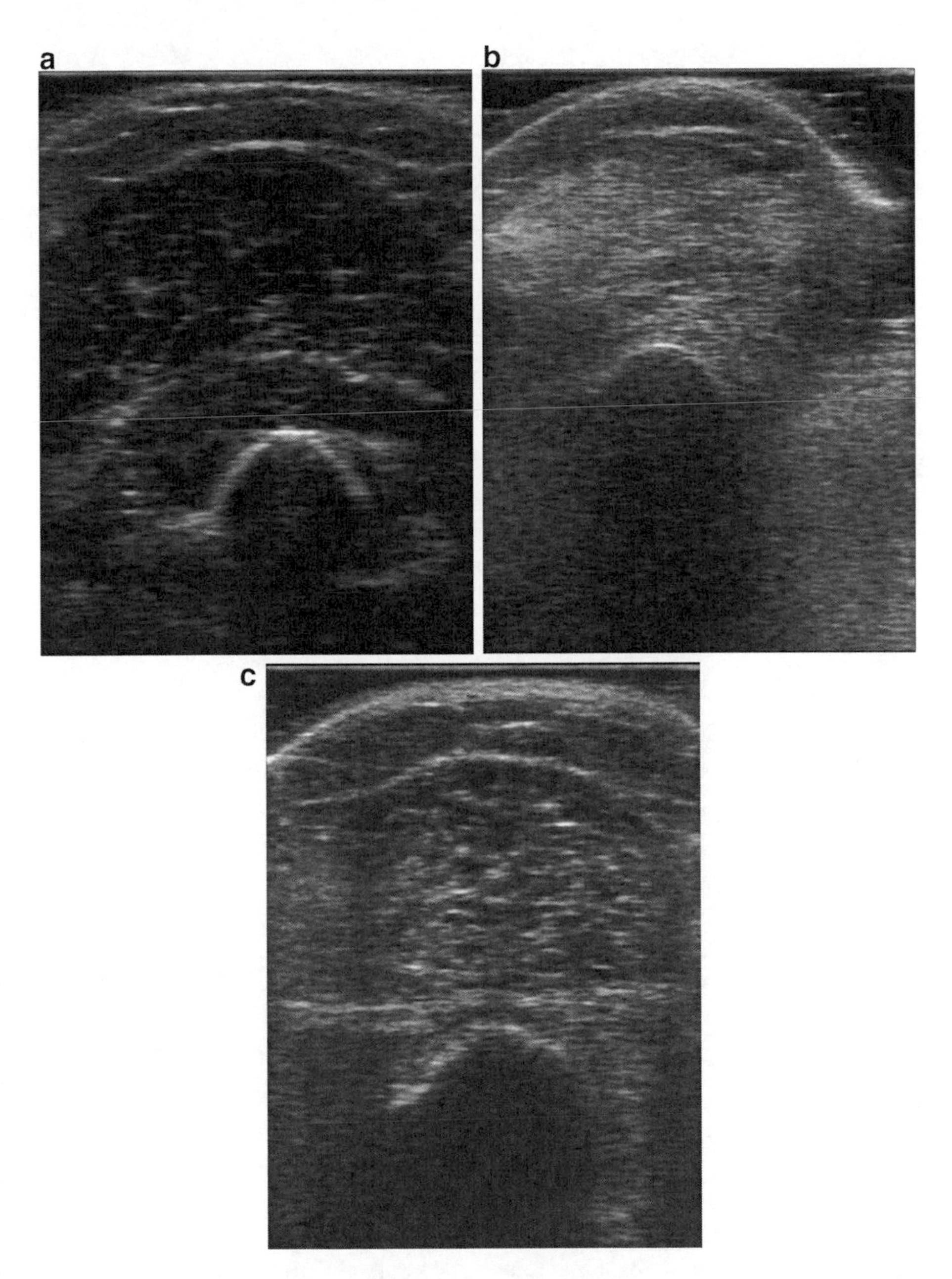

Fig. 9.4 Examples of transversal B-mode ultrasound images of (**a**) a healthy biceps muscle, (**b**) a myopathic muscle, and (**c**) a neuropathic muscle

Box 9.2 Quantitative Heterogeneity Analysis of Muscle Ultrasound Images

A heterogeneous ultrasound image is characterized by the presence of many echogenic patches (whiter areas) and thus by many boundaries between darker and lighter pixels. Therefore, heterogeneity can be calculated (Maurits et al. 2003) by first applying a Sobel filter, which extracts and enhances edges by expressing gradients (grey-value differences) between neighboring pixels in a grey value. This value is calculated from a 3 × 3 neighborhood of the pixel E as follows:

$$\sqrt{(X^2 + Y^2)}, \tag{1}$$

where

$$X = (C + 2F + I) - (A + 2D + G),$$

$$Y = (A + 2B + C) - (G + 2H + I) \tag{2}$$

and the neighborhood of the pixel E is arranged as:

$$\begin{matrix} ABC \\ DEF \\ GHI. \end{matrix} \tag{3}$$

Hence, if there is no difference between neighboring pixels, the grey value is set to 0 (black); if there is maximum difference, the grey value is set to 255 (white). After Sobel filter application, a maximally heterogeneous image would thus consist of black and white pixels alternating in a checkerboard manner. After the Sobel filter has been applied, white areas that are larger than two pixels (to avoid counting noise) are counted. The white areas are determined by all pixels with a value larger than a certain grey value, such that visible contrasts are selected (see Fig. 9.5).

To normalize for the size of the selected part of the ultrasound image, the heterogeneity is then calculated as the number of counted white areas divided by the area of the selection in cm^2. In this way, quantitatively high heterogeneity corresponds to visually high heterogeneity.

Alternative methods to quantify heterogeneity are numerous and here only a few are mentioned. A simple and straightforward method is to calculate properties of the distribution of grey values that provide measures of grey value variability such as the standard deviation or the *skewness*. A disadvantage of these measures is that they do not take the spatial distribution of the grey values into account and are less sensitive for differential diagnosis of neuromuscular disorders.

(continued)

Box 9.2 (continued)

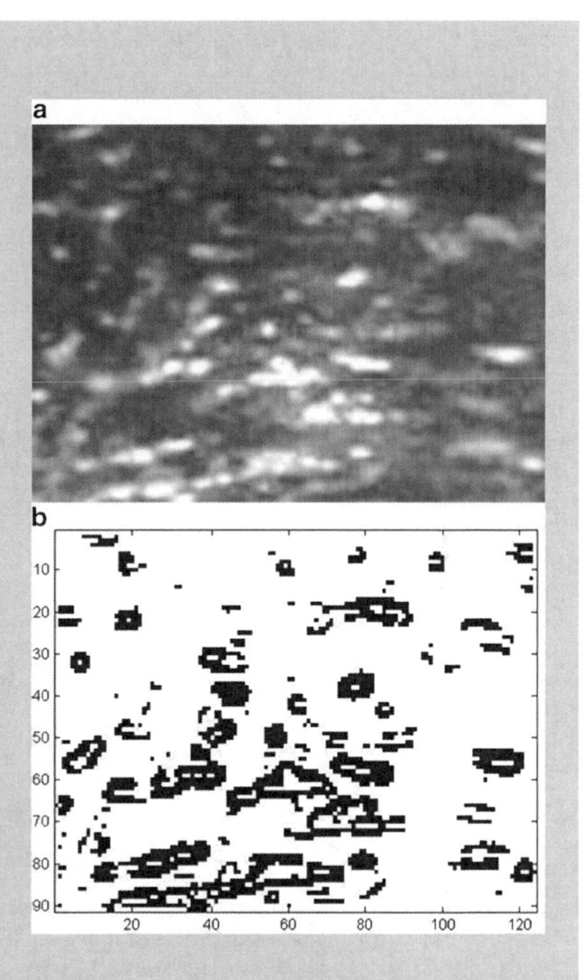

Fig. 9.5 Example of (**a**) selection of the biceps muscle ultrasound image of patient 2 and (**b**) the image after application of the Sobel filter and a grey value cut-off of 197 (for clarity the color scale has been reversed). All *white patches* in the original image selection have been well delineated by the Sobel filter

The Sobel filter method can also be generalized to larger neighboring areas: the larger the neighborhood, the larger the boundaries that will be recognized and emphasized. Other mathematical methods, such as spatial frequency analysis or *fractal* analysis, can also be of use.

Although muscle biopsy (often combined with genetic analysis to detect inherited diseases) is still the gold standard to establish the diagnosis in neuromuscular disease, the quantitative values for echogenicity and heterogeneity can help distinguish between neuropathies and myopathies in a noninvasive and patient-friendly manner.

9.3.4 Tips and Tricks in Muscle Ultrasound Imaging and Analysis

9.3.4.1 Optimizing Image Quality

Higher frequencies are more attenuated than lower frequencies thus providing less penetration depth, while at the same time higher frequencies also provide higher spatial resolution (see Sect. 9.3.2). This means that fine-grained structures can only be visualized when they are located superficially and explains why ultrasound imaging of small structures (such as nerves, see Sect. 9.5.1) is performed with high frequency probes (10–14 MHz) and only for superficial locations. For ultrasound imaging of muscles, a 10–14 MHz transducer usually offers sufficient penetration depth and a reasonably high spatial resolution in children and most adults, whereas 7.5–10 MHz transducers may be needed in adults who have thicker muscles. The focus point should be chosen such that it is in the middle of the area of interest.

To further optimize the image quality, it is important to use a sufficient amount of ultrasound gel without air bubbles, because air will attenuate the ultrasound waves and generate shadows in the image. Gel also aids in getting a good contact between the transducer and the skin without having to exert pressure. Pressure would deform the muscle, thereby making estimates of muscle area or thickness unreliable. For the same reason, muscles should only be imaged when they are relaxed; muscle contraction will increase muscle thickness at the position of the muscle belly.

9.3.4.2 Obtaining Reproducible and Reliable Muscle Ultrasound Images

When ultrasound images are compared between patients or within a patient over time, it is important that differences do not result from the repeated measurement itself, but only from actual changes in muscle composition. As mentioned in Sect. 9.3.3, a first step in improving reliability is to insonate the muscle at the same angle for every measurement, which can be achieved by maximizing the bone echo. Second, the position of the transducer should be the same between measurements and this can be achieved by defining the position with respect to bony external landmarks, such as the lateral epicondyle of the humerus or femur (bony protrusion at the outside of the elbow or knee). Finally, the patient should always be in the same position when measurements are taken.

9.3.4.3 Selection of Area of Interest for Analysis

Knowing that some structures in and around the muscle, such as the epimysium and perimysium, have higher echogenicity than the muscle fibers, the quantitative analysis of a muscle ultrasound image will depend strongly on what part of the image is selected for analysis. Therefore, the procedure for selecting the area of interest should also be standardized to maximize reproducibility. Several approaches exist: some laboratories select a smaller (rectangular or ellipsoidal) part of the innermost part of the muscle, so that the perimysium does not influence the echogenicity calculation. Other laboratories select the entire muscle, following its contours as indicated by the hyperechogenicity of the perimysium. The latter approach will yield higher echogenicity (mean grey values) but consistently so, thereby still providing reliable and reproducible values within the laboratory. These differences between laboratories again emphasize that reference values should be obtained for each laboratory independently.

9.4 Ultrasound Image Analysis in Individual Patients

Patient 1

To confirm or refute the diagnosis of distal myopathy, this patient with symptoms of a distal myopathy was investigated using muscle ultrasound imaging. His m. biceps, m. quadriceps, m. tibialis anterior, and m. gastrocnemius were studied bilaterally, both in transversal and longitudinal orientation using a 14 MHz transducer and a focus at 2–3 cm depth. Both the thickness of the muscles and their echogenicity were evaluated. All muscles had normal thickness and *fasciculations* were not observed. Quantitative values for some muscles are indicated in Table 9.1.

All muscle images were echogenic. In addition, the bone echo had almost disappeared for the image of the m. tibialis anterior. The gastrocnemius muscle was strongly and homogeneously echogenic, as further emphasized by an absent bone echo. The heterogeneity was also increased in all muscles (see Box 9.1), but not in all muscles as strongly as in most neuropathies (Fig. 9.6). In summary, myopathic changes were found in all muscles but particularly in the m. gastrocnemius and to a lesser extent in the m. tibialis anterior, suggesting a distal myopathy. Subsequently, a biopsy was taken from the quadriceps muscle, which showed abnormalities consistent with Miyoshi distal myopathy. This is a hereditary disease leading to muscle weakness in the lower legs, inducing an inability to walk on the toes. In a later phase the upper leg muscles (and arm muscles) may become involved leading to problems with getting up from the floor and walking stairs.

Table 9.1 Quantitatively obtained values for echogenicity and heterogenicity for some muscles in patient 1

Muscle	Side	Echogenicity (mean grey value in selection)	Heterogeneity (# white areas per cm^2 after Sobel filtering)
Biceps	Left	144 {57 ± 20}	24.5 {4.6 ± 2.9}
Tibialis anterior	Left	115	43.3
Gastrocnemius	Left	135	34.4
Gastrocnemius	Right	122	22.3

Normal (mean ± standard deviation) values for his age are indicated between curly brackets when available (see Maurits et al. 2003). Note however that these normal values were obtained with older equipment with lower resolution, which particularly yields lower values for heterogeneity

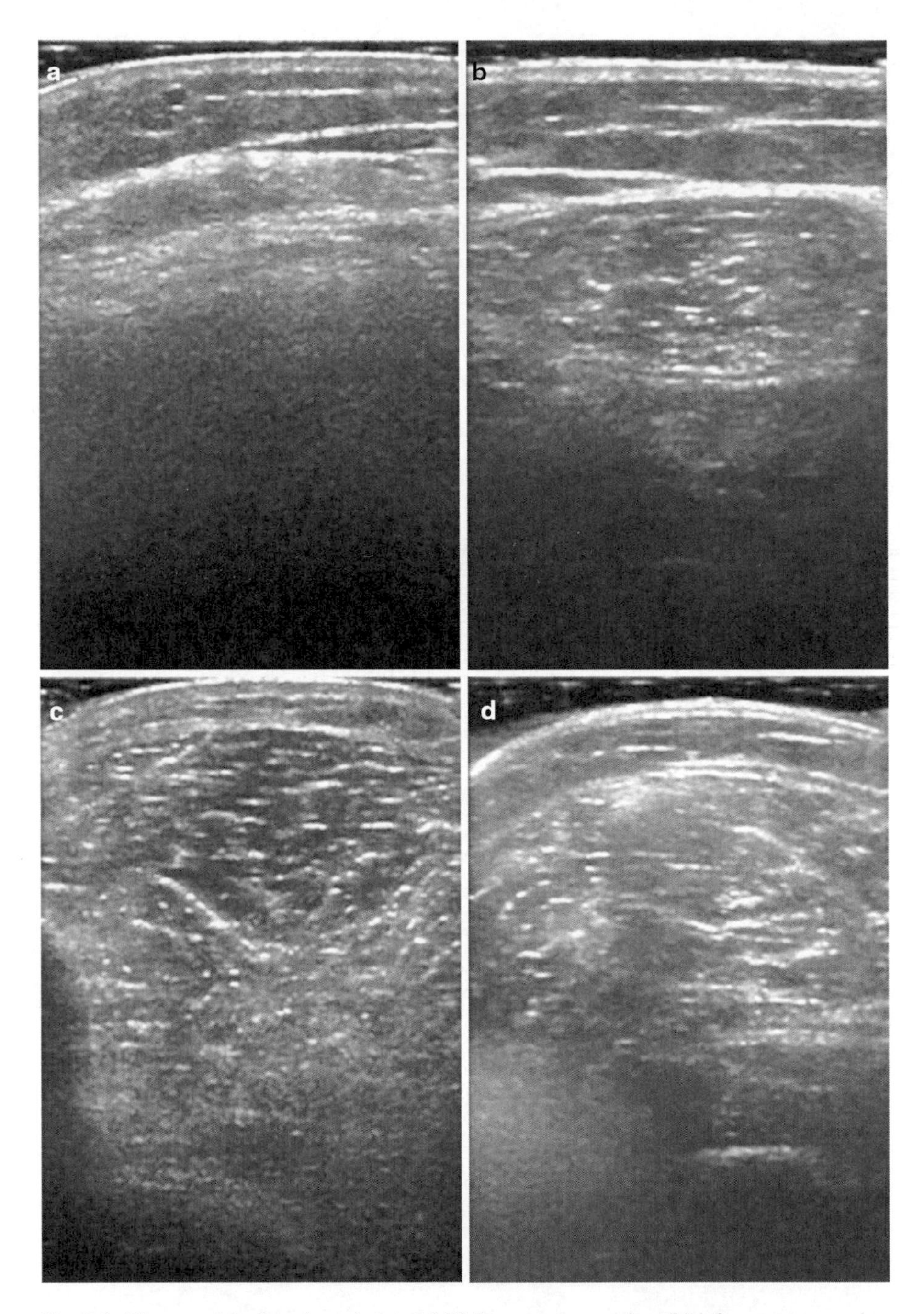

Fig. 9.6 Ultrasound findings in patient 1. (**a**) Right m. gastrocnemius, (**b**) left m. gastrocnemius, (**c**) left m. tibialis anterior, (**d**) left m. biceps. Note that the total image depth was 5 cm for images (**a**) and (**c**) and 6 cm for images (**b**) and (**d**). All images were scanned transversally

Table 9.2 Quantitatively obtained values for echogenicity and heterogenicity for some muscles in patient 2

Muscle	Side	Echogenicity (mean grey value in selection)	Heterogeneity (# white areas per cm^2 after Sobel filtering)
Biceps	Left	106 {30 ± 8}	62.1 {0}
Quadriceps	Left	112 {27 ± 7}	58.6 {0}
Tibialis anterior	Left	91	40.0

Normal (mean ± standard deviation) values for her age are indicated between curly brackets when available (see Maurits et al. 2004). Note however that these normal values were obtained with older equipment with lower resolution, which particularly yields lower values for heterogeneity

Patient 2

To investigate whether this patient with progressive muscle weakness indeed suffers from a problem in multiple nerves, she underwent electrophysiological tests, force recordings using a dynamometer, and muscle ultrasound imaging (14 MHz transducer, focus at 2–3 cm depth). The latter showed that the biceps muscle thickness was normal (2.12 cm thickness; normal 1.60 ± 0.20 cm) and the quadriceps muscle was atrophic (2.06 cm thickness; normal 3.20 ± 0.37 cm). Muscle thickness was compared to weight- and gender-corrected reference values obtained in our own hospital (Maurits et al. 2004). Quantitative values for some muscles are indicated in Table 9.2.

All muscles showed moderate irregularly distributed hyperechogenicity, i.e., neurogenic changes (Fig. 9.7). Note that the values for heterogeneity in Table 9.2 are higher than for patient 1 who was diagnosed with a myopathy. In combination with the force measurements that showed a severe *tetraparesis*, in which the arms were more affected than the legs, the electrophysiological investigations that indicated that this patient suffers from a *demyelinating* polyneuropathy, and the recent onset of symptoms, it was concluded that she suffered from an acquired chronic demyelinating inflammatory polyneuropathy (CDIP) with muscle weakness. She was treated with *immunoglobins* which decreased her symptoms.

Patient 3

To decide in patients as young as patient 3 whether they are suffering from a neurogenic or (hereditary) myopathic disorder, ultrasound imaging is a patient-friendly solution: needle electromyography can often not be performed in very young children. In this patient muscle ultrasound imaging was performed at 14 MHz and at a depth of 2–3 cm. Most muscles had normal thickness for her weight, except for the quadriceps muscle which was atrophic (1.67 cm for the m. rectus femoris, one of the four muscles composing the m. quadriceps. Normally this muscle would be 2.39 ± 0.25 cm for her weight; see Maurits et al. 2004). Fasciculations were not observed. Quantitative values for some muscles are indicated in Table 9.3.

All muscles showed increased echogenicity and particularly increased heterogeneity (Fig. 9.8). The muscle ultrasound findings thus indicated that patient 3 most likely suffers from a neuropathy. Genetic tests confirmed that she has spinal muscular

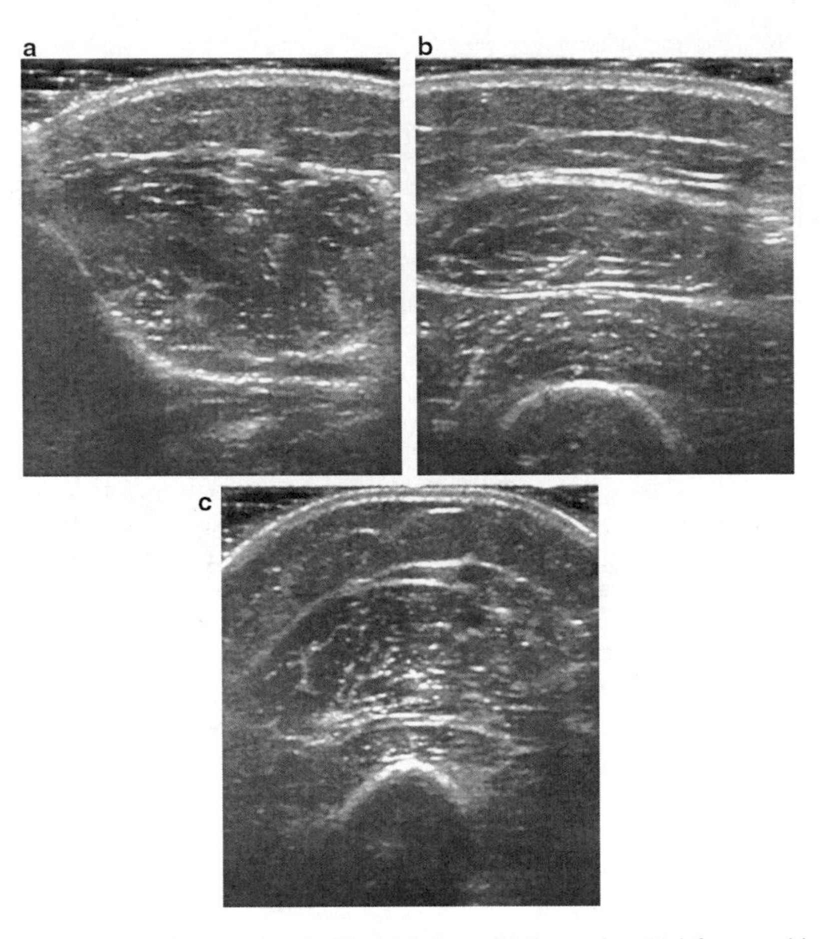

Fig. 9.7 Ultrasound findings in patient 2. (**a**) Left m. tibialis anterior, (**b**) left m. quadriceps, (**c**) left m. biceps. Note that the total image depth was 4 cm for all (transversally scanned) images

Table 9.3 Quantitatively obtained values for echogenicity and heterogenicity for some muscles in patient 3

Muscle	Side	Echogenicity (mean grey value in selection)	Heterogeneity (# white areas per cm^2 after Sobel filtering)
Biceps	Left	86 {30 ± 8}	59.1 {0}
Quadriceps	Left	139 {27 ± 7}	55.9 {0}
Tibialis anterior	Left	74	87.4

Normal (mean ± standard deviation) values for her age are indicated between curly brackets when available (see Maurits et al. 2004). Note however that these normal values were obtained with older equipment with lower resolution, which particularly yields lower values for heterogeneity

atrophy (SMA), a hereditary disease of the *motor neurons* leading to progressive muscle atrophy and weakness. This disease cannot be treated and the patient could only be referred to a rehabilitation clinic to prevent complications of the disease in later life (such as pneumonia due to swallowing problems) and for further improvement of her quality of life.

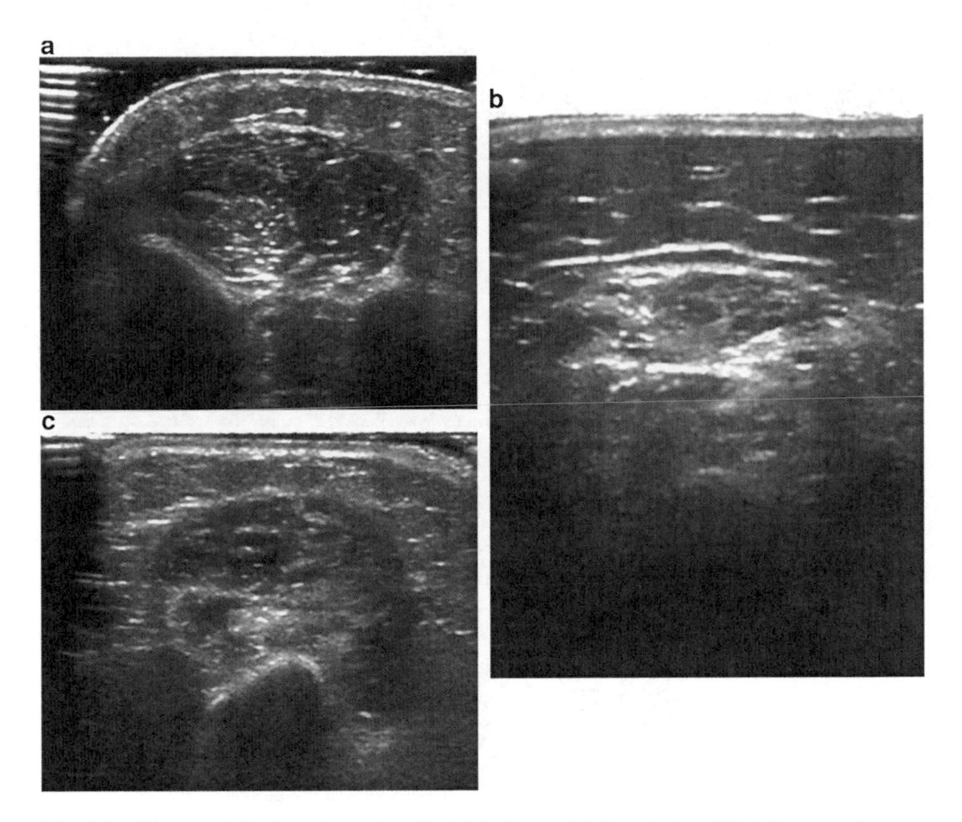

Fig. 9.8 Ultrasound findings in patient 3. (**a**) Left m. tibialis anterior, (**b**) left m. quadriceps, (**c**) left m. biceps. Note that the total image depth was 3 cm for images (**a**) and (**c**) and 5 cm for image (**b**). All images were scanned transversally

9.5 Other Applications of Ultrasound Imaging in Neurology

9.5.1 Nerve Ultrasound

Although evaluation of nerve function through nerve conduction studies has been performed for decades, imaging of peripheral nerves using ultrasound is a relatively recent endeavor, because it can only be achieved with high resolution transducers (preferably >12 MHz). Only when the nerve is located deeper (such as the n. ischiadicus in the upper leg), lower frequency transducers will be used. Nerve ultrasound imaging allows to evaluate the result of nerve decompression surgery (such as in carpal tunnel syndrome, see Chap. 2), to evaluate traumatic damage to nerves and to determine the exact location of nerve entrapment or *neuromas*. As such it can be very helpful in improving diagnosis and treatment of disorders of the peripheral nerves.

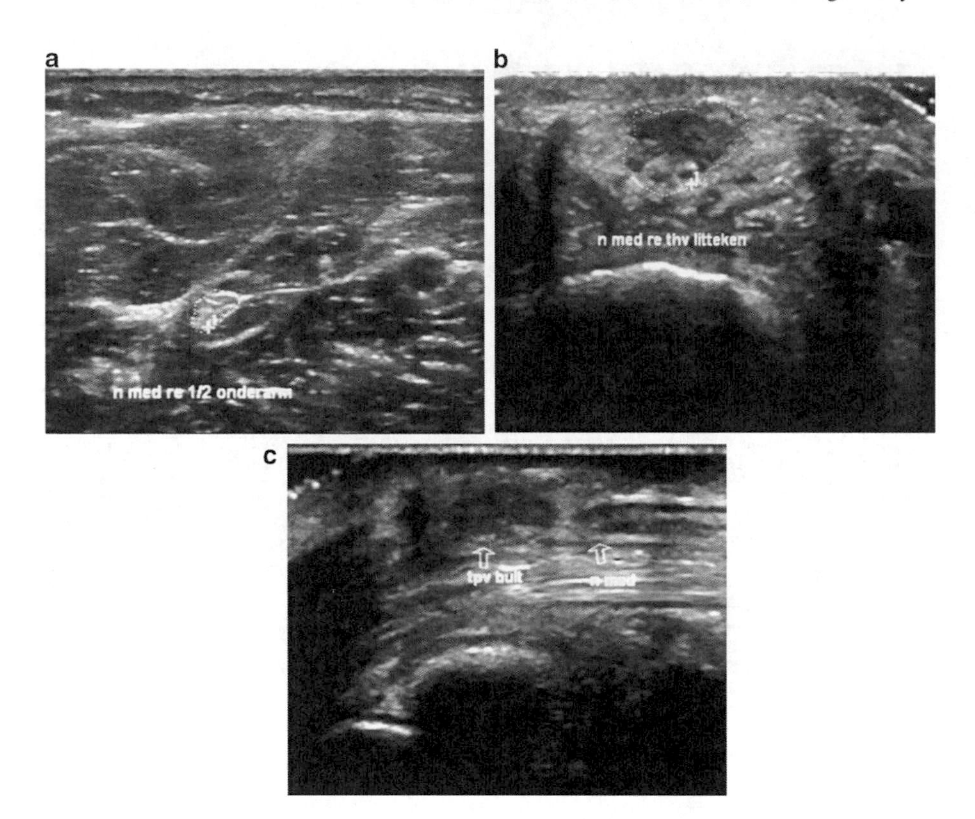

Fig. 9.9 This patient suffered from a traumatic injury to the right wrist 26 years ago for which he underwent surgery. Now he complains about a swelling in his right wrist, and numbness in his second and third finger tips. Electrophysiological tests show that the right median nerve only functions partially; sensory nerve conduction velocities could not be established. Nerve ultrasound shows (**a**) a normal cross-sectional area (0.08 cm^2) of the right median nerve in the lower arm (transversal image), (**b**) an increased cross-sectional area (0.28 cm^2) of the right median nerve at the level of the scar (transversal image), and (**c**) a neuroma of the right median nerve at the location of the swelling (longitudinal image). In images (**a**) and (**b**), the median nerve has been delineated by a *dotted line*. In image (**c**), the *left arrow* indicates the neuroma, the *right arrow* the normal median nerve

In contrast to ultrasound imaging of muscles, the longitudinal orientation of the transducer is generally more helpful in nerve ultrasound imaging. It allows to follow the nerve over a longer trajectory, enabling the identification and measurement of local swelling. The *epineurium* and the nerve *fascicles* are more easily assessed on a transversal image. On a longitudinal image, the nerve is characterized by a fine pattern of parallel lines; on a transversal image, the nerve is oval and has a honeycomb-like internal structure. Nerves can be distinguished from tendons that have a similar structure by the more constant echogenicity along their trajectory and by their immobility: tendons (particularly in the lower arm) can be easily identified on longitudinal images when the fingers are moved. In Figs. 9.9 and 9.10 some examples of nerve ultrasound are given.

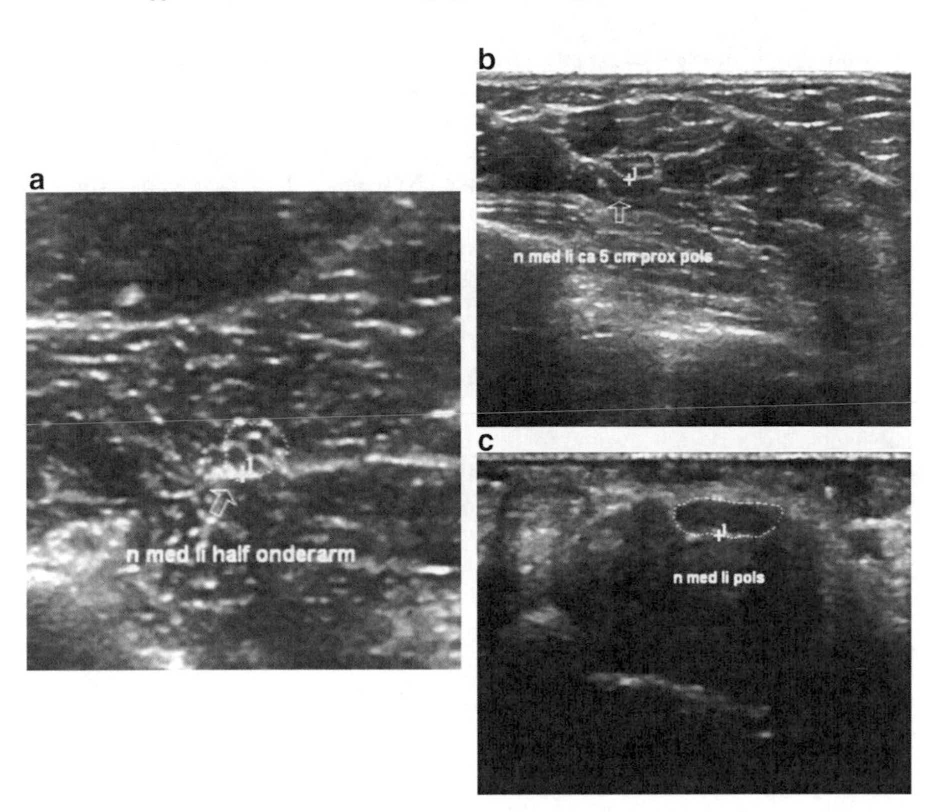

Fig. 9.10 This patient suffers from tingling feelings in both hands. Electrophysiological tests show that she suffers from a carpal tunnel syndrome in her left wrist and has borderline normal conduction velocities in the right median nerve. These findings are confirmed by nerve ultrasound. In her left arm, the median nerve is swollen, compressed, and hypoechogenic at the wrist. In her right arm the pathological changes are present as well, but less severe. (**a**) Normal cross-sectional area (0.07 cm^2) of the left median nerve in the lower arm and (**b**) 5 cm proximal to the wrist. (**c**) Increased cross-sectional area (0.12 cm^2) of the right median nerve at the wrist. All images were taken transversally. In all images the median nerve has been delineated by a *dotted line*

9.5.2 Early Diagnosis of Parkinson's Disease

An interesting recent application of ultrasound imaging is transcranial ultrasound of deep brain structures for the early and differential diagnosis of Parkinson's disease. Early diagnosis of Parkinson's disease is important, because for *neuroprotective* therapies to be of use (when they become available), they must start before the motor symptoms of the disease have become prominent, i.e., before *dopaminergic* neuronal degeneration is already severe. Hyperechogenicity of the *substantia nigra* has been shown to be a sensitive marker for indicating vulnerability for impairment of this brain structure and later development of Parkinson's disease (Berg 2009). Because ultrasound is cheap, noninvasive and patient-friendly, it could be used as a screening tool, in contrast to nuclear imaging (positron emission tomography or PET), although the latter has a good diagnostic yield for Parkinson's disease.

By scanning through the temporal window at a specific angle using a 1.5–3 MHz transducer, the butterfly-shaped hypoechogenic brainstem can be viewed on a B-mode image. The substantia nigra can then be seen as a hyperechogenic structure within the brainstem. It is not possible to reliably quantify the echogenicity of this structure itself, but the area of hyperechogenicity can be determined. Hyperechogenicity of the substantia nigra is then defined by an increased area compared to healthy subjects. Hyperechogenicity of the substantia nigra is usually found most prominently contralateral to the most affected side and is not related to disease stage, but rather stable over the course of the disease, indicating its suitability as a marker of early Parkinson's disease. Additionally, hyperechogenicity of the substantia nigra is very specific for idiopathic Parkinson's disease and is usually not found in patients suffering from atypical Parkinsonian syndromes such as multiple system atrophy (MSA) or progressive supranuclear palsy (PSP), thereby also aiding in the differential diagnosis of these diseases.

9.6 Answers to Questions

Answer 9.1

According to the formula relating the speed of sound, the wave length, and the frequency, a sound wave with a frequency of 10 MHz and a speed of 2 km/s has a wavelength of 2.000/10.000.000 = 2/10.000 = 0.0002 m = 0.2 mm. Thus, the structures that can be visualized need to be larger than (approximately) 0.2 mm or the sound will not be reflected.

Answer 9.2

Time-gain compensation amplifies echoes that are received from deeper structures. When it is switched off deeper structures will thus appear darker in the image.

Answer 9.3

Images in order of increasing echogenicity (mean grey values): healthy, neuropathic, and myopathic. Images in order of increasing heterogeneity (see Box 9.2): healthy, myopathic, and neuropathic.

Glossary

Attenuation Gradual loss of intensity of a signal
Axial Toward the central axis of the body
Caliper Device used to measure distance between two points
Congenital Existing from birth or even before
Creatine kinase An enzyme important for energy storage and transport in the brain, muscles, eyes, and ears
Demyelinating Leading to loss of the insulating myelin layer around nerves
Dopaminergic Related to the neurotransmitter dopamine. Dopamine producing neurons in the substantia nigra are lost in Parkinson's disease
Echogenicity Ability to reflect a sound wave
Epineurium Outermost layer of connective tissue surrounding a nerve
Fascicle Larger bundle of nerve fibers
Fasciculations Involuntary spontaneous contraction of fascicles, visible under the skin from the outside
Fibrillation Spontaneous pathological muscle fiber contraction that cannot be observed under the skin from the outside
Fibrous Containing fibers
Fractal A geometric shape that repeats itself on different spatial scales
Immunoglobins Antibodies
Longitudinal Along the longest axis (of the muscle) or in parallel with the direction of movement (of waves)
Motor neuron Neurons located in the central nervous system with long axons, controlling muscles
Myopathy Disease of the muscles
Neurogenic Originating from the nerves
Neuroma Growth or tumor of nerve tissue
Neuropathy Disease of the nerves
Neuroprotective Protection of neurons from cell death or degeneration
Skewness Measure of asymmetry of a distribution
Substantia nigra Brain structure located in the midbrain
Tetraparesis Paralysis in all limbs
Transducer Device that transforms one form of energy into another: here electrical energy is transformed into sound energy
Transversal Perpendicular to the longest axis (of the muscle) or perpendicular to the direction of movement (of waves)
Trendelenburg gait Abnormal gait caused by weakness of hip muscles

References

Online Sources of Information

http://en.wikipedia.org/wiki/Ultrasound. General overview of ultrasound applications in biomedicine, industry and the animal world

http://en.wikipedia.org/wiki/Medical_ultrasonography. Introduction to ultrasound from a medical perspective

http://en.wikipedia.org/wiki/Edge_detection. Overview of edge detection methods including the Sobel filter

Books

Pillen S (2009) Quantitative muscle ultrasound in childhood neuromuscular disorders. PhD thesis. Available on http://dare.ubn.kun.nl/bitstream/2066/74405/1/74405.pdf

Papers

Berg D (2009) Transcranial ultrasound as a risk marker for Parkinson's disease. Mov Disord 24: S677–S683

Heckmatt JZ, Leeman S, Dubowitz V (1982) Ultrasound imaging in the diagnosis of muscle disease. J Pediatr 101:656–660

Maurits NM, Bollen AE, Windhausen F, de Jager AEJ, van der Hoeven JH (2003) Quantitative muscle ultrasound analysis in healthy adults: weight- and age-related normal values. Ultrasound Med Biol 29:215–225

Maurits NM, Beenakker EAC, Fock JM, van Schaik DEC, van der Hoeven JH (2004) Muscle ultrasound in children: normal values and application to neuromuscular disorders. Ultrasound Med Biol 30:1017–1027

Pillen S, Arts IMP, Zwarts MJ (2008) Muscle ultrasound in neuromuscular disorders. Muscle Nerve 37:679–693

Walker FO, Cartwright MS, Wiesler ER, Caress J (2004) Ultrasound of nerve and muscle. Clin Neurophysiol 115:495–507

Chapter 10
Cerebrovascular Disease, Ultrasound, and Hemodynamical Flow Parameters

After reading this chapter you should:

- Know why ultrasound imaging is useful in the context of cerebrovascular disease
- Understand what factors influence (arterial) flow profiles
- Know how the Doppler effect can be used to determine blood flow velocities
- Be able to discuss the (dis)advantages of using high ultrasound frequencies
- Be able to discuss the (dis)advantages of using high pulse repetition frequencies in pulsed wave Doppler techniques
- Understand how arterial flow parameters can be used to diagnose stenoses

10.1 Patient Cases

Patient 1

This male 65-year-old patient suddenly collapsed. He had been standing for some time, when he unexpectedly started to talk incoherently and fell down. He was unconscious for a few minutes, during which he snored. There were no involuntary movements, nor did the patient bite his tongue. He was incontinent for urine and feces, however. When he came to, he had a grayish color. One week later, he suffered from a minute-long loss of motor function in his left little and ring fingers, but without sensory complaints. Another 3 days later, he had another 1 min episode of loss of motor function in his right lower arm, again without sensory complaints. He never experienced any headache or loss of motor function in his facial muscles. He never suffered from these complaints before. In addition he suffers from *hypertension*, *hypercholesterolemia*, and he has chronic obstructive pulmonary disease (COPD).

N. Maurits, *From Neurology to Methodology and Back: An Introduction to Clinical Neuroengineering*, DOI 10.1007/978-1-4614-1132-1_10,

Patient 2
This female 55-year-old patient suddenly experienced severe headache, which later expanded into neck pain. Simultaneously, she felt severely nauseous, vomited and partially lost her sense of balance. She was sent to the neurologist immediately. Her complaints did not deteriorate during transport.

10.2 Ultrasound Imaging of Extracranial and Intracranial Arteries

The complaints that patient 1 suffers from are not straightforward to diagnose. His first episode seems to be most consistent with an epileptic attack (see Chap. 4), particularly since he had some tell-tale signs such as the tongue bite and incontinence. Yet, the other two episodes are not epileptic in nature and seem to be more consistent with a momentary loss of function in the brain (most likely the motor cortex) which may be caused by a temporary obstruction of blood flow in the supplying arteries. This type of event is referred to as a transient ischemic attack or TIA. Yet, the complaints are not typical for a TIA either, because they were very short-lived, occurred on two sides of the body, and involved a very small part of the body only. Since a TIA is often caused by an *embolus* originating from an *atherosclerotic plaque* in one of the *carotid arteries*, these arteries need to be examined first to check whether the complaints that patient 1 suffered from could be caused by TIAs.

The symptoms that patient 2 shows are rather typical for a subarachnoidal hemorrhage (SAH), a bleeding into the space between the *arachnoid membrane* and the *pia mater* surrounding the brain. An SAH is characterized by severe headache with a sudden onset. In addition, vomiting and seizures can be present. Patients may be confused, have a lower level of consciousness, or may even be in a coma. A few hours after onset of the SAH neck stiffness may be perceived. In most cases of spontaneous SAH the cause is a ruptured aneurysm, a weak spot in one of the intracranial arteries. Generally, a computed tomography (CT) scan is made to exclude or confirm bleeding. If an SAH is indeed confirmed, a CT *angiogram* may be made to determine the origin of the bleeding. A serious complication of SAH is vasospasm, a condition in which the arteries in the brain contract, most probably in reaction to the presence of the excess blood, and changes in the local calcium concentration. The danger of this condition is that the arteries constrict so much that the brain becomes oxygen deprived and a secondary stroke occurs. It is important to detect the development of vasospasm as soon as possible, so that adequate treatment can be given.

A noninvasive, painless, and fast method to investigate the structure and function of both the extracranial (as relevant for TIA as well as for major stroke) and intracranial arteries (as relevant for SAH) is ultrasonography or ultrasound imaging. In the previous chapter, structural ultrasound imaging was introduced. However, ultrasound investigations allow to image anatomical structures as well as to determine blood velocities. When both functions are performed at the same time, the technique is referred to as duplex ultrasound. Since the Doppler effect (see Sect. 10.3.3) is used to determine blood velocities, the latter technique is often referred to as Doppler ultrasound.

Structural ultrasound allows to visualize the (relative) position of the main arteries in the neck (common, internal and external carotid arteries, see Fig. 10.1) and any abnormalities in these arteries, such as thickening of the arterial wall – which may signify the onset of atherosclerotic changes – or the presence and (to some extent) the composition of plaques.

When duplex ultrasound is used, the blood flow can be observed together with any structural changes or abnormalities. An example of duplex ultrasound in the carotid arteries is given in Fig. 10.2.

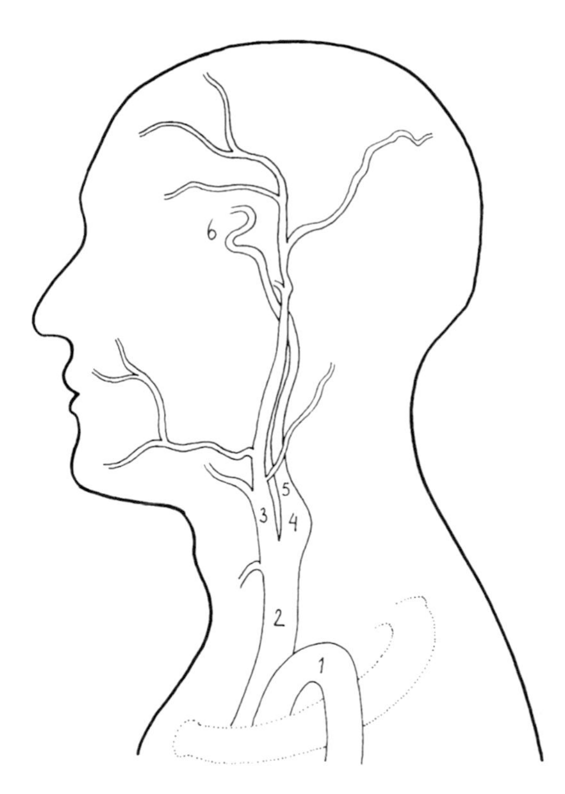

Fig. 10.1 Schematic of the anatomy of the carotid arteries. *1* Subclavian artery; *2* common carotid artery; *3* external carotid artery (recognized by its many branches); *4* bulbus (widening in the internal carotid artery); *5* internal carotid artery; *6* carotid siphon just before the internal carotid artery connects to the arteries of the circle of Willis

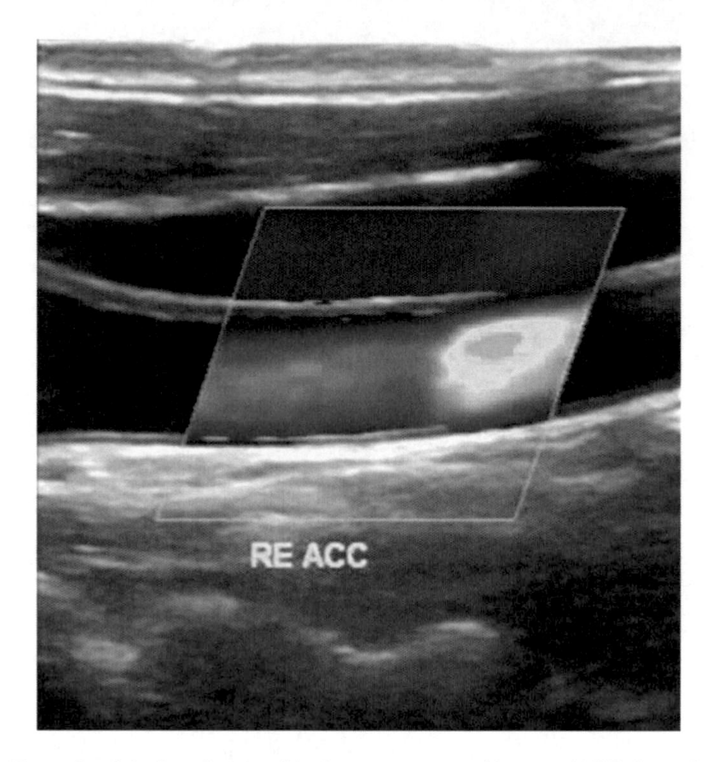

Fig. 10.2 Example of duplex ultrasound in the common carotid artery (ACC; *hot colors*) and the jugular vein (*cold colors*)

10.3 Hemodynamics and Ultrasound Imaging

10.3.1 Normal Flow

To be able to interpret duplex ultrasound results, it is important to have a basic understanding of the behavior of blood flow and the factors that influence it, i.e., of hemodynamics. Blood flow in the human body can be considered as flow in a closed circuit of tubes, in which the blood flows from areas with high pressure (arterial part) to areas with low pressure (venous part). The heart is the pump that maintains the necessary pressure difference. The pulsating flow determined by the beating heart is transformed in a more continuous flow – a necessity for adequate organ functioning – by the elastic arterial walls. The strength of the blood flow depends on both heart function and (peripheral) resistance brought about by the smallest end branches of the arterial tree (arterioles) and the capillaries, where nutrients in the blood are exchanged with the organ tissue. The formal relation between blood flow, pressure difference, and resistance is given by Poiseuille's law (Box 10.1).

Box 10.1 Poiseuille's Law and Bernoulli's Principle

Poiseuille's law is an analog for Ohm's law in fluid dynamics given by:

$$\Delta P = Q \cdot R \tag{1}$$

in which ΔP is the pressure difference over the vessel, Q the flow (the amount of displaced blood per second), and R the resistance given by:

$$R = \frac{8\eta l}{\pi r^4} \tag{2}$$

Here r is the vessel radius, l the vessel length, and η the blood *viscosity*. This equation thus reflects that it is easier for blood to flow through shorter and wider vessels (that have smaller resistance). In blood, viscosity is mainly determined by the hematocrit, the proportion of red blood cells (normally ~40%). In *anemia*, viscosity and thus resistance are lower.

At constant pressure and viscosity, Poiseuille's law states that blood flow will increase by a factor $2^4 = 16$ when the radius of a vessel doubles. Arterioles can thus quickly adapt the blood supply to an organ when required by increasing their diameter.

In an incompressible fluid, the total amount of passing fluid, which is equal to the flow velocity multiplied by the lumen area, must be the same at any cut-plane in the vessel. This implies that the average flow velocity in such a plane must be inversely related to the local lumen, or that the flow velocity doubles when the lumen is halved.

An increase in flow velocity due to a local narrowing in the vessel also influences the local pressure as a consequence of Bernoulli's principle which formalizes that an increase in flow velocity occurs simultaneously with a decrease in pressure by stating that the total energy along a streamline is constant for an incompressible, *inviscid*, time-independent flow:

$$\frac{v^2}{2} + gh + \frac{p}{\rho} = C \tag{3}$$

Here v is the flow velocity of a point on the streamline, g is the gravitational constant, h is the height of the point above a reference plane, p is the pressure at the point, and ρ is the fluid density.

The pattern of blood flow is largely determined by the local anatomy. In a straight vessel blood flow is highly regular and has a so-called laminar or layered pattern. In a laminar flow profile, the speed is highest at the center of the blood vessel and drops to zero at the wall in a parabolic manner (Fig. 10.3 top). The parabolic profile is induced by the friction between particles in the blood. If the

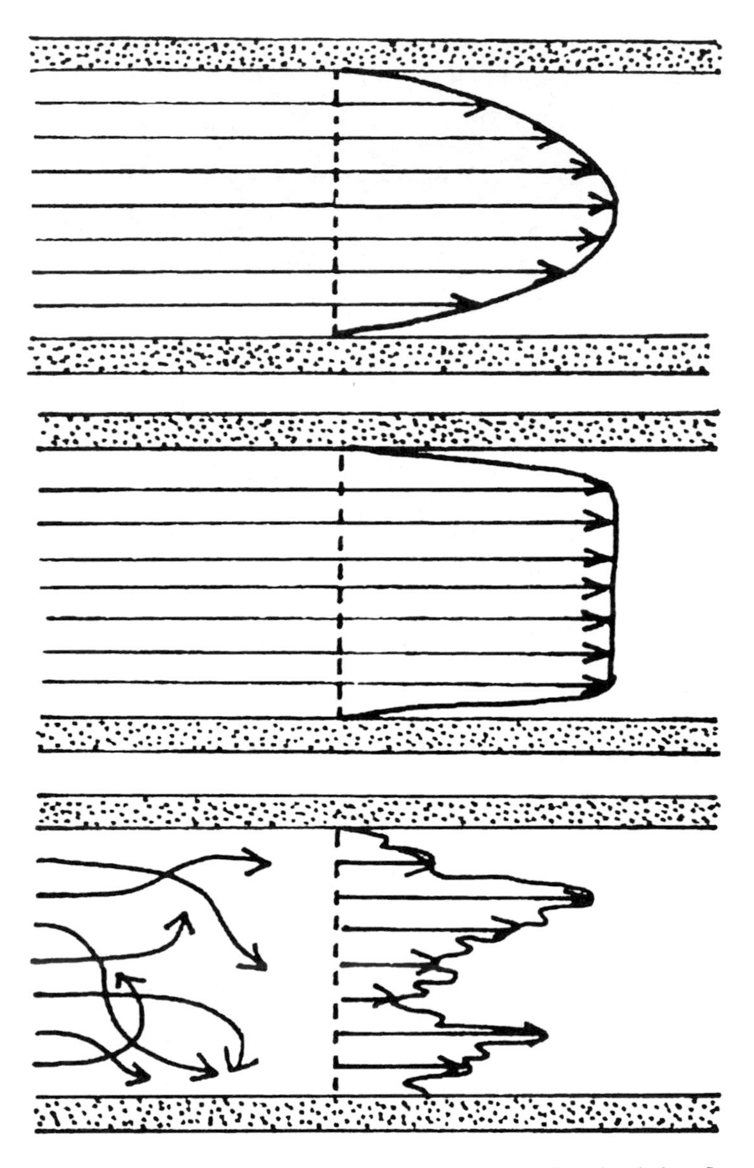

Fig. 10.3 From *top* to *bottom*: laminar (parabolic and plug-shaped) and turbulent flow profiles. The *arrows* indicate flow lines

friction is less, the speed is maximal in the largest part of the *lumen* and a plug-shaped flow profile (Fig. 10.3 middle) results. Near bifurcations and curves or when the vessel widens or narrows the character of the flow pattern changes and the laminar flow profile can become distorted in shape and even transform into a turbulent flow pattern which is characterized by irregularity, inpredictability, and chaos (Fig. 10.3 bottom).

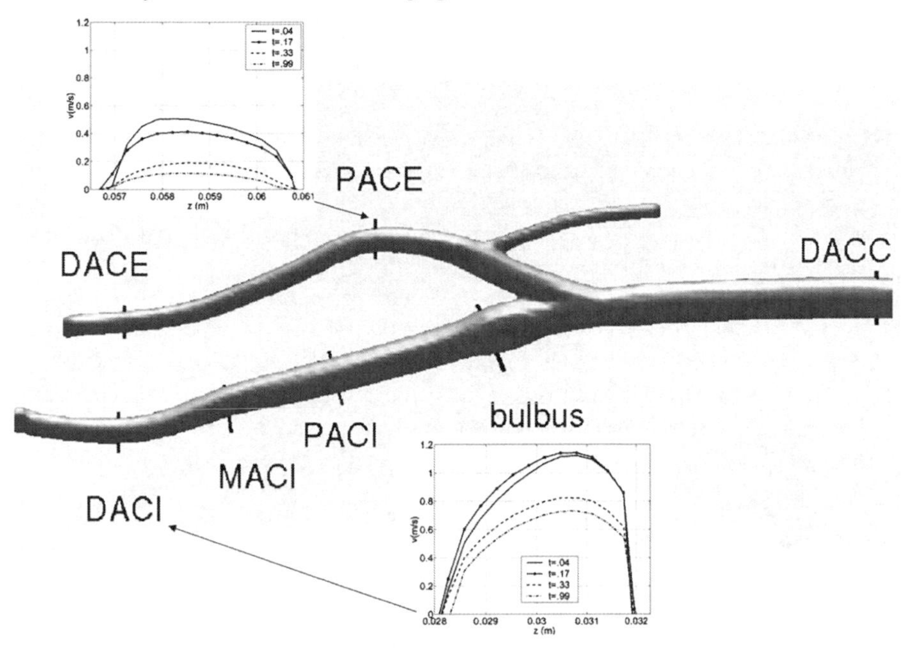

Fig. 10.4 Example of calculated flow profiles in a model of the carotid artery. *D* distal; *P* proximal; *M* mid; *ACC* common carotid artery; *ACI* internal carotid artery; *ACE* external carotid artery

A turbulent flow pattern in a blood vessel does not necessarily indicate pathology. In a healthy carotid artery flow will likely be laminar in the common branch and close to laminar in the external and internal branches (Fig. 10.4), but near the bifurcation and especially in the bulbus – the initial widening of the internal branch – backflow leading to a turbulent flow pattern may occur.

10.3.2 Abnormal Flow

On the other hand, when turbulent blood flow occurs in places where it is not expected, it may be an indication of pathology. Near the carotid bifurcation, the lumen is often narrowed because of atherosclerotic changes. Such an abnormal narrowing is called a stenosis. A stenosis induces hemodynamic changes that can be observed with ultrasound. At the location of the stenosis flow velocity will be increased, simply because the same amount of blood per unit of time has to pass through a smaller opening (see also Box 10.1). Proximally of (before) the stenosis flow will be decreased, whereas distally from (behind) the stenosis flow will become turbulent because of the sudden widening of the vessel (Box 10.2). The degree of turbulence and the relative change in velocity provide information on the extent of vessel diameter reduction.

Box 10.2 Turbulence

Poiseuille's law only holds for vessels with a constant diameter and laminar flow. However, close to a curve, bifurcation, plaque, or stenosis friction induces a loss of flow energy that can no longer be used to transport flow particles, leading to increased flow resistance and an increased pressure drop over this segment of the vessel. This viscous behavior can lead to turbulent flow.

The chances for turbulent flow to occur increase when flow velocity increases, when the lumen is larger, when the relative increase in lumen is larger, when the vessel wall is rougher, and when viscosity is lower. These relations can partially be summarized as follows. Chances for turbulent flow to occur increase when the Reynolds number Re is larger, where:

$$\mathrm{Re} = \frac{2\rho v r}{\eta} \tag{1}$$

in which ρ is the fluid density, v is the flow velocity, r is the vessel radius, and η the blood viscosity.

The critical value of Re for which turbulence develops is further determined by the local vessel anatomy. Near sharp curves, bifurcations, or sudden vessel widenings such as the bulbus, local pressure differences can develop, lowering the critical value for Re compared to that for straight vessels.

Question 10.1 What does the flow profile behind a stenosis look like? Sketch or describe it based on Figs. 10.3 and 10.4.

10.3.3 The Doppler Effect

In Chap. 9 it was already explained how ultrasound can be used to make anatomical images noninvasively. Ultrasound can also be used to determine blood flow velocities noninvasively by employing the Doppler effect. This effect is probably known from what you perceive when an ambulance approaches you and then recedes again. When it approaches its siren is more high-pitched than when it moves away. This is the Doppler effect: when a sound source approaches you, its wave length decreases and thus its frequency increases (see Sect. 9.3.1). By contrast, when the sound source moves away, its wave length increases and its frequency decreases (see Fig. 10.5).

The shift in frequency (the Doppler effect) can be used to determine the velocity of a moving object by sending a sound with a known frequency toward it and

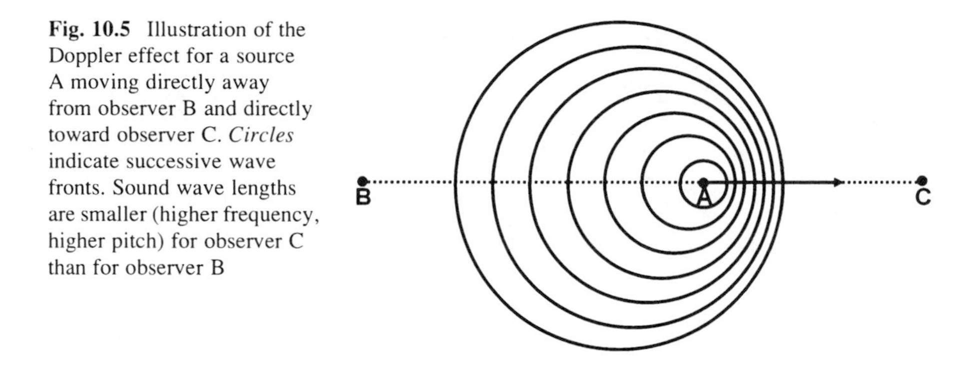

Fig. 10.5 Illustration of the Doppler effect for a source A moving directly away from observer B and directly toward observer C. *Circles* indicate successive wave fronts. Sound wave lengths are smaller (higher frequency, higher pitch) for observer C than for observer B

recording the frequency of the reflected sound wave. The Doppler effect depends on the relative speed difference between the sound source and the observer and the direction in which the sound source moves with respect to the observer (Box 10.3).

As discussed in the context of flow profiles, not all blood particles move at the same speed. Hence, the transducer will receive signals with different frequency shifts that together give a complex signal containing a range of frequencies. Fourier transforms (see Sect. 3.3) can be used to extract the presence of each frequency from the detected complex signal. Each detected frequency corresponds to a particular blood flow velocity. The three-dimensional information (spectral power as a function of frequency and time) can be displayed in a spectrogram (see Fig. 10.6 for an example).

Box 10.3 The Doppler Formula

In Doppler ultrasound measurements, the transducer is both the transmitter (sound source) and the receiver (observer). The Doppler shift mainly originates from the scattering of the sound beam by moving red blood cells. These cells are so small (5–10 μm) that they do not reflect the sound waves that have a larger wave length (O(mm)). The Doppler shift f_D is equal to the difference between the frequencies f_t of the transmitted and f_r of the received sound:

$$f_D = f_t - f_r = f_t \frac{2v \cos \alpha}{c} \tag{1}$$

Here v is the blood velocity, c the speed of sound in the body (~1,540 m/s for soft tissue), and α the angle between the transducer sound beam and the direction of the blood flow. Note that $v \cos \alpha$ is simply the component of the blood velocity in the direction of the ultrasound beam.

This formula also shows that the Doppler shift depends strongly on the angle between the incident sound beam and the direction of blood flow. When the sound beam is perpendicular to the blood flow direction, a Doppler shift and thus velocity cannot be measured ($\cos 90° = 0$).

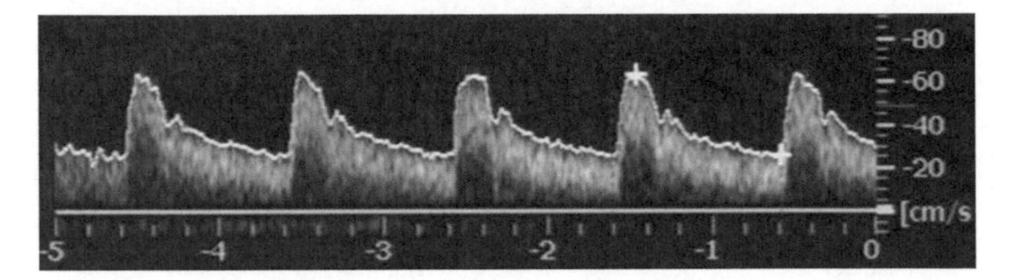

Fig. 10.6 Example of a Doppler spectrogram as measured in the internal carotid artery. Time in seconds is on the horizontal axis; the Doppler shift (proportional to the flow velocity in cm/s) is on the vertical axis. Grey values code for the spectral power, a measure for the number of particles with that particular velocity

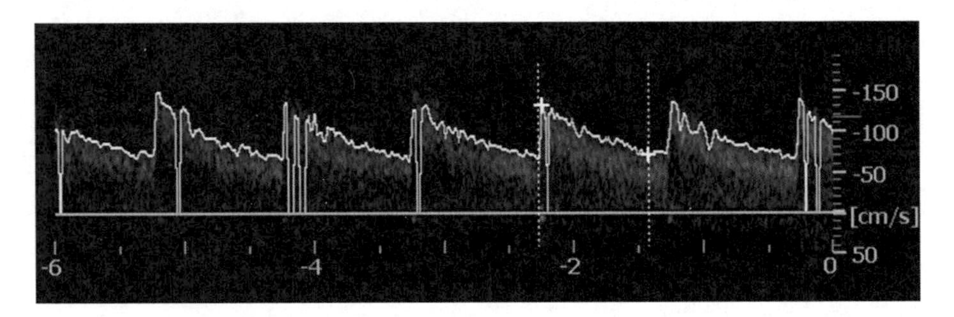

Fig. 10.7 Example of a Doppler spectrogram in an internal carotid artery with a stenosis. Compare to the shape and velocities of the spectrogram in Fig. 10.6

The lowest value on the vertical axis in the spectrogram thus represents the lowest velocity, the highest value the highest velocity. In normal arteries, the flow profile will be parabolic or plug-like (see Fig. 10.3), implying that most particles move at (close to) maximum velocity and the range in flow velocities in the sampled volume of blood is small. This yields a narrow spectrum. When the flow becomes turbulent, particles with lower velocities will also be present. This is reflected in a broader spectrum; the range of flow velocities is then much larger (see Fig. 10.7) and can even include negative velocities.

The spectrogram (and particularly its *envelope*) yields several blood flow parameters that can be used for diagnosis. The peak *systolic* velocity (V_s) is enlarged in a stenosis. The end *diastolic* velocity (V_d) provides information on the peripheral resistance (distal to the measurement site). High end diastolic velocity implies that even when the pressure is at its minimum, the flow is still large, i.e., low peripheral resistance. Such a low resistance profile is normally seen in the internal carotid artery, which is one of the main providers of blood to the brain. Vice versa, high systolic flow, but no diastolic flow (or even backflow), is called a high resistance profile and is normal for the external carotid artery. Examples of normal spectral flow profiles in the three branches of the carotid artery are given in Fig. 10.8.

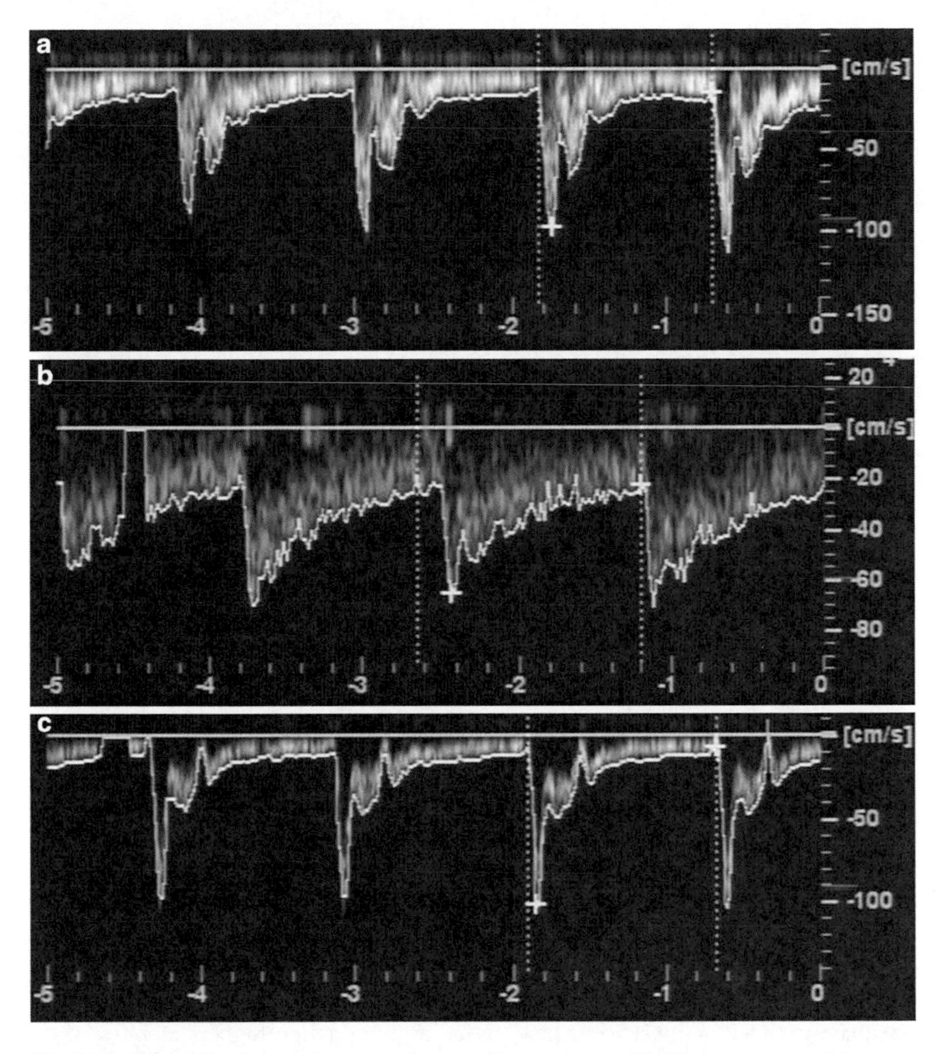

Fig. 10.8 Examples of normal spectral flow profiles in the (**a**) (distal) common carotid artery, (**b**) (distal) internal carotid artery (low resistance profile), and (**c**) (proximal) external carotid artery (high resistance profile). Note that, because the probe was reversed with respect to the orientation used for Figs. 10.6 and 10.7, velocities are plotted in the negative direction

The mean velocity (V_m) over a heart cycle is mostly calculated to be used in the pulsatility index:

$$\mathrm{PI} = \frac{V_s - V_d}{V_m} \tag{10.1}$$

When proximal disease dampens the flow waveform, the pulsatility index will be much lower than in a healthy elastic artery. When using Doppler ultrasound only, the direction of blood flow is not always known and the velocities cannot be corrected for the angle of insonation (the angle between the ultrasound beam and the blood flow). The advantage of the PI is that it does not depend on this angle, because it appears in both the nominator and denominator of the index. A disadvantage is that V_d is lower when the heart rate is lower, incorrectly giving higher values for the PI. In clinical practice other indices, such as the resistance index (similar to PI but with V_m replaced by V_s), are also employed.

Normally, the arterial walls expand during the systolic blood pressure rise and bounce back during the diastolic phase. Because of the peripheral resistance, the rate of blood entering the arteries surpasses that leaving them. This induces a net storage of blood during the systolic phase which is ejected again during the diastolic phase. This so-called Windkessel effect helps dampen blood pressure fluctuations. When the arteries stiffen due to atherosclerotic disease, the systolic pressure wave is not dampened anymore and this can lead to further damage to arteries and tissues.

10.3.4 Pulsed Wave Doppler Ultrasound

In the pulsed wave Doppler technique, short microsecond ultrasound pulses are emitted by the transducer, which also receives the echoes. The deeper the source of the echo, the longer the time between the sent pulse and the received echo will be. By only incorporating echoes that are received within certain temporal boundaries, information on the flow velocities at a certain depth can be obtained selectively. In pulsed wave Doppler, the Doppler signal must be reconstructed based on the known pulse repetition frequency (PRF) and the echoes that are received in between pulses. Since echoes that are generated at a greater depth need more time to return, the PRF must be reduced for deeper vessels. On the other hand, Nyquist's theorem (see Sect. 3.3.2) demands that to be able to reconstruct the Doppler frequencies adequately, the PRF must be at least twice as high as the highest Doppler shift. When the received sounds are sampled at a lower frequency, aliasing (see Sect. 3.3.2) may occur, presenting high Doppler shifts belonging to high velocities wrongly as low or even negative velocities. In a duplex ultrasound image, aliasing can often be recognized as a "turbulent" image in a position where this is not expected based on the anatomy (see Fig. 10.9), but this is often impossible for Doppler measurements only.

Thus, the higher the flow velocities the higher the PRF should be, whereas the deeper the vessel, the lower the PRF should be, implying that high velocities in deep vessels cannot be measured reliably with pulsed wave Doppler. In these cases an alternative technique (continuous wave Doppler) can be used, which has the important disadvantage that depth information is not available at all. For most cerebrovascular ultrasound applications, pulsed wave Doppler suffices.

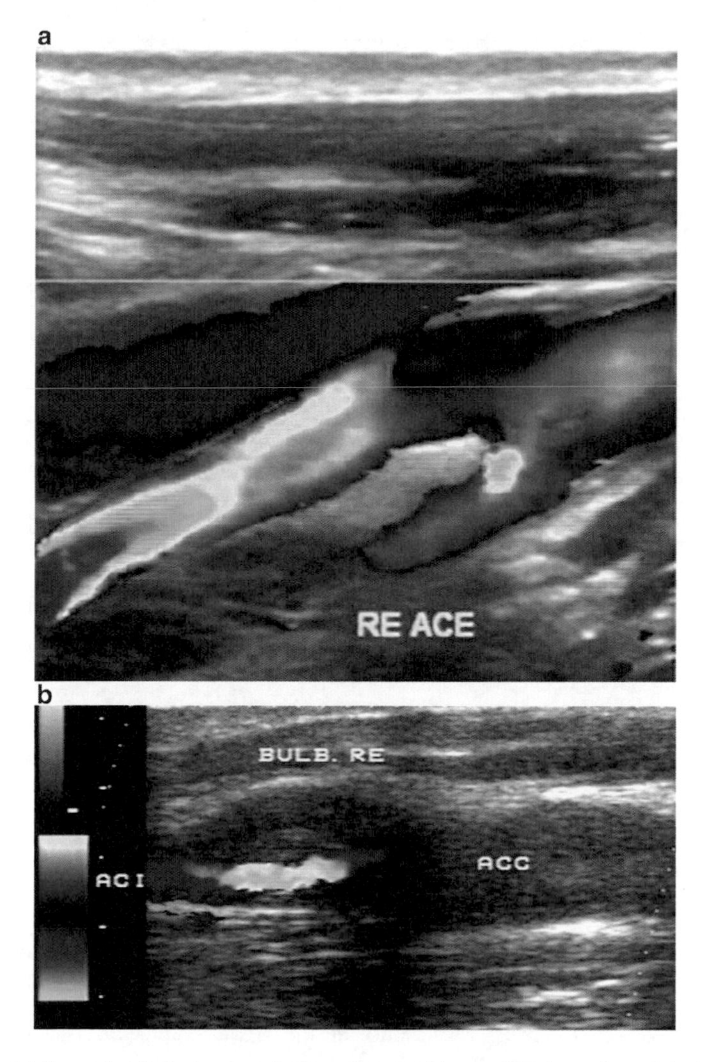

Fig. 10.9 (**a**) Example of aliasing in a duplex ultrasound image. The lower branch is the external carotid artery (ACE); the upper branch is the internal carotid artery. Very high velocities cannot be represented in the *hot color* scale and are aliased and displayed in *cold colors*. Turbulence is not expected here, because the arterial branch maintains its width. For comparison, image (**b**) displays turbulent flow in a stenoted bulbus

10.3.5 Tips and Tricks in Duplex Ultrasound Imaging

10.3.5.1 Optimizing the Doppler Recording

When Doppler ultrasound is used without B-mode (anatomical) imaging, the direction of the blood flow is unknown and the estimated velocity cannot be

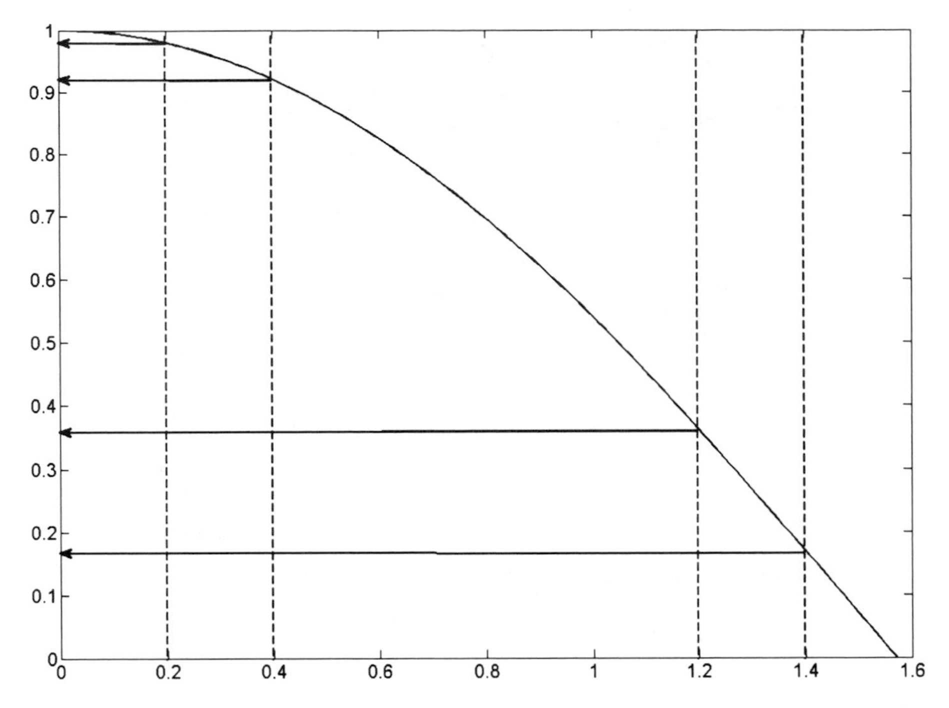

Fig. 10.10 Effect of error in insonation angle on velocity estimation. The cosine is plotted for angles between 0 and $\pi/2$ (90°). As can be seen, the same difference in angle (distance between *dashed lines*) results in a smaller difference in velocity estimation (as represented by the distance between the *arrows*) for smaller angles than for larger angles

corrected for the angle of insonation. When anatomical information is present however, the insonation angle can be optimized so that the incident sound beam is as parallel to the blood flow as possible. When the angle is larger, the effect of a small error in the angle estimation has a large influence on the estimation of the flow velocity (because the cosine varies more when the angle is closer to 90° than when it is closer to 0°; cf. (1) in Box 10.3 and see Fig. 10.10).

Theoretically, it would be best to make the angle as small as possible, but this is often anatomically impossible.

Question 10.2 Why is it anatomically impossible to make the insonation angle zero?

Furthermore, at lower angles the vessel wall strongly reflects sound and only a limited part of the sound will reach the blood particles. At large angles, the reflection from the (moving) vessel wall will also give a large component in the direction of the received Doppler signal from the blood particles, thereby masking the echoes of interest. Therefore, a practical compromise is an angle between 30° and 60°. Yet, even with this approach the vessel walls will still generate echoes themselves. Luckily, the walls move slowly compared to the blood velocities and the resulting lower frequency echoes can be filtered out with a high-pass filter (see Sect. 4.3.4).

Sound does not penetrate through bone. Therefore, the intracranial arteries can only be investigated through windows in the skull where the bone is absent or thinner. The temporal window (in front and above the ear) can be used to investigate flow in the branches of the circle of Willis, the occipital window (foramen magnum; just below the skull at the back of the head) can be used to investigate flow in the vertebral arteries, and the orbital window (through the eye) can be used to investigate flow in the ophthalmic artery (see Fig. 10.11).

For transcranial recordings, relatively low frequencies (1–2.5 MHz) are used, because they are less attenuated by bone and allow deeper recordings.

10.3.5.2 Recognizing Arteries

The advantage of using duplex ultrasound is that both the anatomical and the flow information can be used for recognizing arterial branches. Typically, in a duplex ultrasound investigation of the carotid arteries, the entire carotid arterial trajectory is screened first by sliding the transducer along the trajectory in a perpendicular orientation to the vessel, to get an impression of the anatomy. In the neck, arteries can be distinguished from veins by the direction of the blood flow and because veins can be compressed from the outside. It may be more difficult to distinguish the internal and external carotid arteries, particularly when the bulbus is not so obvious. Anatomically, the external carotid artery typically has more branches than the internal carotid artery. Furthermore, the flow profiles differ between the internal and external carotid arteries; the first has a low resistance profile whereas the latter has a high resistance profile (see Sect. 10.3.3).

To recognize the different intracranial arteries in the circle of Willis, anatomical information cannot be used. For orientation within the skull, first the diameter of the head can be measured. This provides a reference for the depth at which the different branches (posterior, media, and anterior) of the circle of Willis can be found when approached from the temporal window. Additionally, the orientation of the transducer with respect to the window and the flow direction with respect to the transducer can be used for identification. For instance, when the transducer is held perpendicular to the temporal window and a flow toward the transducer is observed at a depth of 40–50 mm, the signal likely originates from the a. cerebri media.

Question 10.3 What arteries can the signal originate from when there is a flow away from the transducer at a depth of approximately 60 mm?

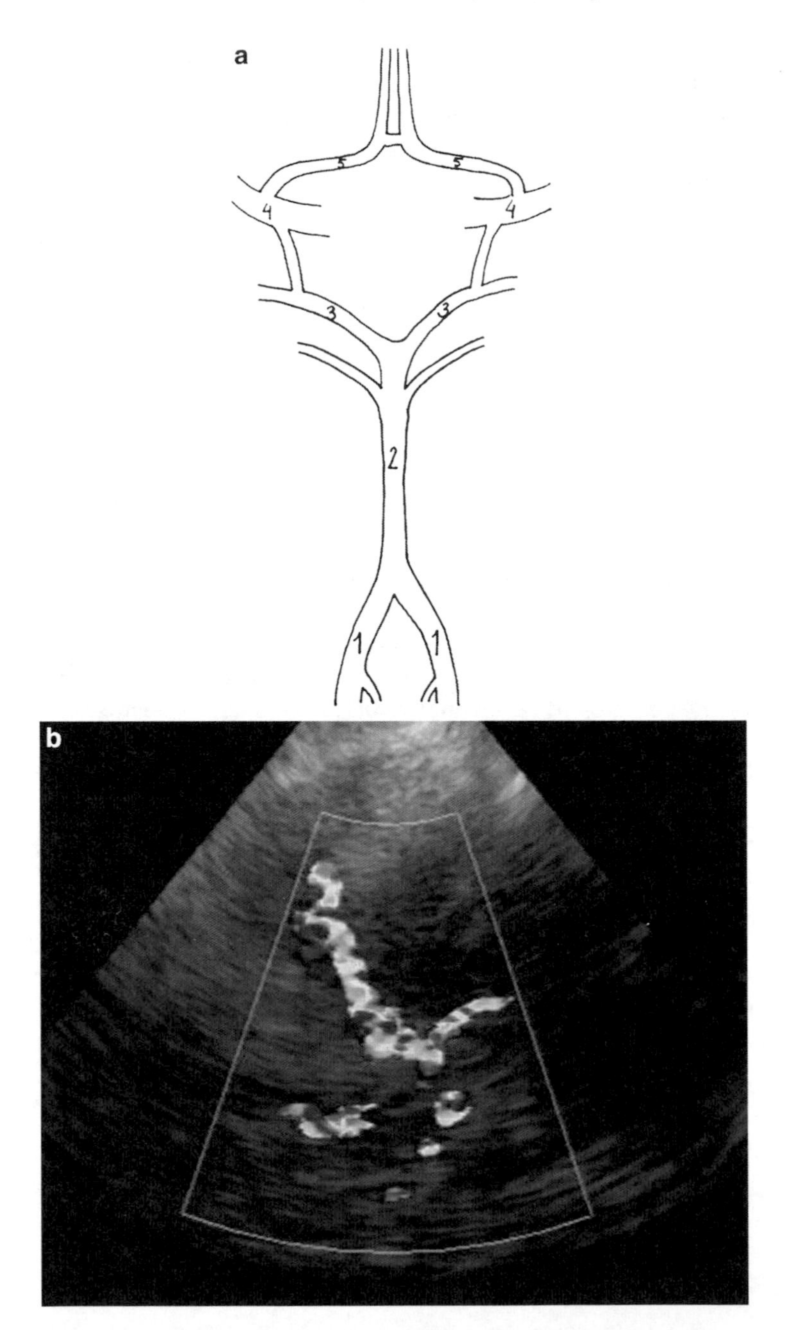

Fig. 10.11 (**a**) Overview of most common anatomy of the intracranial arteries. *1* Vertebral arteries; *2* basilar artery; *3* posterior cerebral arteries; *4* medial cerebral arteries; *5* anterior cerebral arteries. 3–5 are part of the circle of Willis. For comparison an example of an intracranial TCD recording of part of the circle of Willis (viewed from the *right* in the *top image*) is displayed in image (**b**). The medial cerebral artery is extending to the *top left*

Table 10.1 Doppler criteria for stenosis as used in our hospital

	PSV ratio ICA/CCA	PSV	EDV
0–50%	<2.0	<130	<50
51–70%	>2.0	>130	>50
71–99%	>3.0	>200	>90

PSV peak systolic velocity; *EDV* end diastolic velocity; *ICA* internal carotid artery; *CCA* common carotid artery

Furthermore, the relative velocities in the branches (media > anterior > posterior) and the flow profiles (more pulsatile for the a. cerebri media) can be taken into account. In case of pathology and anatomical variants recognition can be difficult, however.

10.3.5.3 Estimating Stenosis Severity

When patients have suffered from a TIA or stroke that may have been caused by a plaque in the carotid artery, *carotid endarterectomy* may be considered as a treatment, but only in cases of severe (>70% lumen reduction) stenosis. In principle, anatomical imaging can be very helpful to estimate the amount of reduction in vascular lumen due to a plaque. Yet, the percentage stenosis is conventionally determined indirectly based on flow velocities, mainly because older ultrasound equipment did not allow anatomical imaging. By comparing Doppler ultrasound recordings to digital subtraction *angiography* (DSA) images (the *gold standard*), cut-off values have been determined for significant stenoses (i.e., >70% lumen reduction) in many laboratories.

In our own hospital, we have a very low risk of complications due to carotid endarterectomy (3%). Therefore, we studied the available literature (e.g., Jahromi et al. 2005) and chose slightly lower cut-off values (Table 10.1), so as not to miss patients who need to be operated on, at the cost of some patients who will undergo surgery even though it would not have been strictly necessary (higher sensitivity at the cost of a lower specificity).

Ideally, because of differences in ultrasound equipment and local standards, each laboratory should determine its own cut-off values by comparing DSA with ultrasound results in a large group of patients, but this is often not very practical.

10.4 Vascular Ultrasound Imaging in Individual Patients

Patient 1

To examine whether this patient's complaints could be explained by TIAs, his extracranial (carotid, basilar and vertebral) and intracranial arteries were examined by duplex and Doppler ultrasound, respectively.

Table 10.2 Velocities in cm/s as detected by Doppler ultrasound in patient 1

	Left				Right			
	PSV	EDV	Mean	PI	PSV	EDV	Mean	PI
CCA prox	78	16	33	1.88	74	24	38	1.32
CCA dist	82	19	41	1.54	61	21	37	1.08
ICA prox	454	234	304	0.72	472	247	339	0.66
ICA dist	22	13	18	0.50	60	31	44	0.66
ECA	238	54	108	1.70	225	47	96	1.85
Vertebral	62	28	42	0.81	52	16	30	1.20

CCA common carotid artery; *ICA* internal carotid artery; *ECA* external carotid artery; *prox* proximal; *dist* distal; *PSV* peak systolic velocity; *EDV* end diastolic velocity; *mean* mean velocity; *PI* pulsatility index

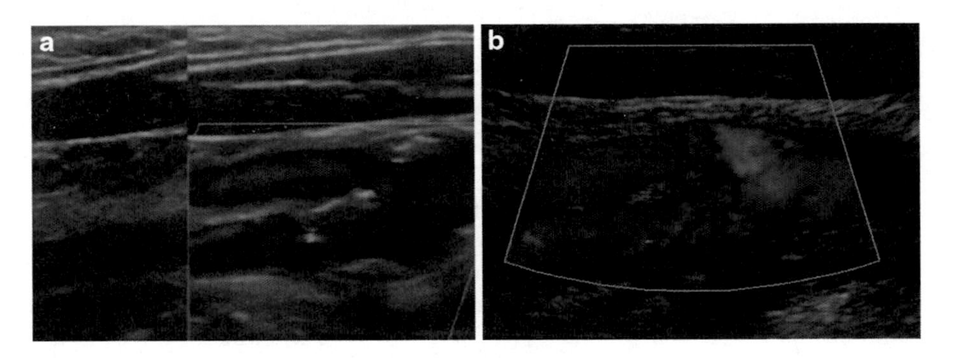

Fig. 10.12 Duplex ultrasound images of the left ICA (**a**) before and (**b**) after surgery

The systolic, diastolic, and mean velocities that were detected in his left and right carotid arteries are given in Table 10.2.

The ratio between the systolic and diastolic velocities in the ICA and CCA was 5.5 (left, PSV), 12 (left, EDV), 7.7 (right, PSV), and 12 (right, EDV). The flow velocities are normal in the CCA bilaterally. In the proximal ICA however, flow velocities are extremely high bilaterally, with concurrent very low velocities in the left distal ICA. The flow velocities in the ECA are also strongly increased bilaterally. Flow in the vertebral artery is slightly increased on the left and normal on the right. The ratio between the ICA and CCA velocities is strongly increased bilaterally as well. The PI is slightly decreased in the ICA bilaterally, particularly distally. By itself, these results strongly suggest stenoses in the ICA and ECA which is confirmed by the Duplex ultrasound images (Fig. 10.12a).

In the left CCA, the vessel wall is irregular with multiple *echogenic* plaques. On the level of the bifurcation, the lumen of the left proximal ICA is strongly decreased because of the presence of large mixed plaques. On the right, irregular atherosclerotic abnormalities are also found. In the proximal ICA and the proximal ECA, the lumen is strongly decreased due to local plaques.

Table 10.3 Velocities in cm/s as detected by Doppler ultrasound in patient 1

	Left				Right			
	PSV	EDV	Mean	PI	PSV	EDV	Mean	PI
Media prox	196	113	145	0.57	90	53	69	0.54
Media dist	43	30	35	0.37	50	27	38	0.61
Anterior	36	16	26	0.77	103	48	76	0.72
Posterior	115	46	74	0.93	67	30	49	0.76

Media: a. cerebri media, anterior: a. cerebri anterior, posterior: a. cerebri posterior, *prox* proximal; *dist* distal; *PSV* peak systolic velocity; *EDV* end diastolic velocity; *mean* mean velocity; *PI* pulsatility index

Table 10.4 Mean blood velocities (± standard deviation in cm/s) according to Ringelstein et al. (1990)

Age	MCA	ACA	PCA
10–29	70 ± 16.4	61 ± 14.7	55 ± 9.0
30–49	57 ± 11.2	48 ± 7.1	42 ± 8.9
50–59	51 ± 9.7	46 ± 9.4	39 ± 9.9
60–70	41 ± 7.0	38 ± 5.6	36 ± 7.9
Insonated depth	50–55	65–70	60–65

Based on these observations of bilateral extensive atherosclerotic abnormalities and the Doppler criteria for stenosis as used in our hospital (Table 10.1), patient 1 is diagnosed with bilateral ICA stenoses of 70–99% and stenoses of the bilateral ECA.

The systolic, diastolic, and mean velocities that were detected in his left and right intracranial arteries are given in Table 10.3.

To interpret these velocities, reference values are required. In our hospital the values determined by Ringelstein et al. (1990) are employed (see Table 10.4).

Thus, the velocities in the a. cerebri media are slightly increased on the right side and strongly increased proximally on the left side. More distally, the velocities decrease. The internal carotid artery was also assessed intracerebrally, and a strongly increased peak systolic velocity of 196 cm/s was found. In the a. cerebri anterior, velocities are increased on the right and decreased on the left. Finally, in the a. cerebri posterior, flow velocities are increased, left more than right. PI values are generally low. Summarizing, there is a local stenosis intracerebrally in the left internal carotid artery, probably continuing into the proximal a. cerebri media.

Based on the extracranial duplex ultrasound recording and the intracranial Doppler ultrasound recording, the patient underwent a carotid endarterectomy on the left (see Fig. 10.12b) and 3 months later a carotid endarterectomy on the right, from which he recovered well.

Patient 2

To determine the origin of the bleeding, both CT angiography and DSA were performed in this patient, but an aneurysm or arteriovenous malformation (AVM) was not found. It was concluded that she had suffered from a nonaneurysmal *perimesencephalic* hemorrhage, and the origin of the bleeding remained uncertain.

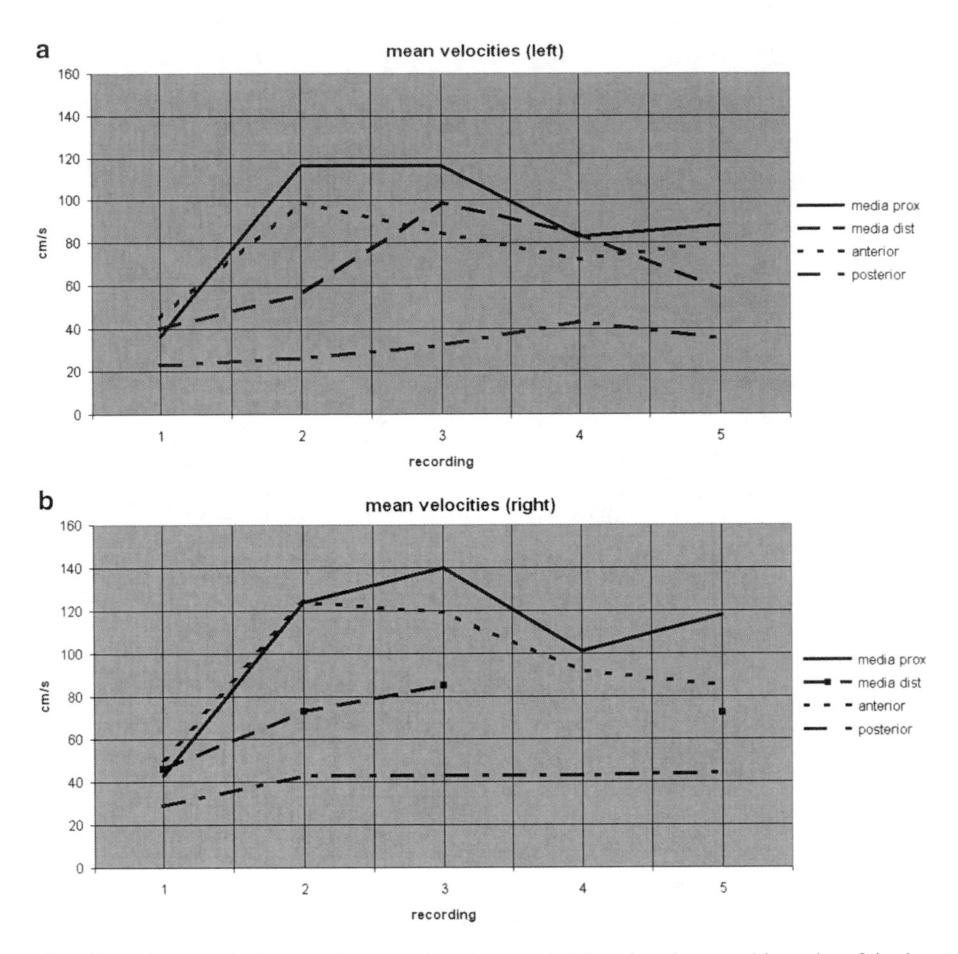

Fig. 10.13 Mean velocities as determined by intracranial Doppler ultrasound in patient 2 in the (**a**) left and (**b**) right hemispheres for the proximal and distal a. cerebri media, the a. cerebri anterior, and the a. cerebri posterior at five separate recordings (2, 8, 10, 14, and 17 days after SAH). The value for the right distal a. cerebri media is missing for recording 4

To screen for vascular spasm in this patient, a transcranial Doppler investigation was performed 2, 8, 10, 14, and 17 days after the SAH, knowing that spasms typically do not occur until a few days after the hemorrhage. The mean velocities in the proximal and distal a. cerebri media, the a. cerebri anterior, and the a. cerebri posterior are indicated in Fig. 10.13 for both hemispheres.

Comparing the results to the reference values in Table 10.4, it can be seen that although initially there are no spasms, from the second recording the patient suffers from spasm, particularly on the right side in the a. cerebri media and a. cerebri anterior, for which she was treated. Although spasm severity decreases after day 10, the spasms persist until the last recording when the patient is discharged from the hospital in clinically good condition.

In both patients in this chapter, duplex or Doppler ultrasonography helped gain more insight in the severity of the vascular disease. In patient 1, the results of the duplex ultrasonography were of crucial importance for deciding on further treatment.

10.5 Other Applications of Duplex and Doppler Ultrasound in Neurology

10.5.1 Intraoperative Neuromonitoring: Carotid Endarterectomy

Duplex ultrasound is not only used to estimate the percentage stenosis in case of symptomatic plaques in the carotid arteries, it can also be used to monitor brain functioning during subsequent carotid endarterectomy (i.e., intraoperatively). As part of this surgery, the thickened *intima* layer has to be removed over a certain length of the artery. This can only be done when the common, internal, and external carotid arteries are clamped for a short time, so that the surgeon has a clear view of the arterial walls. This procedure has the risk of inducing ischemia in the part of the brain being supplied by oxygenated blood through the carotid artery. To estimate this risk during surgery, the EEG could be monitored while shortly clamping the artery as a trial. When the EEG shows that the brain will likely suffer from ischemia when the carotid artery is clamped, the surgeon might choose to put a shunt between the proximal common carotid artery and the distal internal carotid artery, so that the area in between can be clamped and operated on without risks. However, it takes some time before the EEG responds to ischemia in the brain and irreversible damage may have been done already. Furthermore, the EEG can be difficult to interpret as a result of the centrally active anesthetics. Transcranial Doppler is therefore used as an additional tool for more direct monitoring of the cerebral blood supply. An ultrasound probe is placed on the temporal window preoperatively and aimed at the a. cerebri media. The probe can be fixated by a headband, and some state-of-the-art systems even have robots repositioning the probe when it has moved during surgery. This allows to monitor the blood flow in the a. cerebri media continuously. When the blood flow drops too much (more than 70%) during trial clamping, the surgeon will generally choose to use shunting. Blood velocity is also monitored during the rest of the surgical procedure, allowing the anesthetist to react adequately to sudden drops in blood velocity as a result of changes in blood pressure. Furthermore, monitoring allows to check whether blood flow in the a. cerebri media has recuperated sufficiently at the end of surgery. Sometimes monitoring is continued after surgery to monitor for reocclusion of the vessel or the occurrence of cerebral emboli (see Sect. 10.5.2) originating from the wound. Because of these monitoring techniques, carotid endarterectomy has become safer resulting in fewer serious complications.

10.5.2 Embolus Detection

A TIA can be caused by a temporary occlusion of an artery in the brain due to an embolus. When a patient has suffered from a TIA, it is thus important to find out if there are possible sources of emboli, such as a damaged area on one of the arterial walls. In such an area a local *thrombus* consisting of blood platelets may develop. When part of this thrombus breaks off, it will be carried by the blood flow until it gets stuck in a smaller artery. In general, emboli may originate from the carotid arteries, but also from the heart or the arteries in the legs.

> *Question 10.4* What is the most likely source of an embolus in the a. cerebri media? And what is the most likely source of an embolus in the a. cerebri posterior?

An embolus can be detected in the Doppler signal due to its size and due to its movement with the flow. Since the embolus is large compared to the blood particles, it yields a relatively strong echo and even more so when it is a gaseous embolus. Since an embolus moves with the flow at a speed maximally equal to the blood flow velocity, the embolus signal appears at one side of the spectral flow profile only (see Fig. 10.14), in contrast to artifacts that typically appear on both sides of the spectral flow profile.

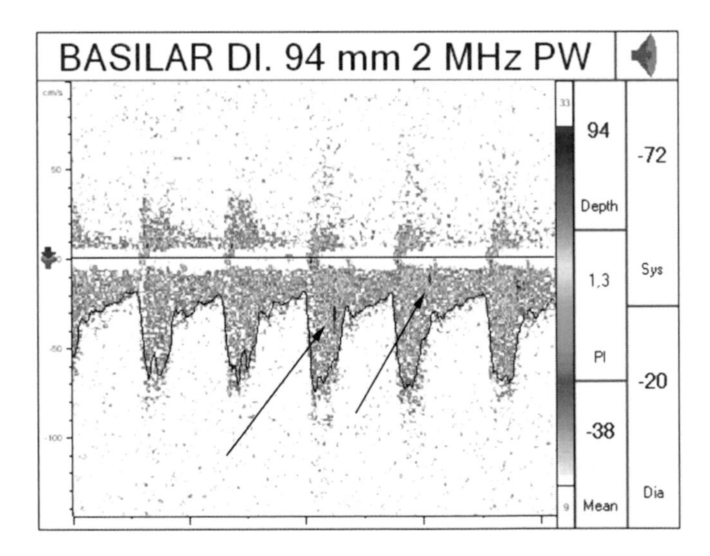

Fig. 10.14 Example of high intensity embolus signals (indicated by *arrows*) in the basilar artery in a patient who recovered from an SAH due to an aneurysm in the basilar artery

10.5.3 Confirmation of Brain Death

As explained in Sect. 4.6.3, an isoelectric EEG is an important criterion to confirm brain death. However, the EEG is unreliable for this purpose in patients who have received medication that suppresses neural activity such as sedatives or anesthetics. Since a few years some countries have adopted a protocol to confirm brain death that includes an important role for transcranial Doppler ultrasonography. This technique can be used to confirm cerebral circulatory arrest. Two specific patterns may be observed as a result of cerebral circulatory arrest. A reverberating pattern is the result of oscillating forward and backward flow reflected in the area under the systolic peak (positive flow) being the same as the area under the flow curve during the diastolic phase (negative flow), resulting in net zero flow. Systolic spikes are short (<200 ms) unidirectional peaks during early systole with a maximum flow velocity of 50 cm/s, resulting in net zero flow, as well. When either of these patterns is observed, brain death can be confirmed with more certainty.

10.6 Answers to Questions

Answer 10.1 There will be backflow directly behind the stenosis. In contrast to the flow in the bulbus, the backflow will not be limited to one side, but will be present along the entire circumference.

Answer 10.2. An insonation angle of zero is only possible when the incident sound beam is parallel with the vessel. This implies that to actually receive echoes, the transducer should be in the vessel, which is anatomically impossible when vessels are aligned with the skin. If the vessel would curve away from the skin such that its orientation becomes almost perpendicular to the skin, an insonation angle of zero would be possible. However, such an anatomy does not typically occur in the carotid arteries.

Answer 10.3. Depending on the orientation of the probe, the signal can originate from the (proximal) a. cerebri anterior or (proximal) a. cerebri posterior. In both cases the velocity will be lower than in the a. cerebri media.

Answer 10.4. The a. cerebri media is generally supplied by the carotid arteries, the a. cerebri posterior by the vertebral and basilar arteries, indicating the most likely source of emboli in these cerebral arteries. However, in a small group of patients, anatomical variants may exist, allowing carotid pathology to be the cause of emboli in the a. cerebri posterior. It is therefore important to always investigate all extracranial arteries.

Glossary

Anemia Deficiency in red blood cells.

Angiography Imaging technique used to visualize the lumen of part of the vasculature, e.g., in an organ such as the brain.

Arachnoid membrane Second (spiderweb-like) membrane covering the brain.

Arteriovenous malformation Abnormal connection between veins and arteries.

Atherosclerotic plaque Accumulation of cholesterol, fatty acids, calcium, and fibrous tissues on arterial walls.

Carotid arteries The main arteries in the neck supplying the brain and the face with blood.

Carotid endarterectomy Removal of atherosclerotic plaque in the carotid artery by surgery.

COPD Combination of bronchitis (inflammation in larger airways) and emphysema (secondary destruction of lung tissue in final branches of the airways (alveoli)). Mostly caused by prolonged tobacco smoking.

Diastolic Phase of the heart cycle when arterial pressure is minimal, occurring near the beginning of the cardiac cycle when the *ventricles* are filled with blood.

Echogenic High ability to return an echo.

Embolus Small mass transported by the circulation; can be solid, liquid, or gaseous.

Envelope Here: the curve joining the maximum velocities of the spectrogram at each moment in time.

Gold standard Benchmark or test that is regarded as perfect, to compare other tests to.

Hypercholesterolemia High level of cholesterol in the blood.

Hypertension High blood pressure.

Intima Innermost layer of an artery or vein. It is directly in contact with the blood flow.

Inviscid Nonviscous.

Lumen Interior of a blood vessel.

Pia mater First membrane covering the brain.

Perimesencephalic Around the mesencephalon (midbrain).

Systolic Phase of the heart cycle when arterial pressure in the arteries peaks, occurring when the ventricles are contracting.

Thrombus Blood clot.

Transducer Device that transforms one form of energy into another: here electrical energy is transformed into sound energy.

Ventricle Lower heart chamber.

Viscosity Measure of resistance of a material to deformation. A viscous fluid is "thicker" than an inviscid fluid.

References

Online Sources of Information

http://en.wikipedia.org/wiki/Stroke. Extensive overview of etiology, pathophysiology, prevention, treatment and epidemiology

http://www.stroke.org.uk/information/index.html. Accessible information on stroke

http://www.bbc.co.uk/health/physical_health/conditions/in_depth/stroke. Accessible information on stroke

http://en.wikipedia.org/wiki/Transient_ischemic_attack. Information on transient ischemic attacks

http://en.wikipedia.org/wiki/Subarachnoid_hemorrhage. Information on subarachnoid hemorrhages

http://en.wikipedia.org/wiki/Bernoulli%27s_principle. Bernoulli's principle as used in fluid dynamics

http://en.wikipedia.org/wiki/Poiseuille%27s_law. Includes derivation of this law also known as Hagen-Poiseuille equation

http://sydney.edu.au/science/uniserve_science/school/curric/stage6/phys/medphys.html#ultrasound. Extensive list of useful links

Books

Aburahma A, Bergan JJ (eds) (2010) Noninvasive cerebrovascular diagnosis. Springer Verlag, London. Also (partially) on http://books.google.com

Allen PL, Dubbins PA, Pozniak MA, McDicken WN (2006) Clinical Doppler ultrasound, 2nd edn. Elsevier, Philadelphia. Also on http://books.google.com

Arger PH, DeBari Iyoob S (2004) The complete guide to vascular ultrasound. Lippincott, Williams and Wilkins, Philadelphia. Also on http://books.google.com

Papers

Jahromi AS, Cinà CS, Liu Y, Clase CM (2005) Sensitivity and specificity of color duplex ultrasound measurement in the estimation of internal carotid artery stenosis: a systematic review and meta-analysis. J Vasc Surg 41(6):962–972

Maurits NM, Loots GE, Veldman AEP (2007) The influence of vessel wall elasticity and peripheral resistance on the carotid artery flow wave form: a CFD model compared to in vivo ultrasound measurements. J Biomech 40(2):427–436

Ringelstein EB, Kahlscheuer B, Niggemeyer E, Otis SM (1990) Transcranial Doppler sonography: anatomical landmarks and normal velocity values. Ultrasound Med Biol 16(8):745–761

Chapter 11
Spinal Dysfunction, Transcranial Magnetic Stimulation, and Motor Evoked Potentials

After reading this chapter you should:

- Understand the basic electromagnetic principles underlying magnetic stimulation of biological tissue
- Know why a motor evoked potential is suited to investigate the integrity of the corticospinal pathway
- Know how a motor evoked potential is evoked, both practically and physiologically
- Know what measures can be derived from the motor evoked potential to assess corticospinal pathway functioning
- Be able to mention at least two applications of repetitive transcranial magnetic stimulation

11.1 Patient Cases

Patient 1

Since 5 years, this 45-year-old female patient occasionally suffers from a tingling sensation and weakness in her left arm and leg. After some initial improvement, the muscle weakness in her left leg has slowly, but progressively worsened. Her leg is dragging, particularly after she has walked for a while. The leg also feels unstable although sensibility is normal. She currently has no complaints about her right leg or her arms. The neurologist finds that she has normal muscle force in her arms and a slight weakness in her legs (proximal iliopsoas and hamstring muscles *MRC* 5−). Although she can still walk on her toes and heels and can stand on one leg, hopscotching on her left leg is impossible. Her reflexes are bilaterally normal in her arms, but the knee jerk and Achilles tendon reflexes in her legs are asymmetrically increased (left stronger than right with extinguishing *clonus*). Her footsole (plantar) reflexes cannot be evoked. These findings indicate that this patient suffers

N. Maurits, *From Neurology to Methodology and Back: An Introduction to Clinical Neuroengineering*, DOI 10.1007/978-1-4614-1132-1_11,

from a slight *pyramidal syndrome*, which results from damage to or dysfunction of the upper *motor neurons*. Since these complaints are rather aspecific and can still reflect several diseases, the neurologist decides to initiate further investigations.

Patient 2

A few months ago, this 60-year-old female patient woke up with a weak right hand, disabling her fine motor skills. Although a CT scan did not show any abnormalities, the patient was thought to have suffered from a cerebrovascular accident and she was started on blood thinning and *lipid* lowering medication. The weakness in her right hand persisted however, and since 1 month she additionally suffers from increasing stiffness in her right hand and arm. In the last 2 weeks she has noticed a weakness in her right leg, which she now drags along. Since a couple of days, this progressive motor weakness now also involves her left hand. She has no sensory disturbances and no cramps, but she does have *fasciculations* in both hands. The distal muscle force in her arms has deteriorated considerably: wrist extensors MRC 3/4+, finger extensors MRC 1/4, wrist flexors MRC 3/4+, finger flexors MRC 2–3/4+ (all left/right). Proximal arm and leg muscle force appear to be normal. Standing on her right toes or right heel is difficult. Her muscle tone is increased in her right arm and hand. Her biceps tendon and knee tendon reflexes are lively, and slightly high on the right side. Her Achilles tendon reflex is absent on the left and lively on the right. All other reflexes are normal. The patient is admitted to hospital for further investigations.

11.2 Magnetic Stimulation of the Central Nervous System: Assessing the Pyramidal Tract

Both patients 1 and 2 suffer from complaints that are, at least partially, consistent with a dysfunctioning pyramidal tract. The pyramidal or corticospinal tract originates from pyramidal neurons in the outer layer of the cerebral cortex – mostly in the primary motor cortex – and involves their axons that run down into the spinal cord. The corticospinal tract is mainly involved in voluntary movement, particularly of the distal parts of the limbs. The neuronal cell body in the motor cortex and its axon traveling through the brain stem to the spinal cord are referred to as upper motor neuron. In the spinal cord, the upper motor neurons connect, both directly and indirectly (through interneurons), with the lower (alpha) motor neurons that originate mostly from the brainstem and innervate voluntary muscles. When the upper motor neurons are damaged, this can result in a variety of symptoms, such as *spasticity*, decreased and less easily evoked superficial reflexes, increased deep tendon reflexes, difficulty performing fine motor skills, and an abnormal plantar reflex, the *Babinski* response.

To investigate the integrity of central motor pathway functioning noninvasively, magnetic stimulation can be used. The basic principle of magnetic stimulation is

Box 11.1 Principles of Magnetic Stimulation

In 1820, a first relation between electricity and magnetism was found by Hans Christian Oersted, who discovered that a compass needle placed near an electric wire deflects as soon as the current in the wire starts to flow. He thus discovered that an electric current produces a magnetic field. Ten years later, Faraday discovered that the inverse is also true; an electric field is produced by a changing magnetic field. Both fields are perpendicular to each other. The magnetic field produced by a current or changing electric field can be calculated using the Ampère-Maxwell law:

$$\oint_{\partial S} \mathbf{B} \cdot d\mathbf{l} = \mu_0 \mathbf{I}_S + \mu_0 \varepsilon_0 \frac{\partial \Phi_{\mathbf{E},S}}{\partial t} \tag{1}$$

in which $\mathbf{B}$ is the produced magnetic field, μ_0 the magnetic constant (electric permeability of vacuum), $\mathbf{I}_S$ the current passing through the integrated surface, ε_0 the dielectric constant (permittivity) of vacuum and

$$\Phi_{\mathbf{E},S} = \iint_S \mathbf{E} \cdot dA \tag{2}$$

the electric flux through any surface S. In this equation, $\mathbf{E}$ is the electric field. Similar to (1), the electric field produced by a changing magnetic field can be calculated using Faraday's law:

$$\oint_{\partial S} \mathbf{E} \cdot \mathbf{dl} = -\frac{\partial \Phi_{\mathbf{B},S}}{\partial t} \tag{3}$$

in which

$$\Phi_{\mathbf{B},S} = \iint_S \mathbf{B} \cdot dA \tag{4}$$

is the magnetic flux through any surface S. Note that the direction of the induced electric field is opposite to the direction of the change in flux that caused the electric field.

The coil used for TMS stimulation consists of a long wire, looped multiple times. Through the coil, a powerful, short-lived, and rapidly changing current is sent. This changing current produces a magnetic field perpendicular to the plane of the coil that penetrates the cranium. Because this magnetic field changes rapidly, this induces an electric field in the brain perpendicular to the magnetic field, and thus in the same plane as that of the coil. The voltage of the electric field itself may excite neurons directly. However, because the brain is a conducting medium, the spatial changes in the electric field also cause small currents which can excite neurons by themselves. As Lenz's law (an equivalent of Newton's third law: "action = −reaction") states that an induced current produces a magnetic field that opposes the original change in magnetic flux, the direction of the induced currents in the brain is opposite to the direction of the current in the coil.

that a magnetic field that changes over time induces an electric field (and thus electric current) in nearby conducting tissue. In clinical applications, this is done by sending a strong current through a coil, thereby evoking a strong magnetic field perpendicular to the coil. This magnetic field then induces an electric current in the tissue, in the direction opposite to the current direction in the coil (see Box 11.1 for details).

Thus, even though the mode of stimulation is magnetic, the cell membrane is depolarized through electrical stimulation on a cellular level. An important difference with direct electrical stimulation (see e.g. Sect. 2.2) is that the field strength is maximal near the stimulation points in electrical stimulation, causing the pain fibers in the skin to be stimulated as well. Magnetic stimulation only causes a weak electrical current in the skin, making it a less painful experience. Another important difference between the two stimulation modalities is that for transcranial magnetic stimulation (TMS) the magnetic field travels easily through scalp and skull to the superficial brain layers, in contrast to an electric field which is strongly disturbed by, especially, the skull.

The central motor pathway can be assessed by stimulating magnetically over the motor cortex and recording the response in the muscle corresponding to the stimulated area in the primary motor cortex. This motor response is called the motor evoked potential, or MEP. Since the representations in the motor cortex of the hand and thumb are so large, it is quite common to assess central motor pathway functioning by stimulating the motor cortex such that a motor response can be recorded in the thumb muscle (m. abductor pollicis brevis). Because the induced current of a round stimulation coil is maximal near its outer edge (whereas, the magnetic field is maximal under the center of the coil), the coil may be positioned over the *vertex* to stimulate the hand area of the motor cortex. Note that activation of the coil in this position will stimulate both motor cortices. To assess upper motor neuron functioning for the legs, the coil may be repositioned to have its maximum induced current more medially and a response can be recorded from the m. tibialis anterior. However, other parts of the cortex may be stimulated depending on the clinical question.

11.3 Motor Evoked Potentials

11.3.1 MEP Physiology

Physiologically, when the motor cortex is stimulated by a single magnetic pulse during TMS, repetitive discharges of action potentials are evoked along the descending motor pathway (pyramidal tract). Two types of waves are discerned; the direct or D-waves and the indirect or I-waves. The D-waves have the shortest latency and are the result of direct stimulation of the proximal part of the axon of the pyramidal neuron. These D-waves are typically evoked as a result of electrical

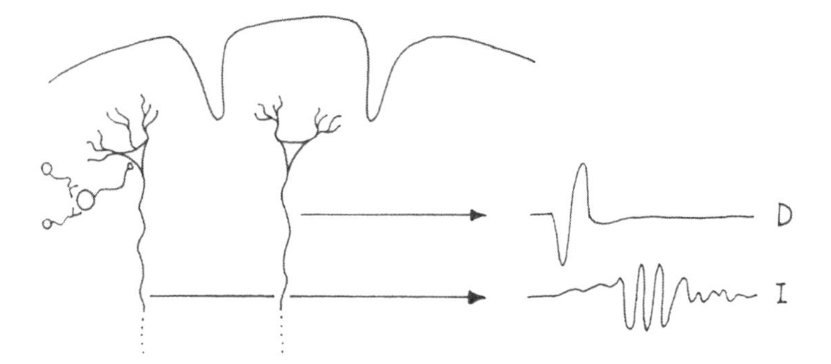

Fig. 11.1 Schematic overview of D- and I-waves evoked by a magnetic pulse over the motor cortex. Two schematic pyramidal cells are indicated: the D-wave is evoked by direct stimulation of the cell, the I-wave is evoked indirectly through trans-synaptic connections with other excitatory cells

stimulation and also appear at higher magnetic stimulation intensities. I-waves develop when excitatory cells in the deeper layers of the motor cortex are activated which then indirectly activate the pyramidal cells through trans-synaptic connections. This results in a variable delay in latency with respect to the D-wave of one to several milliseconds. One magnetic pulse thus yields a complex, *multiphasic* response in the pyramidal neurons (see Fig. 11.1).

When this response reaches the alpha motor neurons in the *anterior horn* cells in the spinal cord, depending on the threshold for firing of these neurons, a MEP will be evoked in the muscle. The multiphasic response in the pyramidal neurons implies that the response in the muscle will vary in latency and amplitude from stimulus to stimulus, depending on when the threshold to generate an action potential is reached by the alpha motor neuron. The number and intensities of the D- and I-waves and the initial excitability of the alpha motor neuron determine its activation. In addition, the direction of the induced current in the cortex also determines the threshold at which an MEP is evoked, since pyramidal neurons are more easily activated by currents in posterior-anterior direction, perpendicular to the central sulcus. Thus, even though both motor cortices will be stimulated when a circular coil is placed over the vertex, the MEP response will be larger on one side.

Since the magnetic field penetrates the tissues quite deeply, it can also be used to stimulate the peripheral nervous system and deeply positioned nerves, the *plexus brachialis* and the ventral roots (*efferent* motor roots of the spinal nerves), in particular. In the latter situation, the upper edge of the coil is placed above the root, while the center of the coil is placed over the middle of the spine. In contrast to TMS, the MEP amplitude, as a result of ventral root stimulation, is highly variable and not maximal. Therefore, only the latency is used for clinical interpretation. Latency depends on the peripheral motor conduction velocity of the nerve fibers, and thus on age and temperature (see also Sect. 2.2.1.3).

11.3.2 MEP Outcome Measures

Besides the MEP amplitude and latency, other measures can be derived from TMS and peripheral magnetic stimulation that are of value in clinical diagnostic procedures. The central motor conduction time (CMCT) – the time it takes the evoked action potentials to travel from the motor cortex to the spinal cord – can be derived by subtracting the peripheral latency, which is obtained by stimulating the ventral root innervating the target muscle, from the MEP latency. Alternatively, the peripheral latency may be derived from the F-response (see Sect. 2.2.1.1).

Question 11.1 Why would it be useful to record the peripherally evoked compound motor action potential (CMAP, see Sect. 2.2.1.1) as part of a clinical MEP investigation?

For clinical diagnostic applications in which spinal cord integrity is assessed, it is most important to obtain a reliable estimate of the MEP latency and the MEP amplitude is less important. However, in research applications, the MEP amplitude is often used to indicate if a certain repetitive transcranial magnetic stimulation (rTMS) protocol (see Sect. 11.4) has had any effect on the excitability of the cortical neurons, and thus the intensity is much more important. At low stimulation intensities, smaller populations of cortical neurons are recruited than at higher intensities. Different rTMS protocols can influence different populations of neurons. Thus, to measure the effect of a protocol, it is important to measure the MEP amplitude at different intensities. The measurement of the MEP amplitude as a function of the stimulation intensity is known as the MEP recruitment curve. In a recruitment curve, the MEP amplitude is plotted as a function of stimulation intensity, which would typically vary between 90% and 150% of the resting motor threshold (RMT), with steps of 10%. The RMT is defined as the minimum stimulation intensity at which five out of ten TMS pulses result in an MEP response with an amplitude of more than 50 μV in a relaxed muscle. To get a more reliable estimate of the recruitment curve, the stimulations are repeated several times at each stimulation intensity, in random order. Examples of recruitment curves are given in Fig. 11.2.

Other measures of the excitability of interneuronal circuits in the motor cortex are intracortical facilitation (ICF) and intracortical inhibition, first described by Kujirai et al. (1993). They showed that if a TMS stimulus is preceded by a conditioning stimulus, the MEP response to the second test stimulus can be altered in amplitude. If the response to the conditioned test stimulus increases in amplitude compared to the test stimulus alone, this is known as ICF. If the response decreases, this is known as (short-interval) intracortical inhibition (SICI). Whether the response increases or decreases in amplitude depends on the interstimulus interval (SICI: 1–5 ms, ICF 10–15 ms) and the stimulation intensities of the conditioning and the test stimulus.

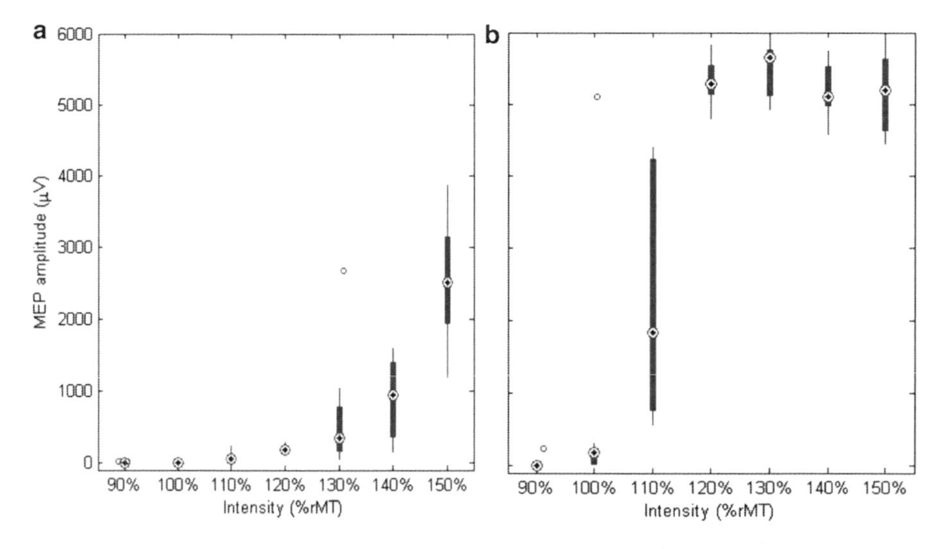

Fig. 11.2 Two recruitment curves in different subjects. The curve in (**a**) has not reached a plateau at the highest intensity of 150% of the RMT, while the curve in (**b**) already reached the plateau at an intensity of 120% of the RMT. The results are displayed for ten stimulations at every intensity

11.3.3 Tips and Tricks for TMS

11.3.3.1 Safety

Whether a neuron is activated by stimulation depends on the strength and duration of the stimulation: it can be activated both by strong and short stimulation or by weak and long stimulation. However, if a current is sent through a stimulation coil for a longer time, it may overheat, so from a technical point of view, shorter pulses are better. Thus, a strong, short pulse is needed to optimally stimulate the neurons and avoid overheating of the coil. In practice, a trade-off between these two requirements must be made. Most magnetic stimulators consist of a circuit in which a *capacitor* is used to store energy. When the capacitor is suddenly discharged, a short and strong current flows through the coil evoking the magnetic field (see Box 11.2).

To generate a 10 T magnetic field during 100–150 μs, currents of 4–10 kA and voltages up to 3 kV are involved. These specifications require particular safety measures in the stimulator itself, but also regarding its application in humans. Patients with implanted stimulators (such as a cardiac pacemaker or deep brain stimulation electrodes) are not allowed to participate in an MEP investigation. Other metal implants in or near the area that will be stimulated (such as vascular clips or metal fragments) are also a contraindication for magnetic stimulation, since the magnetic field can exert strong mechanical forces on these materials and may heat

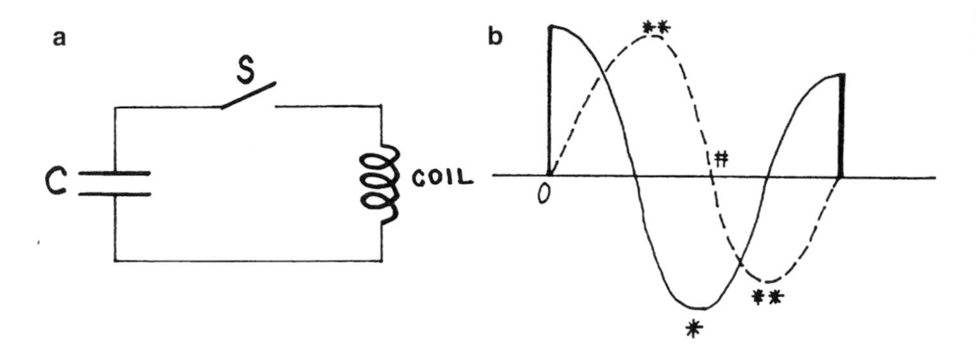

Fig. 11.3 (**a**) A simplified TMS circuit. *C* capacitor; *S* switch; *COIL* TMS coil (inductor) and (**b**) the biphasic pulse that can be generated by it. *Solid line*: voltage, *dashed line*: current. *, **, #: see Box 11.2

Box 11.2 TMS Circuit and Biphasic Pulses

In its simplest form, a TMS circuit consists of a power source, a capacitor, the stimulation coil, and a switch (see Fig. 11.3a). When there is an electric charge between the two plates of the capacitor, energy (charge) is being stored. When the capacitor is discharged by closing the switch, a current flows through the circuit and energy is transferred from the capacitor to the coil. This is the start of the biphasic magnetic pulse in Fig. 11.3b. In the next phase of the cycle, the current decreases again, the voltage decreases and the capacitor is charged again, but at a voltage that is reversed compared to the initial value (*). In the second half of the cycle, the procedure is repeated, but with the current flowing in the other direction. Thus, in general, when the current is maximum (**), all energy is in the coil and when the current is zero (#), all energy is in the capacitor. At the end of the cycle, the capacitor is charged again at the same polarity as in the beginning, but because of energy dissipation in the circuit, the voltage is lower. In principle, the cycle could start again at this moment, but typically the circuit is switched off here to avoid oscillatory behavior. When the voltage over the stimulation coil is maximal, (either positive or negative) the induced voltage (and current) is also maximal in the underlying tissue.

It is also possible to generate monophasic pulses by slowly dissipating the coil current instead of returning it to the capacitor. Because biphasic pulses are generally shorter than monophasic pulses, they are more often used, especially when stimulating repetitively.

them up. Since the strong magnetic fields have evoked epileptic seizures in some people with a history of epilepsy, epilepsy is a contraindication for magnetic stimulation as well, except when the importance of the MEP investigation is larger than the relative risk of epilepsy (just like flash stimulation is used in

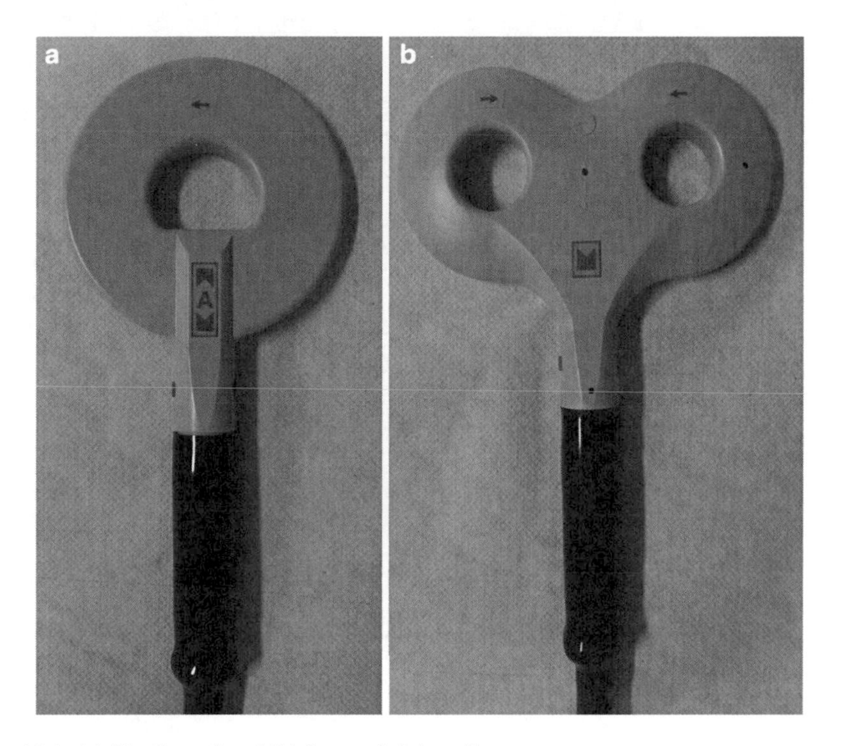

Fig. 11.4 (**a**) Circular coil and (**b**) figure-of-eight coil

electroencephalography (EEG) investigations, even though it may evoke an epileptic seizure). Finally, the magnetic field can also discharge magnetic information carriers such as credit cards and computer disks, so that they should be kept out of the magnetic field as well.

11.3.3.2 Localization

For optimal stimulation (maximal MEP amplitude), the maximum of the evoked current should be over the motor area representing the target muscle. For a circular coil, this maximum is close to its outer edge, making it difficult to precisely identify the position on the scalp, where the MEP is maximal. In other words; circular coils are not very focused since the coil edge overlies a large part of the brain and the extent of the strongest field is not precisely known. When two circular coils are positioned next to each other in a configuration known as a figure-of-eight or butterfly coil (see Fig. 11.4), the currents in these two coils can be made to flow in the same direction at the point where they meet and the result is an induced electric field that adds up, and is maximal below the junction of the two coils and more focused.

When the goal is to stimulate the hand area in the primary motor cortex, typically the EEG location C3 (left hemisphere) or C4 (right hemisphere) is taken as a

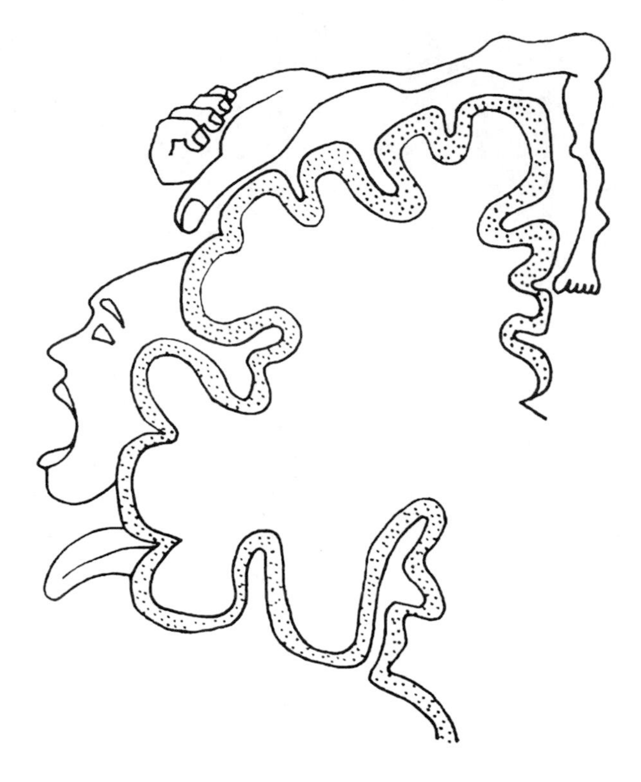

Fig. 11.5 Motor homunculus according to Penfield

starting point. This is a well-defined point with respect to the bony landmarks on the skull (see Sect. 4.2) and is thought to overlie the central sulcus near the *hand knob*. In practice, however, holding a figure-of-eight coil over C3 does not necessarily yield the maximum MEP response in the right hand, because the anatomy and *somatotopy* of the motor cortex are slightly different in every individual. To find the position where the MEP amplitude is maximal, the motor homunculus can be used. This is a map of the primary motor cortex derived by Wilder Penfield in the 1950s by direct electrical stimulation of the brain. In this map, hands and face have much larger representations in the motor cortex than e.g., arms and legs. From medial to lateral positions along the motor strip, the feet, legs, arms, hands and thumb, face and tongue are represented (see Fig. 11.5).

By closely observing the evoked responses in the muscles close to the target muscles, the position of the coil can be adapted to more lateral or more medial sites to get closer to the representation of the target muscle. Furthermore, it may be necessary to relocate the coil slightly more anteriorly or posteriorly, to get the best MEP response.

Question 11.2 In which direction should the coil be moved over the scalp when a contraction in the lower arm muscles is observed instead of in the target muscle, the m. abductor pollicis brevis?

11.3.3.3 Optimizing the MEP Response

As discussed in Sect. 11.3.1, the initial excitability of the alpha motor neuron is one of the factors determining its activation. Since the resting potential of an inactive alpha motor neuron is closer to the threshold at which it will fire when the target muscle is contracted voluntarily, an MEP will be easier to evoke when the patient is asked to contract the target muscle during TMS (so-called pre-activation). A single descending volley may then already provide sufficient input to the alpha motor neuron to discharge it. Thus, MEPs are facilitated by voluntary muscle contraction; their latency is reduced, their amplitude increased and the threshold at which they are evoked is lower. Since prior contraction of the target muscle causes an irregular baseline, making MEP latency determination more difficult, the corresponding muscle on the other side may alternatively be contracted strongly for facilitation. However, facilitation cannot be employed when the RMT needs to be determined. The RMT, compared to the active motor threshold (AMT), has better reproducibility, because it does not depend on prior activation of the alpha motor neurons and thus is a better reference for used stimulation intensities.

11.4 Repetitive TMS

So far, the use of single pulse TMS to assess the integrity of the corticospinal tract has been discussed. However, the magnetic pulses can also be applied repeatedly, at a certain frequency or rhythm. This application of TMS is called rTMS. With rTMS, effects of stimulation can be generated that last longer than the stimulation itself, making it interesting for therapeutic applications. Since the development of rTMS, TMS research has expanded. rTMS allows to temporarily "switch off" certain brain areas and thereby examine the function of each brain area separately in healthy subjects. The advantage of this examination of single brain areas over imaging techniques such as functional magnetic resonance imaging (fMRI) is that rTMS can demonstrate causality. If a brain area is switched off and the subject performs worse on a task as a result, this is strong evidence that the region is essential for good task performance. Furthermore, rTMS can be used to upregulate (activate or excite) or downregulate (inhibit) brain activity in certain brain areas, and it is thought that this type of brain modulation can also influence entire brain networks.

Because of these properties of rTMS, many disorders and syndromes that involve altered activity in cortical areas can be a target for rTMS therapy. Severe depression, *tinnitus* and movement disorders are promising areas for application of rTMS therapy. Yet, results in different studies vary, probably because rTMS protocols have many variables. Stimulation intensity, frequency, rhythm, target area, coil design, and coil positioning can all potentially influence the effect of the stimulation. Furthermore, factors not directly related to the stimulation protocol such as the target muscle, attention of the subject, subject's age, individual genetic susceptibility to rTMS, and background muscle activity, can all be of influence. As a result, the effect of apparently the same protocol may vary from inhibitory to excitatory through different studies.

11.4.1 rTMS Outcome Measures

Typically, the MEP amplitude is used to indicate the type of effect of an rTMS protocol on cortical excitability. If the MEP amplitude post-stimulation is increased compared to the pre-stimulation amplitude, the protocol is said to be excitatory, while the protocol is inhibitory when the change in amplitude is in the opposite direction, i.e. when the MEP amplitude is smaller post-stimulation. All TMS outcome measures as discussed in Sect. 11.3.2 can also be used as outcome measures for rTMS protocols.

Outcome measures that are more specific for rTMS protocols are functional brain measures such as fMRI or EEG. When rTMS is used to inhibit or increase the activity in a certain brain area, fMRI can be used to measure the induced change in brain activity. Ideally, fMRI is executed immediately after rTMS, so that short-lasting effects can be captured. However, magnetic stimulation inside a magnetic field poses many difficult technical challenges and is only performed in a few labs worldwide. In practice, rTMS is mostly performed close to, but outside the scanner and fMRI is performed as soon as possible after stimulation. A disadvantage of using fMRI to measure the effect of an rTMS protocol is that it only provides an indirect metabolic measure (the blood oxygen-level dependent [BOLD] effect) of a change in neuronal activity. An advantage of fMRI is that, in principle, changes in activity elsewhere in the brain, outside the directly stimulated area, can also be detected.

Even though EEG is strongly disturbed by TMS, special amplifiers exist that allow concurrent TMS and EEG recording.

> *Question 11.3* Why does TMS disturb the EEG?

Because of the thickness of the electrodes on the scalp, the coil needs to be held further away from the scalp, which causes the magnetic field to penetrate less deeply (the magnetic field decreases in strength with distance), allowing only more superficial stimulation. An advantage of EEG-based outcome measures is that they have high temporal resolution, allowing e.g., to monitor the spread of activation from the stimulated cortical site via intra- and interhemispheric *corticocortical* fibers.

11.4.2 rTMS Protocols

The oldest rTMS protocols are continuous: stimuli are delivered at a constant frequency and a constant stimulation intensity. These protocols are divided into low frequency protocols (stimulation frequency $\leq$ 1 Hz) and high frequency protocols (stimulation frequency mostly $\geq$ 5 Hz). In general, low frequency protocols have an inhibitory effect; the MEP amplitude after application of such a protocol is lower than before. High frequency protocols generally have an excitatory effect; after these protocols, the MEP amplitude is generally higher than before. Mostly, the effect of these continuous protocols does not last long after the end of stimulation, although the effect can be lengthened by longer stimulation. Yet, stimulation cannot be lengthened indefinitely, because epileptic seizures may be evoked and safety guidelines have been developed to prevent this (Wassermann 1998).

With the advancement of TMS hardware technology, alternative protocols have become available with longer effect durations. One of these protocols – theta burst stimulation – was found to have different, long lasting effects depending on the exact rhythm at which it is delivered (Huang et al. 2005). The theta burst stimulation (TBS) protocols consist of short bursts of three subthreshold TMS pulses at a frequency of 50 Hz, repeated at inter-train intervals of 200 ms. Continuous application of these blocks (continuous or cTBS) leads to inhibitory effects. When blocks of 2 s are followed by 8 s rest before the next block (intermittent or iTBS), the effects are found to be excitatory. Because the stimulation duration with theta burst protocols is very short (between 40 s and 3 min) and the effects are promising (lasting at least 30 min), many recent studies have used theta burst stimulation protocols.

11.4.3 Tips and Tricks of rTMS

11.4.3.1 Neuronavigation

As described in Sect. 11.3.3.2, general knowledge about motor cortex somatotopy may be used to find the optimal stimulation point for (r)TMS, but only when the aim is to stimulate part of the motor cortex, of course. As rTMS may also be used to selectively activate or inhibit brain areas that are involved in psychiatric or neurological disorders, or that are of interest from a fundamental point of view, other methods are needed to localize brain areas that are outside the primary motor cortex. One of these methods is to use prior anatomical or functional knowledge about brain areas as obtained from (f)MRI, magnetoencephalography (MEG), or positron emission tomography (PET). To maneuver the coil to a position right above an fMRI "hot spot", neuronavigation tools can be used. A neuronavigation system can determine the position of the coil in three-dimensional space with respect to the subject's brain area of interest as obtained from an anatomical or functional MR image, using an optical tracking system which records the positions

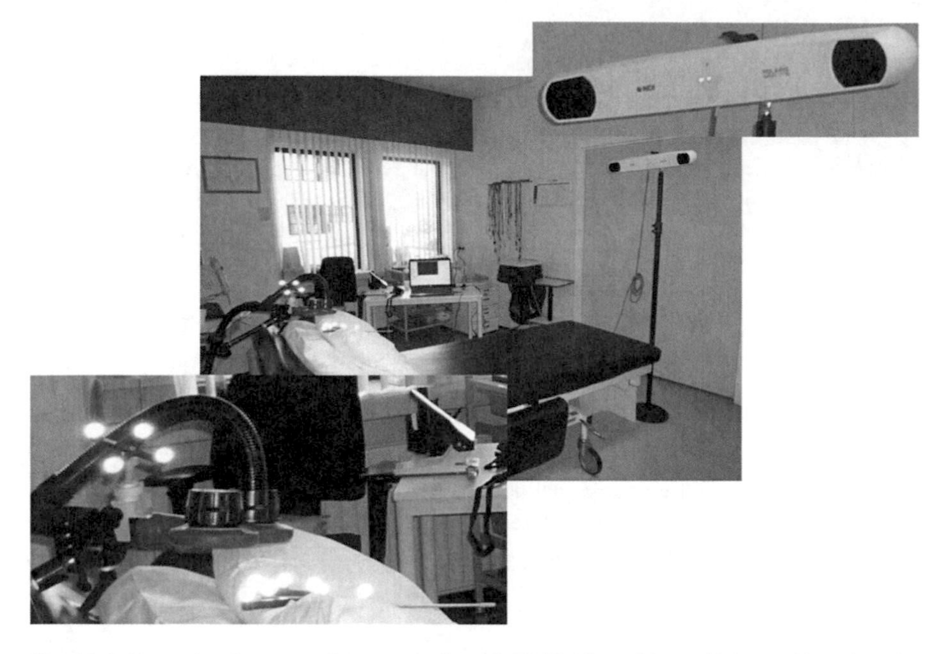

Fig. 11.6 Example of a set-up for neuronavigated rTMS. The subject will be positioned on the bed. The TMS coil with the attached cooling system and the pointer (enlarged on the *bottom left*) at the head-end of the bed are both fitted with reflecting spheres that can be recognized by the camera (enlarged on the *top right*) at the foot-end of the bed. Note that since the photo was taken with a flash, the spheres actually reflect in the photo

of the subject's head and the coil in 3D space continuously and a co-registration procedure to link the subject's head with the subject's (f)MRI (see Fig. 11.6).

Additionally, the induced electric field may be calculated using mathematical procedures (cf. Sect. 8.3) and projected onto the (f)MRI to verify the focus of the rTMS. Neuronavigation systems were first applied for neurosurgical applications to determine the position of surgical instruments with respect to anatomical structures in the brain. These techniques also allow to hold the position of the coil constant with respect to the subject's head, even during longer stimulation protocols and even without the necessity to fixate the subject's head in case a neuronavigation robot is used. These robots reposition the coil in real time, in reaction to a movement of the subject. Another advantage of navigated rTMS for therapeutic applications is that the exact same coil position and orientation can be maintained between sessions, thereby improving its reliability and repeatability.

11.4.3.2 Sham Stimulation

To be able to evaluate the effect of rTMS for therapeutic applications, it is important to have a placebo condition, which does not induce physiological effects, for comparison. In the context of rTMS, this is called sham stimulation. Ideal sham rTMS would resemble real rTMS both visually and acoustically and create a comparable cutaneous sensation without producing cortical stimulation. Furthermore, sham stimulation should ideally be double-blind, i.e. unrecognizable to both the subject and the technician. It is very difficult to meet all these requirements at once, which has led to multiple sham TMS approaches being used. A simple approach is to lower the stimulation to an intensity which has no physiological effects. However, at lower stimulation intensities, the acoustic intensity and the cutaneous sensation are also less. Another simple method is tilting and/or moving the coil with respect to the scalp, which preserves the acoustic experience, but may still have physiological effects and there is no cutaneous sensation. Furthermore, neither approach allows blinding to the technician. Commercial sham coils, that look just like real coils but deliver lower magnetic fields, typically allow blinding to the technician but lack the cutaneous sensation and sound different, as well. Some sham coils have been developed that deliver scalp sensations, usually by electrical stimulation synchronized to the sound of the stimulation (e.g. Rossi et al. 2007). Even though these types of coils cannot be used in a double-blind manner, they seem to be the best solution at the moment.

11.5 MEPs in Individual Patients

Corticospinal tract integrity is assessed by magnetic stimulation over the (ipsi- and contralateral) motor cortex. Furthermore, the spinal cord is stimulated cervically to assess motor conduction to the arms and lumbally to assess motor conduction to the legs. The MEP is then derived from the m. abductor pollicis brevis (APB) for the arms and from the m. tibialis anterior (TA) for the legs. Before cortical and spinal stimulation, the CMAP in the APB and TA muscles is also determined for reference purposes by stimulating electrically at the wrist and knee, respectively.

Patient 1

The MEP latencies and amplitudes resulting from the protocol described above are indicated in Tables 11.1 and 11.2 and Fig. 11.7 for patient 1, who suffers from a slight pyramidal syndrome in particularly her left leg.

Table 11.1 Results of MEP investigation in patient 1 for the arms

	Left arm		Right arm	
Stimulation position	Latency (ms)	Amplitude (mV)	Latency (ms)	Amplitude (mV)
Wrist	3.3	10.7	3.7	8.1
Cervical spine	13.5	5.7	15.0	2.6
Cortical (contra)	24.4	0.2	24.4	2.1
Cortical (ipsi)	20.8	2	23.1	4.4

Contra contralateral pre-activation; *ipsi* ipsilateral pre-activation

Table 11.2 Results of MEP investigation in patient 1 for the legs

Stimulation position	Left leg		Right leg	
	Latency (ms)	Amplitude (mV)	Latency (ms)	Amplitude (mV)
Knee	3.3	6.0	3.5	7.9
Lumbal spine	12.9	0.3	13.8	1.3
Cortical (contra)	36.4	1.9	32.4	0.3
Cortical (ipsi)	37.5	1.9	27.2	1.9

Contra contralateral pre-activation; *ipsi* ipsilateral pre-activation

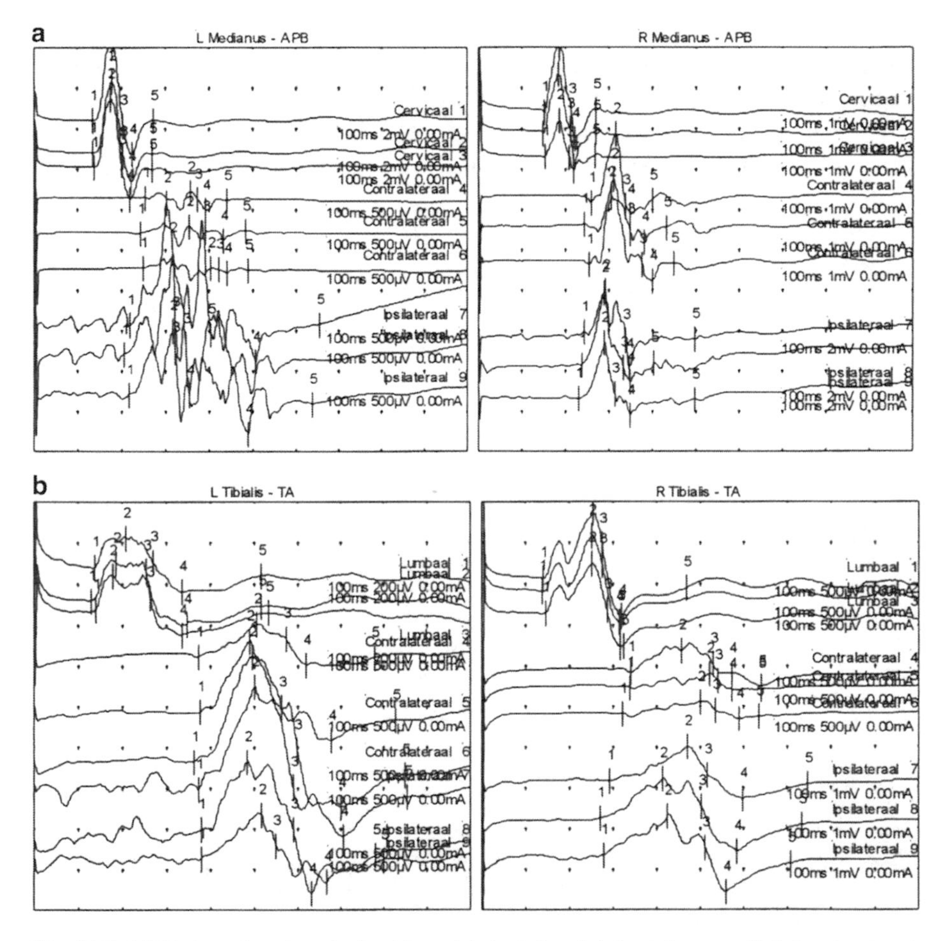

Fig. 11.7 MEPs in patient 1 as obtained from (**a**) the left and right abductor pollicis brevis (thumb abductor) muscles and (**b**) left and right tibial anterior (ankle flexor) muscles after cervical or lumbar stimulation (*top three traces*), after cortical stimulation with contralateral pre-activation (*middle three traces*) and cortical stimulation with ipsilateral pre-activation (*bottom three traces*)

Table 11.3 Results of MEP investigation in patient 2 for the arms

	Left arm		Right arm	
Stimulation position	Latency (ms)	Amplitude (mV)	Latency (ms)	Amplitude (mV)
Wrist	2.3	11.7	2.3	3.9
Cervical spine	12.4	1.0	13.5	1.7
Cortical (contra)	–	–	–	–
Cortical (ipsi)	–	–	–	–

Contra contralateral pre-activation; *ipsi* ipsilateral pre-activation; – MEP not obtainable

Table 11.4 Results of MEP investigation in patient 2 for the legs

	Left leg		Right leg	
Stimulation position	Latency (ms)	Amplitude (mV)	Latency (ms)	Amplitude (mV)
Knee	3.2	9.8	3.3	10.3
Lumbal spine	12.5	1.5	13.3	1.5
Cortical (contra)	–	–	–	–
Cortical (ipsi)	–	–	–	–

Contra contralateral pre-activation; *ipsi* ipsilateral pre-activation; – MEP not obtainable

For the arms, the MEP shapes and latencies were found to be normal for peripheral (wrist) and root (cervical) stimulation. However, when the cortex was stimulated the (contralateral) CMCT was delayed for the left arm (24.4−13.5=10.9 ms: +5.1 SD compared to a normal population as derived in our hospital). In addition, the MEP was polyphasic, also indicating impaired motor conduction. The MEP for cortical stimulation was slow, but still within normal ranges and the appearance was more biphasic. For the legs, the MEP shapes and latencies were also normal for peripheral (knee) and root (lumbal) stimulation. Again, the CMCT for contralateral and ipsilateral cortical stimulation was found to be abnormally increased for the left leg (36.4 − 12.9 = 23.5 ms (+5.5 SD) and 37.5 − 12.9 = 24.6 (+6.2 SD), respectively). In summary, these findings indicate that patient 1 suffers from a central motor conduction disorder, both to the left arm and left leg. Combining these results with the findings of other clinical investigations that the neurologist ordered (multiple white matter lesions on MRI, consistent with demyelination) and the patient's complaints, it is most likely that the patient suffers from a progressive form of multiple sclerosis (see also Chap. 5).

Patient 2

In this patient, who suffers from progressive motor weakness, corticospinal integrity was assessed in the same manner as for patient 1. MEP latencies and amplitudes are indicated in Tables 11.3 and 11.4 and Fig. 11.8.

The MEP shapes and latencies were normal for peripheral and spinal stimulation, for both arms and legs. Despite maximal stimulator output, MEPs could not be obtained for cortical stimulation, however, neither for the arms nor for the legs.

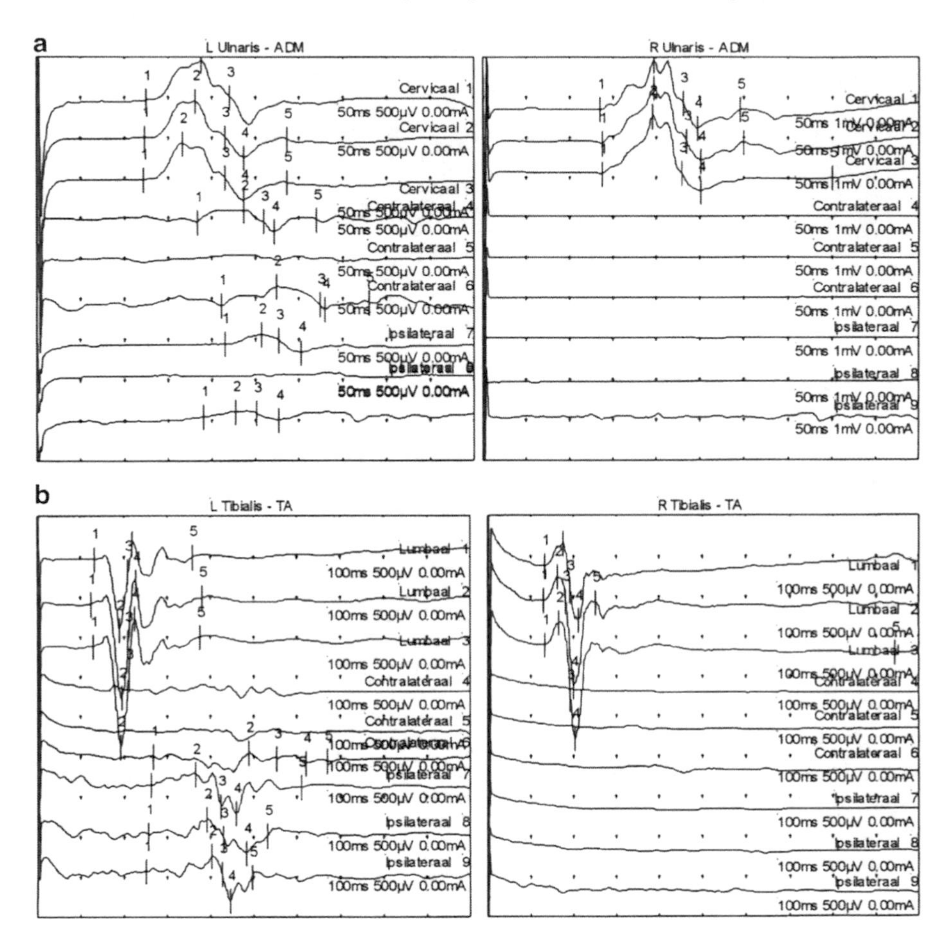

Fig. 11.8 MEPs in patient 2 as obtained from (**a**) the left and right abductor digiti minimi (little finger abductor) muscles and (**b**) left and right tibial anterior (ankle flexor) muscles after cervical or lumbar stimulation (*top three traces*), after cortical stimulation with contralateral pre-activation (*middle three traces*) and cortical stimulation with ipsilateral pre-activation (*bottom three traces*)

These findings indicate that patient 2 suffers from a severe central motor conduction disorder. These results combined with the other clinical (e.g., progressive muscle weakness and fasciculations) and EMG findings (not described here) indicate that patient 2 suffers from a progressive upper motor neuron disease (with – initially – less involvement of the lower motor neuron), which is most likely amyotrophic lateral sclerosis (ALS, also known as Lou Gehrig's disease). ALS involves progressive degeneration of both the upper and lower motor neurons, causing muscle weakness and atrophy. Depending on which motor neurons are affected first, initial symptoms may vary, but the disease is inevitably fatal. Because all muscles become involved eventually, most patients die of pneumonia or respiratory failure, within 2–5 years of diagnosis.

11.6 Other Applications of TMS and MEP in Neurology

11.6.1 Preoperative Motor Cortex Mapping

As discussed in Sect. 11.4.3.1 neuronavigation can be used to localize targets for TMS based on anatomical or functional information as obtained from (f)MRI or PET. Vice versa, neuronavigation can also be used to localize the primary motor cortex when it can no longer be recognized on anatomical images, e.g., when a tumor in the *Rolandic region* has displaced the precentral gyrus, such as in patient 1 in Chap. 8.

However, the spatial accuracy of motor cortex mapping using TMS is limited due to the inherent error of the tracking system which is in the order of a few milliseconds and the limited spatial focus of the TMS itself. This accuracy can be improved by shifting the (figure-of-eight) coil over the scalp in small increments to obtain the MEP in the target muscle at each location. The location of the representation of the target muscle in the primary motor cortex can then be determined by the coil focus position at which the MEP response is maximal. In this manner, preoperative motor cortex mapping can serve to plan brain surgery, but the actual position of the central sulcus will always be verified intraoperatively by direct cortical stimulation (using Penfield's methods; see Sect. 11.3.3.2).

11.6.2 Intraoperative Monitoring

Another application of TMS during neurosurgery is intraoperative monitoring of the MEP response. In Sect. 5.6.1 it was already discussed how SEPs can be used to monitor spinal cord functioning during surgery to correct a scoliotic spine. During such a procedure, MEPs are also routinely recorded with the same purpose. As for SEPs, MEPs will be obtained preoperatively to be sure that spinal cord functioning was initially intact and to have a reference value for the intraoperative MEP amplitude. An advantage of MEPs compared to SEPs is that only one stimulation is required. Because of the anesthetics that suppress I-wave production but allow more painful procedures at the same time and the spatial limitations in the operating room, MEPs are often obtained by electrical (needle) stimulation instead of magnetic stimulation. Furthermore, multipulse stimulation is used to facilitate the MEP response. Intraoperative MEP monitoring helps determine the maximal correction of the scoliotic spine, without damage to the nerves or nerve roots or obstruction of blood flow. An example is given in Fig. 11.9.

11.6.3 Experimental Treatment of Movement Disorders

From a neurological point of view, one of the most interesting potential applications of rTMS is the treatment of movement disorders associated with altered brain network activity, such as Parkinson's disease and dystonia. In 2008, Wu et al.,

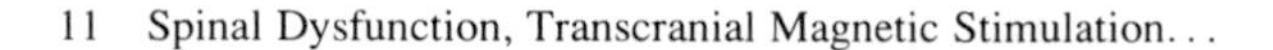

a

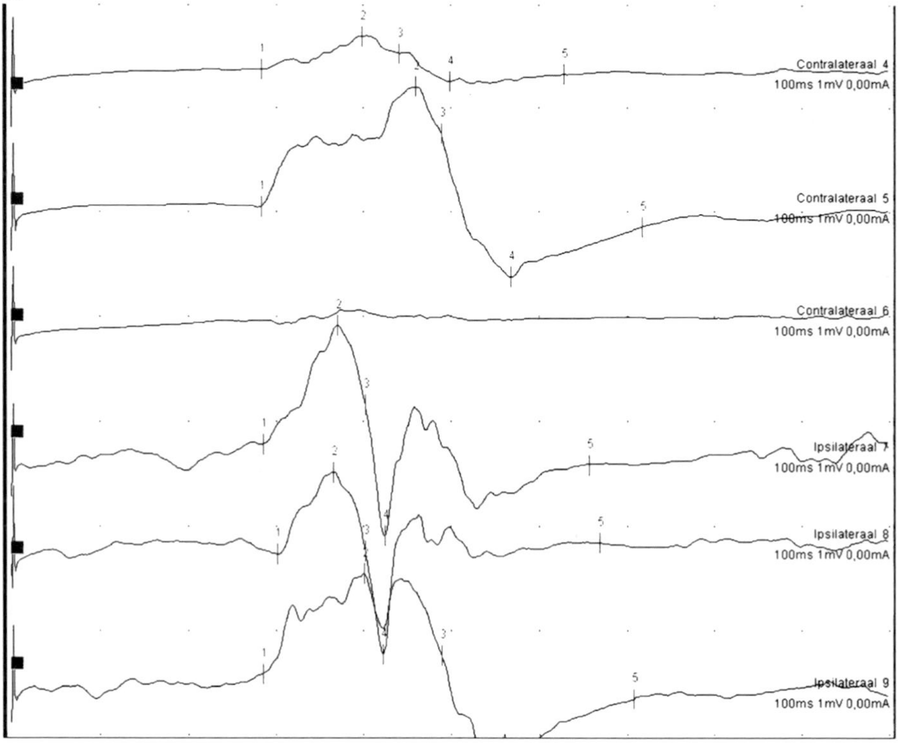

b

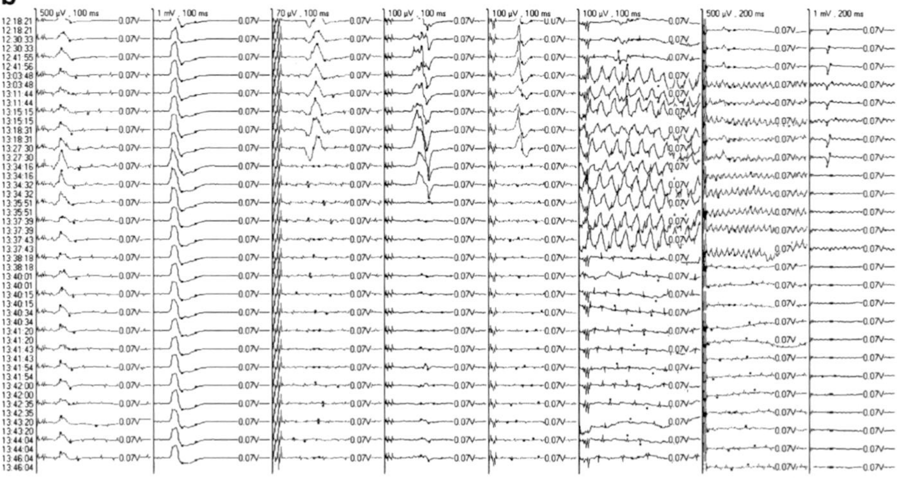

Fig. 11.9 Example of MEPs (**a**) pre-operatively, (**b**) intraoperatively during scoliotic correction in a 14-year old female patient. Pre-operatively, MEPs could be recorded from the tibial anterior (TA) muscles both with pre-activation of the TA muscles on the other side (*top three traces*) and with pre-activation of the TA muscles on the stimulated side (*bottom three traces*), indicating a preserved spinal cord functioning. In (**b**) the following muscles are displayed from left to right: arm flexor

published an overview of rTMS studies in these disorders. In general, rTMS treatments aim at normalizing brain network activity (neuromodulation) and usually base their protocols on results from metabolic neuroimaging (fMRI and PET) studies in these disorders. Excitatory protocols are then chosen to enhance brain activity, inhibitory protocols are chosen to decrease network activity. Although fMRI and PET studies indicate both cortical and subcortical hyper- and hypoactive brain areas in movement disorders, only superficial (cortical) areas can be directly reached by (r)TMS. Yet, it is thought that stimulation of superficial brain areas also influences activity in other brain areas in the rest of the network. Since there is no consensus yet on the duration and number of rTMS treatment sessions needed for a positive effect, various studies in patients with Parkinson's disease have used protocols stimulating the primary motor cortex, the vertex, premotor cortex, supplementary motor cortex, or (dorsolateral pre-) frontal cortex at frequencies between 0.2 and 25 Hz for durations anywhere between minutes (single session) and months (multiple sessions). Outcome is usually evaluated by clinical scores such as the *UPDRS*. In Parkinson's disease, most studies have shown beneficial effects on clinical symptoms, but due to the variability in rTMS protocols used, it is still very difficult to determine rTMS parameters for the best effect in individual patients. In dystonia, available studies have mostly focused on inhibiting activity, aiming for the primary or pre-motor cortex using stimulation frequencies of maximally 1 Hz. Although many studies have shown physiological changes in dystonia after rTMS, clinical benefits have not been clearly established yet and additional studies systematically investigating the effect on clinical symptoms of different rTMS protocols are required.

11.7 Answers to Questions

Answer 11.1 Since the peripherally, electrically, evoked CMAP only depends on the most distal part of the motor pathway, its amplitude and shape may serve as a reference for the maximal cortically stimulated MEP amplitude, when lesions more proximally in the motor pathway are expected.

Fig. 11.9 (continued) muscles (lower arm; left and right), TA muscles (front lower leg; left and right), gastrocnemius muscles (back lower leg; left and right), abductor hallucis muscles (foot sole; left and right). Channels 6 and 7 display 50 Hz noise due to the electrical tools used by the surgeons. From top to bottom subsequent MEPs are plotted. It can be observed that after some time, coinciding with the moment that the surgeons started to correct the spinal curvature, the MEPs from the leg muscles disappeared, whereas the MEPs from the arm muscles were preserved. This event was immediately notified to the surgeons and the corrective procedure was reversed. The patient recovered her spinal cord function some time after the surgery and the scoliotic correction was performed successfully at a later occasion

Answer 11.2 Since the thumb (m. abductor pollicis brevis) is represented more laterally than the arm in the motor cortex according to Penfield's homunculus, the coil should be moved more laterally.

Answer 11.3 TMS involves a rapidly changing and strong magnetic field. Since this changing magnetic field not only induces a current in the brain but also in any other conducting material, TMS will cause currents in the conducting wires connecting the electrodes to the amplifier, thereby causing large artifacts in the EEG.

Glossary

Anterior horn Here: frontal grey matter of the spinal cord, containing the alpha motor neurons

Babinski sign Abnormal plantar (footsole) reflex: upward response of the big toe (dorsiflexion) and spreading of other toes

Capacitor Electric component consisting of a pair of conductors separated by an insulator; when there is a potential difference across the two conductors, the capacitor can store energy

Clonus Repetitive relatively large movement resulting from a reflex

Corticocortical Between two cortical areas

Efferent Away from the center; e.g. nerve fibers running from the spinal cord to the hand. Opposite to afferent

Fasciculations See Glossary Chap. 9

Hand knob Area of the precentral gyrus representing hand motor function, recognizable on MRI by its knob-like shape

Hot spot Here: area of increased activation in an fMR image

Lipid Fatty molecule

Motor neuron See Glossary Chap. 2

MRC See Glossary Chap. 2

Multiphasic Here: response curve exhibiting multiple positive and negative peaks

Plexus brachialis Bundle of nerve fibers, originating from the spine and proceeding through the neck and armpit to the arm

Pyramidal syndrome The complaints associated with damage to the pyramidal or corticospinal tract (axons originating in the cerebral cortex and running down into the spinal cord); *spasticity*, muscle weakness, *Babinski sign*

Rolandic region Brain area around the central sulcus, separating the frontal and the parietal lobe

Somatotopy Preservation of relative spatial location throughout the central nervous system (e.g., hand and arm sensory information is processed in adjacent areas in the primary somatosensory cortex)

Spasticity Increased resistance to passive movement at increasing speed of antigravity muscles

Tinnitus Perception of sound in the ear without an external sound source being present

UPDRS Unified Parkinson's disease Rating Scale used for longitudinal assessments of the progression of Parkinson's disease. The scale incorporates questions related to motor functioning, activities of daily life and mental problems

Vertex Top of the head, coincident with the EEG position Cz, which is halfway the inion and nasion and the two pre-auricular points (see Glossary Chap. 4)

References

Online Sources of Information

http://en.wikipedia.org/wiki/Transcranial_magnetic_stimulation. Short introduction to the basics of TMS

http://emedicine.medscape.com/article/1139085-overview. Overview of TMS and MEP, protocols and applications

Books

Wassermann EM, Epstein CM, Ziemann U, Paus T, Lisanby SH (eds) (2008) The Oxford handbook of transcranial stimulation. Oxford University Press, Oxford (Available on books.google.co.uk)

Papers

Huang YZ, Edwards MJ, Rounis E, Bhatia KP, Rothwell JC (2005) Theta burst stimulation of the human motor cortex. Neuron 45(2):201–206

Kujirai T, Caramia MD, Rothwell JC, Day BL, Thompson PD, Ferbert A, Wroe S, Asselman P, Marsden CD (1993) Corticocortical inhibition in human motorcortex. J Physiol 471:501–519

Rossi S, Ferro M, Cincotta M, Ulivelli M, Bartalini S, Miniussi C, Giovannelli F, Passero S (2007) A real electro-magnetic placebo (remp) device for sham transcranial magnetic stimulation. Clin Neurophysiol 118(3):709–716

Ruohonen J, Karhu J (2010) Navigated transcranial magnetic stimulation. Neurophysiol Clin 40:7–17

Wassermann EM (1998) Risk and safety of repetitive transcranial magnetic stimulation: report and suggested guidelines from the international workshop on the safety of repetitive transcranial magnetic stimulation, june 5–7, 1996. Electroencephalogr Clin Neurophysiol 108(1):1–16

Wu AD, Fregni F, Simon DK, Deblieck C, Pascual-Leone A (2008) Noninvasive brain stimulation for Parkinson's disease and dystonia. Neurotherapeutics 5:345–361

Index

N. Maurits, *From Neurology to Methodology and Back: An Introduction to Clinical Neuroengineering*, DOI 10.1007/978-1-4614-1132-1,

T

U

V

W

Z